VOSMAER'S
BIBLIOGRAPHY OF SPONGES

BIBLIOGRAPHY OF SPONGES
1551–1913

BY THE LATE

G. C. J. VOSMAER

PROFESSOR OF ZOOLOGY IN THE UNIVERSITY OF LEIDEN
MEMBER OF THE ROYAL ACADEMY OF SCIENCE
IN AMSTERDAM

EDITED BY

G. P. BIDDER, M.A., Sc.D.

AND

C. S. VOSMAER-RÖELL

CAMBRIDGE
AT THE UNIVERSITY PRESS
1928

CAMBRIDGE UNIVERSITY PRESS
Cambridge, New York, Melbourne, Madrid, Cape Town, Singapore,
São Paulo, Delhi, Dubai, Tokyo

Cambridge University Press
The Edinburgh Building, Cambridge CB2 8RU, UK

Published in the United States of America by Cambridge University Press, New York

www.cambridge.org
Information on this title: www.cambridge.org/9780521141499

First published 1928
This digitally printed version 2010

A catalogue record for this publication is available from the British Library

ISBN 978-0-521-06715-7 Hardback
ISBN 978-0-521-14149-9 Paperback

ERRATA

(See footnote to Author's Preface.)

p. 19. Lamouroux.   *For* 1825, 1825 (*a*), 1825 (*β*) *read* 1824 (*γ*), 1824 (*δ*), 1824 (*ε*).

p. 19. After the footnote: *add* —G.P.B.

p. 29. Ehrenberg.   *For* 1837 (*γ*) *read* 1837 (*δ*).

             ,, 1837 (*δ*) ,, 1837 (*γ*).

p. 30. Ehrenberg.   ,, 1838 (*γ*) ,, 1838 (*δ*).

             ,, 1838 (*δ*) ,, 1838 (*γ*).

p. 34. Hogg.   ,, 1840 (*β*) ,, 1840 (*δ*).

             ,, 1840 (*δ*) ,, 1840 (*β*).

p. 38. Ehrenberg.   ,, 1842 (*a*) ,, 1842 (*β*).

             ,, 1842 (*β*) ,, 1842 (*a*).

p. 98. Thomson.   ,, 1870   ,, 1870 (*a*).

             ,, 1870 (*a*) ,, 1870.

p. 104. Pagenstecher. ,, Votrag ,, Vortrag.

p. 106. Dawson.   ,, (see p.) ,, (see p. 7).

AUTHOR'S PREFACE

THIS list of titles of books and articles on Sponges is as complete as possible and contains the literature on recent as well as on fossil sponges. Nevertheless I am thoroughly convinced that the list is not really complete; especially with regard to the fossils perhaps many are omitted. I did not include quite popular accounts of no scientific use whatever; if there are such, it is because I knew of the title, without having been able to see the paper. Ordinary text-books on Zoology are as a rule not included in the list, unless there was some special reason for which I thought it advisable. For instance: 1858 Schlegel, *Handleiding beoef. dierkunde* (an obscure but very good little text-book), because here the generic name *Poterion* is established. The papers are arranged chronologically within the years, in alphabetical order of the authors. If there is more than one paper of an author in the same year, Greek letters are added to the second, third, and so on, without any special order. Thus a paper marked 1820κ may have been published before one marked 1820η, and *vice versa**.

This well known system is doubtless the best existent, but a drawback of it is that all primarily depends on the date. But there are examples enough that one author gives a date of "publication" different from that given by another author writing of the very same paper. I have, therefore, most scrupulously tried to verify the real date of appearance of a paper in the volume of a periodical which bears on the title-page simply one year, whereas the various papers were published in different years. This could be done for several cases, but in at least as many cases I was unable to find out the truth.

As far as I know no more or less complete list of papers on recent and fossil *Porifera* has ever been published. Both Lendenfeld and Rauff published rather extensive lists of the most important papers. I have thought it advisable to include in my list also less important papers. On the other hand I did not include abstracts, translations, reviews, etc., as so many quite superfluous numbers. I only mentioned them under the title in some rare cases where the knowledge of their existence might be of use to somebody.

* This preface was not found until the text of several sheets had been printed, and in the early pages (see "Errata") some of the Greek letters were exchanged to bring the papers into their true chronological order within the year. It was then realised that this would falsify references in the Monograph, and after p. 38 Vosmaer's lettering has been left unaltered even when one of the titles has been dropped.—G. P. B.

Lendenfeld's list contains about 1400 entries, not counting the abstracts, etc.; Rauff's list about 600 titles. Since these papers appeared (1889 and 1893) many others have been published; all together the following list contains about 3800 books and papers, in which something about sponges is to be found. It is a considerable number, alas not quite in adequate relation to our knowledge of sponges. Without doubt there are numerous very valuable papers, but I am afraid, that on several papers Erasmus' famous dictum may still be applied: "What a thing it is to cultivate literature! Better far grow cabbages in a garden*."

* Vosmaer's MS ends with the colon after the word "applied:". The above extract from the letter of Erasmus to Cardinal Wolsey (Froude's translation) may be the quotation intended; alternatives are from Ep. ccxli (to Dr Pace), "the empty babble of educated ignorance," or (cited Enc. Brit., "Erasmus," p. 730), "*crescit scribendo scribendi studium.*" But possibly the "dictum" of Erasmus was itself the quotation, from an earlier seeker after truth, which lies deep in the heart of every bibliographer:—"*Faciendi plures libros nullus est finis.*"—G. P. B.

EDITOR'S PREFACE

GUALTHERUS CAREL JACOB VOSMAER, eldest son of Carel Vosmaer (1826–1888, the poet, and author of the great monograph on Rembrandt), was born August 29th, 1854, in Nieuw-Beijerland, Netherlands. A collateral ancestor was the eighteenth-century zoologist Arnout Vosmaer (1720–1799), best remembered for his description and drawings of the orang-outang; Jacob N. Vosmaer, the explorer of Celebes, and Jacob Vosmaer, Professor of Medicine, were great-uncles.

After school in the Hague, G. C. J. Vosmaer entered the University of Leiden in 1873. He studied sponges with F. E. Schulze in Graz during 1879–1880, and in the latter year took his doctor's degree at Leiden. In 1882 he was appointed a zoologist on the staff of the Zoological Station at Naples, a position which he held for six years; during this time he laid the foundation for his *magnum opus*, the Monograph for the *Fauna und Flora* of the Zoological Station, and in 1882 he published the first part of the volume on Sponges in Bronn's *Klassen und Ordnungen*, completed in 1886. Returning to Holland after the death of his father, he edited first a novel, left by his father nearly finished, and in 1889 the translation of the *Odyssey* into Dutch hexameters, left finished but unpublished. In 1889 he became Assistant for Zoology, and in 1895 Lector on Zoology, in the University of Utrecht; he was made a member of the Royal Academy of Science in Amsterdam in 1900, and Professor of Zoology in the University of Leiden in 1903: he died unexpectedly, after an operation, on September 23rd, 1916. Vosmaer married in 1906 Catalina S. Röell, and leaves one son, Carel J. J. G. Vosmaer.

It is the exception when a young man's doctoral dissertation remains a classic, but it is not too much to say that no one ever writes of the canal-system of sponges, and rarely does anyone describe the anatomy of a sponge, without alluding consciously or unconsciously to Vosmaer's *Leucandra Aspera* (1880(a), 1881(γ)). That paper showed for the first time the differing schemes of sponge-structure as parallel series, such as the evolutionist had recognised in other groups of organisms. It also showed, which was then of no less importance, that Haeckel's *Kalkschwämme* (1872) was not only fallible in its conclusions, but untrustworthy in its dogmatic anatomical descriptions; while Vosmaer's following paper from Naples on the same *Leucandra aspera* (1884(α)) attacked successfully Haeckel's conception of a sponge-species.

Perhaps the attempt to determine what it is that may logically be called a species among sponges was the problem which mainly occupied Vosmaer's mind and ran through all his work. After a useful preliminary training in description of the sponges in Leiden Museum (1880) he wrote his valuable Report on the Sponges of the Willem Barents Expedition (1882); then came his great text-book (1882–1886), which obliged him to survey the whole field and weigh the work of all his predecessors and contemporaries. Topsent (1892(ε)) wrote: "Vosmaer résumait d'une façon magistrale l'état des connaissances acquises sur les Spongiaires. A peine publié, cet ouvrage se trouve avoir considérablement vieilli, parce que...des Spongiaires recueillis par le *Challenger*....Vosmaer n'a pu profiter que des Rapports de Poléjaeff." This is true, but now that the Challenger Monographs, for which we watched so eagerly, are also antiques, Vosmaer's *Bronn* remains a mine of clear thought as well as of facts; and its historical review of previous literature increases in value as the originals are less and less studied.

At Naples he plunged into work for a Monograph of Naples Sponges in the *Fauna und Flora* series, and at this he continued laboriously to the end of his life, for thirty-four years. Occasional papers appeared in the *Mittheilungen* of the Naples Station, and then from Holland he published (with Pekelharing) the fine histological description of the collar cells (1893, 1898), which was a *tour de force* fitting for one who had been a disciple of F. E. Schulze. Half a dozen other papers, principally on spicules, culminated in the masterly Monograph on *Spirastrella* (1911 (β)). Two sentences (p. 3) may serve as its summary: "There are described, so far as I know, 36 species of *Spirastrella*, about 8 species, placed into other genera, ought to be included.... Some of these are quite insufficiently described...others cannot remain under *Spirastrella*....Of the remaining 34 there are two... about which it is open to discussion....The rest, according to me, represent specimens of one species, *S. purpurea*." This is not a mere *ipse dixi*; it is proved by exhaustive examination and comparison of 125 specimens; and, in proving it, something of the nature of a sponge-species is shown (as I think) for the first time.

I have been privileged to read the manuscript of the Naples Mono-graph, and what Vosmaer did for *Spirastrella* in the Siboga volume, he has done for genus after genus of the Naples sponges. All the alleged species are patiently compared, with the correlation of their alleged specific characters, and the degree of transition through various examples from the one type-specimen to the other type-specimens. When Madame Vosmaer-Röell has edited this Monograph, it will be no longer necessary for the conscientious spongologist to multiply species, on single specimens, because of the narrow artificial limits which have prevailed in specific definition.

Such exhaustive comparison, however, meant work, and the in-numerable measurements and the thousands of drawings meant work beside and far beyond straightforward description and anatomy. Vosmaer was blessed and cursed with the hatred of imperfection. This made so good and thorough the work of his which we have; this made him again and again turn back the Monograph for revision, in order that it should reach a state of completeness that even his great industry could not compass in a lifetime. When he was leaving Naples in 1888 he showed me the manuscript "practically ready for press," he said, needing only a few additions and some verbal revision and text references. After his death Madame Vosmaer-Röell showed me the manuscript in 1920—still "practically ready for press," apparently needing only a few additions and some verbal revision and text references. Yet in the interval twenty-nine years of hard work had been sucked into its leaves.

Had he not so over-weighted himself with detailed labour, we should have had much more fruit from the clear zoological insight which Vosmaer shows in his writings, published and unpublished. It is my belief that no one has ever known more than he about sponges, and no one has ever done more work upon them. But the artist in him always craved for perfection in what he did, and the scholar for per-fection in what he wrote; for he was by nature artist and scholar as well as zoologist.

I will only add that his high ideals of honour demanded also per-fection, and that he was a loyal friend.

This Bibliography was begun in 1880, when he published a notice in the *Zoologischer Anzeiger* asking for contributions. The idea ap-pealed to Lendenfeld, who successfully anticipated him (1889).

Vosmaer was indignant both with the characteristic opportunism of the author and with the inaccuracies and omissions which he found in the work; but it clearly became useless to publish his Bibliography as it stood, and he told me he was going to recast it with a greater degree of completeness.

I should perhaps warn the reader that I think this completeness is in some ways overdone. Among the earlier references, to geological journals in particular, there are many which will not repay the trouble of taking the volume from the shelf: it will be found merely recorded that "several sponges" were found in a given stratum. Again with regard to accuracy of transcript, the word "by," or the full stop before it, is intended to be inserted after a title only if either occur in the original. I have seen little significance in the point, and, though the attempt to observe it has cost considerable labour, sins of omission and commission will be found in this respect.

Vosmaer has taken trouble, where he could prove that a paper appeared earlier or later than the date on the title-page, to catalogue it under the year of its actual publication. This is important in cases of priority, especially of names of species, but (a) it can only be carried out where the fact happens to have come under notice, in probably a minority of instances of the occurrence; (b) it means that an author who writes his reference conscientiously from an actual title-page may find that in the Bibliography the year given is that of a different paper.

With regard to initials, the attempt has been made to give the name as found on the title, as "Haeckel, E.," adding in brackets "(H. P. A.)," additional initials used by the author elsewhere. Here also inconsistencies will be found, partly because a number of sheets had been passed for press before the need for such a rule was recognised.

I must apologise very much for the imperfection of my editing. When Madame Vosmaer-Röell showed me the MSS, I suggested that the Bibliography, being actually ready in fair copy, should appear at once, and should be printed independently of the Monograph in a handy octavo volume, the honour of editing which I accepted, in anticipation of correcting merely printers' errors and obvious clerical errors. On the proofs, however, Madame Vosmaer's great knowledge enabled her to supply a number of important corrections, largely necessitated by the typing of the "fair copy"; and, while on the first sheet my investigations were confined to entries which interested or attracted, it then became clear that, where possible, every entry must be checked.

An attempt to do this from p. 17 to p. 146 occupied three months, after which the work suffered two long interruptions; and in the later pages the proof has been compared in a less degree with the actual books. The difference in accuracy may not be large. Verification of the earlier titles could not be made even at the Cambridge University Library without aid from the British Museum (Natural History), the Zoological Society, and the Linnean Society; but after 1875 the greater part of the works catalogued are fairly accessible.

Some "curiosities of literature" may be noted. Systematic zoologists may find attention necessary to the late editions of Jonston's *Natural History* (see 1650 and 1769). They do not exist in any English libraries accessible to me, but I have inserted references to them on the faith of Agassiz's Bibliography (Ray Society, 1848–1854), since it is conceivable that the French and German versions, with their new titles and language, may have been recast in a sufficiently binomial form to give authority for the names of species. If so, the result would possibly confirm a number of traditional names, which might otherwise be displaced for combinations used by obscure writers at about the transition of the eighteenth to the nineteenth century.

Those who have to deal with early references to the *Annals and Magazine of Natural History* or its congeners may be saved some trouble by the footnote I have added on p. 34. In the rarer event of anyone wishing to follow a reference to "*Bonn, Sitz. Ber. Niederrhein. Gesell.*" he may find useful the note on p. 101. This gives opportunity to say that my indebtedness to the Royal Society Catalogue of Scientific Papers is great, and my admiration for its astonishing accuracy is unlimited.

The important, though neglected, classification of sponges in S. F. Gray's *Botany* (1821) had been assigned already by Vosmaer to J. E. Gray (Vosmaer, 1883, p. 44). Professor Dendy, in conversation with me, doubted if we have justification for this attribution, but the learned General Secretary of the Linnean Society, Dr Daydon Jackson, pointed out that J. E. Gray claims the work in his own catalogue of his own works.

Bowerbank's Monograph, vol. I, 1866, reprints 1858(α) and 1862 with few changes; and with no acknowledgment, although, of 381 figures, 371 are inferior copies from the Philosophical Transactions. Zittel's Monograph (1877) appeared in three journals and in book-form.

In spite of five months spent in correcting a Bibliography which Vosmaer had already been correcting for thirty years, and in spite of Madame Vosmaer's expert assistance, I am confident that a new editor with some months to spare could find correction to make on every page. For that those who have done similar work will measure the blame. But, if perfection be unattained, yet, thanks to Vosmaer's great learning and untiring diligence, the Bibliography adds a new instrument to the apparatus of the few who investigate these multicellular organisms, interesting if only because they claim their own genealogy from the flagellates, and their own evolution.

Some younger biologists incline now to cut themselves loose from the lengthening chain of literature, and to read nothing that has appeared more than twenty years ago. Whatever they gain in time for research, they lose much. They lose pleasure, because, for the recreation of a biologist in an idle hour, the biology of 100 years ago affords most delightful and charming reading. They lose knowledge, because forgotten observations of great value lie hidden in papers which excited little interest among contemporaries, and consequently have not been passed through the generations on the stream of second-hand quotation. They lose perspective, because we can none of us appreciate the relative value of our own conceptions and beliefs unless we know the history of their growth. We are all unconsciously tied in our reasoning by a number of unconfessed superstitions—traditional beliefs without reasonable foundation. These are at their strongest if we believe that our own generation has made all its own thought, and that no brave man lived before Agamemnon.

Surveying the old literature in the light of modern knowledge, we are warned of the fettering power of unreviewed postulates: such were the dogmas crystallised about the cell-theory thirty years ago, and earlier the assumption that living things must be animals or vegetables. Yet we find in every year writers whose acute reasoning compels our admiration, and we recognise that the intellects of the men who stumbled up the lower rungs of our ladder were as powerful and adept as any which now survey them, pityingly, from a momentary supremacy.

GEO. P. BIDDER

Cambridge
April 16th, 1928

VOSMAER'S BIBLIOGRAPHY OF SPONGES

1551 **Entzel, C.** De re metallica, hoc est, de Origine, Varietate et Natura Corporum Metallicorum, Lapidum, Gemmarum, atq, aliarum, quae ex fodinis eruuntur, rerum ad Medicinae usum deseruientium, libri iii. Auctore Christophoro Encelio...Franc. apud Chr. Egenolphum. 271 pp. With woodcuts*. Ed. ii, 1557.

1552 **Wotton, E.** De differentiis animalium libri decem...Lutetiae Parisiorum apud Vascosanum...Fol. 22, 220 and 26 pp. Vide lib. x, cap. 248, § 1, cap. 251.

1553 **(Belon).** Petri Bellonii cenomani De aquatilibus, Libri duo... Parisiis...448 pp. With several woodcuts. French transl. 1555.

1554 **(Rondelet).** Gulielmi Rondeletii...Libri de Piscibus Marinis, in quibus verae Piscium effigies expressae sunt...Lugduni... Fol. 583 and 37 pp. With numerous woodcuts. French transl. 1558.

1555 **(Rondelet).** Gulielmi Rondeletii...Universae aquatilium Historiae pars altera...242 and 19 pp. With numerous woodcuts. French transl. 1558.

1558 **Gesner, C.** Historiae animalium lib. iiii...Tiguri apud Christoph. Froschoverum...Fol. 20 and 1297 pp. Numerous figures. Ed. ii, 1620. Transl. German by C. Forer, Zürich, 1563, ed. ii, 1575, id. by J. Saur, Frankfurt, 1598.

1583 **Cesalpini, A.** De plantis libri xvi...Florentiae apud Georgium Marescottum. 4°. 621 and 50 pp.

1583 **Dodoens, R.** Stirpium historiae pemptades sex, sive libri xxx. Antverpiae, ex officina Christophori Plantini...Fol. 860 pp. With woodcuts. Ed. ii, 1616.

1598 **Bauhin, J.** Historia novi & admirabilis Fontis Balneique Bollensis in Ducatu Wirtembergico ad acidulas Goepingenses:...a... Montisbeligardi...4°. German transl. M. D. Förtèr, 1602, 4°. Ed. ii, 1605.

1599 **Imperato, F.** Dell' historia naturale di Ferrante Imperato Napolitano. Libri xxviii...In Napoli. Fol. 2 and 791 pp. With several illustrations. Ed. ii, 1672. Latin transl. 1695, 4°.

1605 **Écluse, C. de l'.** Caroli Clusii...Exoticorum Libri Decem...Ex officina Plantiniana Raphalengii...Fol.

1606 **(Aldrovandi).** Ulyssis Aldrovandi...De Reliquis Animalibus exanguibus libri quatuor, post mortem eius editi...Bononiae...Fol. 593 pp. With numerous woodcuts. Ed. ii, 1618; ed. iii, 1623; ed. iv, 1642.

* No date on title-page, but dated 19 Aug. 1551 in preface.

1609 **Boodt, A. B. de.** Gemmarum et Lapidum Historia...Hanoviae...
   4º. 294 and 33 pp. With woodcuts. Ed. ii, 1636.

1612 **Franz, W.** Historia Animalium sacra...Wittembergae...

1623 **Bauhin, K.** Πιναξ theatri botanici...Basiliae...4º.

1631 **Donati, A.** Trattato de semplici, pietre, et pesci marini, che nascono
   del lito di Venetia...Venetia...4º. 120 pp. With illustr.

1633 **Gérarde, J.** The Herball, or general historie of plantes, gathered
   by...Very much enlarged and amended by Thomas Johnson.
   London...Fol. 1630 pp. With woodcuts*.

1635 **Nieremberg, J. E.** Historia Naturae, maxime peregrinae, Libris 16,
   distincta...Antverpiae...Fol. 502 and 108 pp. With woodcuts.

1640 **Parkinson, J.** Theatrum botanicum...London...Fol.·· 9 and 1755 pp.
   With woodcuts.

1642 **Besler, M. R.** Garophylacium Rerum Naturalium...Fol.

1643 **Forsius, S. A.** Minerographia...190 pp.

1650 **Jonston, J.** Historiae naturalis De exanguibus aquaticis libri iv.
   Cum figuris aeneis...Francofurti...Fol. Many editions, reprints
   and versions, including Rouen 1768 and Heidelberg 1769, q.v.

1651 **Bauhin, J. & J. H. Cherler.** Historia plantarum Universalis...
   Auctoribus...Tomus iii. Ebroduni...Fol.

1655 **Worm, O.** Museum Wormianum...Lugduni Batavorum...Fol. 389
   and 13 pp. 2 Pl. and woodcuts.

1656 **Moscardo, L.** Note overo memorie del Museo di Ludovico Moscardo
   ...Padoa...Fol. 306 and 1 pp. With illustr.

1658 **(Piso, W.)** Gulielmi Pisonis...Historiae Naturalis & Medicae Indiae
   occidentalis libri quinque...332 pp. Illustr.

1661 **Jonston, J.** Notitia regni mineralis,...Lipsiae...12º. 101 pp.

1665 **Hooke, R.** Micrographia...London...Fol. 246 pp. Illustr.

1666 **Merret, C.** Pinax rerum naturalium Britannicarum...Londini...
   12º.

1677 **Plot, R.** The Natural History of Oxfordshire...By...Oxford...
   Fol.

1680 **Wagner, J. J.** Historia Naturalis Helvetiae curiosa Tiguri...12º.

1681 **Grew, N.** Musaeum Regalis Societatis...London...Fol.

1684 **Sibbald, R.** Scotia Illustrata...Edinburgi...Fol.

1686 **Ray, J.** Historia plantarum...Tomus primus. Londini...Fol. 983
   and 22 pp.

1688 **Ray, J.** Historia plantarum...Tomus secundus. Londini...Fol.

1690 **Ray, J.** Synopsis methodica stirpium britannicarum...Londini.

1691 **Plukenet, L.** Phytographia...Pars altera. Londini...4º.

* This is the second edition, which is generally quoted.

1694 **Faber, J. M.** De Spongite lapide observatio. In: Miscellan. curios. (3) i, p. 196–200, Pl. III.

1694 **Pomet, P.** Histoire générale des drogues...Paris...Fol.

1694 **Tournefort, J. P.** de. Élémens de botanique, ou méthode pour connaître les plantes. Vol. i. Paris.

1696 **Cupani, F.** Hortus Catholicus...Neapoli...4º.

1696 **Plukenet, L.** Leonardi Pluc'netii Phytographiae Pars quarta cui Nil nisi praemia desunt...Londini...4º.

1696 (α) **Plukenet, L.** Almagestum botanicum...Londini...4º.

1696 **Sloane, H.** Catalogus plantarum quae in Insula Jamaica Sponte proveniunt...Londini...12º.

1697 **Boccone, P.*** Museo di Fisica e di Esperienze variato, e decorato di Osservazioni Naturali...Con una Dissertazione dell' Origine ...e della Prima Impressione delle Produzzioni Marini, come Fucus, Coralline, Zoophite, Spongie...Venezia...4º.

1699 **Cunningham, J.** A Catalogue of Shell, etc. gathered at the Island of Ascention, by...In: Phil. Trans. xxi, p. 295–300.

1699 **Lhwyd, E. Eduardi Luidii.** ...Lithophylacii Britannici Ichnographia ...Londini.

1700 **Tournefort, J. P.** ...Institutiones rei herbariae. Editio altera, gallica longe auctior, quingentis circiter Tabulis aeneis adornata... Parisiis...4º. Ed. i, 1694.

1702 **Scheuchzer, J. J.** Specimen Lithographiae Helveticae curiosae,... Tiguri.

1703 **Loeselius, J.** Flora prussica...Regiomonti...4º.

1704 **Petiver, J.** Garophylacii naturae et artis decades x...Vol. i... Londini.

1704 **Ray, J.** Historiae Plantarum Tomus tertius...Londini...Fol.

1705 **Plukenet, L.** ...amaltheum botanicum...Londini...4º.

1705 **Rumpf, G. E.** D' Amboinsche rariteitkamer...Amsterdam...4º.

1706 **Leeuwenhoek, A. van.** Several Microscopical Observations on the Pumice-Stone, Coral, Spunges etc. In: Phil. Trans. xxiv, p. 2158–2163, Pl. I.

1707 **Geoffroy, C. J.** Analyse chimique de l'éponge de la moyenne espèce. In: Hist. Acad. Paris, 1706, mém. p. 507–508.

1707 **Sloane, H.** A voyage To the Islands Madera, Barbados, Nieves, S. Christophers and Jamaica...By...Vol. i. London...Fol.

1708 **Baier, J. J.** Ορυκτογραφια Norica...Norimbergae...4º.

1708 **Lange, C. N.** Historia Lapidum Figuratorum Helvetiae...Venetiis-Lucernae...4º.

1709 **Buonanni, F.** Musaeum Kircherianum...Romae...Fol.

***

* P. Later changed into Sylvius.

1—2

1709 **Geoffroy, E. F.** Observations sur les analyses du Corail, et de quelques autres plantes piereuses, faites par M. le Comte Marsigli. In: Hist. Acad. Paris, 1708, p. 102–105.

1709 **M(ylius), G. F. M.** Memorabilium Saxoniae subterraneae Paris I... Leipzig, 4º.

1711 (?) **Marsigli, L. F.** Osservazioni naturali intorno al Mare, ed alla Grana detta Kermes.—Brieve Ristretto del Saggio Fisico intorno alla Storia del Mare scritta alla Regia Academia delle Scienze di Parigi...Venezia. 4º. 72 and 4 pp. ?Transl. 1725, 1786.

1712 **Helwing, G. A.** Flora quasimodogenita...4º.

1712 **Petiver, J.** Pterigraphia Americana icones...London...Fol.

1714 **Büttner, D. S.** Coralliographia subterranea...Lipsiae...4º.

1716 **Lochner, J. H.** Rariora Musei Besleriani...Nürnberg...Fol.

1717 **Helwing, G. A.** Lithographia Angerburgica...Regimonti...4º.

1717 **Mercati, M.** Metallotheca Vaticana...Romae...Fol.

1717 **Reneaume, M. L.** Spongia fluviatilis ramosa, fragilis et piscem olens ...Par...In: Hist. Acad. Paris, 1714, mém. p. 231–239, Pl. 9.

1718 **Ruppius, H. B.** ...Flora Jenensis...Francofurti & Lipsiae.

1718 **Scheuchzer, J. J.** Helvetiae Historia Naturalis...Zürich...4º.

1720 **Boerhaave, H.** Index alter Plantarum quae in Horto Academico Lugduno—Batavo aluntur. Lugduni Batavorum...4º.

1720 **Volkmann, G. A.** Silesia Subterranea...Leipzig,..4º.

1722 **Vallisnieri de Vallisnera, A.** De Stella marina discoide. In: Acad. nat. curios. ephem. ix–x, p. 345–348.

1725 **(Marsigli, L. F.)** Histoire Physique de la Mer. Ouvrage enrichi de figures dessinées d'après le Naturel. Par Louis Ferdinand Comte de Marsilli...Amsterdam. Fol. 17 and 173 pp., 52 Pl. Is supposed to be a translation (by P. Le Clercq) from Marsigli 1711. It is provided with a preface of Boerhaave. From this French edition a Dutch translation appeared in 1786, w. t. Natuurkundige beschryving der zeën...'s Gravenhage. Fol. xxvi and 216 pp. Ed. ii hereof appeared 1787 in Leiden.

1729 **Buxbaum, J. C.** Nova plantarum genera. In: Comment. acad. petropol. ii, p. 343–...

1729 **Woodward, J.** An attempt towards a Natural History of the Fossils of England...Tome 1. London.

1734 **Krigel, A.** Dissertatio de Spongiarum apud veteres usu...Lipsiae ...4º. 32 pp.

1734 **Henkel, J. F.** Idea generalis de Lapidum origine per observationes experimenta & consectaria succincte adumbrata. Dresdae & Lipsiae.

1735 **Linné, C. von.** ...Systema Naturae...Lugduni Batavorum...Fol. 14 pp.

1737 **Kundman, J. C.** Rariora Naturae & Artis, item in Re Medica...
von...Breslau und Leipzig. Fol.

1737 **Linné, C. von.** Hortus Cliffortianus...Amstelaedami...Fol.

1737 **(α) Linné, C. von.** Flora Lapponica...Amstelaedami.

1739 **(Bianchi, G.)** Jani Planci ariminensis de conchis minus notis Liber
cui accessit specimen aestus reciproci maris superi ad littus
portumque arimini. Venetiis. 4º. 88 pp., 5 Pl.

1740 **Buxbaum, J. C.** Plantarum minus cognitarum Centuria ιv com-
plectens Plantas circa Byzantium & in oriente observatas per
...Petropoli...4º.

1740 **Royen, A. van.** Florae Leydensis Prodromus, exhibens Plantas quae
in Horto Academico Lugduno-Batavo aluntur. Lugduni-
Batavorum.

1742 **(Bourget, L. & P. Cartier).** Traité des Pétrifications. Avec Figures.
A Paris...4º. Reprinted (?) as: Mémoires pour servir à
l'Histoire Naturelle des Pétrifications dans les quatre parties
du Monde...A la Haye...1742.

1743 **Hebenstreit, J. E.** Museum Richterianum...Lipsiae...Fol.

1744 **Morandi, G.** Historia botanica practica...Mediolani...Fol.

1745 **Jussieu, B. de.** Examen de quelques productions marines qui ont
été mises au nombre des Plantes et qui sont l'ouvrage d'une
sorte d'Insectes de mer. In: Hist. Acad. Paris, 1742, mém.
p. 290–303.

1745 **Linné, C. von.** ...Flora Suecica...Stockholmiae.

1748 **Linné, C. von.** Materia medica, Liber i...Holmiae.

1749 **Brückmann, F. E.** ...Centuria ιι. Epistolarum Itinerariarum.
Accedit Museum Closterianum. Wolffenbuttelae...4º.

1750 **Donati, V.** Della storia naturale marina dell' Adriatico. Saggio del
...In Venezia, Apresso Francesco Storti...4º. Transl. into
German 1775; into French 1758.

1750 **Rumpf, G. E.** ...Herbarium amboinense...Pars sexta...Amste-
laedami...

1751 **Argenville, A. J. D. d'.** Enumerationis fossilium quae in omnibus
Galliae provinciis reperiuntur, tentamina. Paris.

1751 **Gleditsch, J. G.** Systeme des plantes, fondé sur la situation et la
liaison des etamines. In: Hist. Acad. Berlin, 1749, p. 109–136.

1751 **Linné, C. von.** ...Philosophia botanica...Stockholmiae.

1752 **Peyssonel, J. A.** An Account of a manuscript treatise, presented to
the Royal Society intituled, Traité du corail, contenant les
nouvelles découvertes, qu'on a fait sur le corail, les pores, madre-
pores, scharras, litophitons, éponges, et autres corps et productions, que la mer fournit, pour servir à l'histoire naturelle
de la mer...Extracted and translated...by Mr William Watson,
...In: Phil. Trans. xlvii, p. 445–469.

1752 Pontoppidan, E. Det förste forsög paa Norges natuurlige historie forestillende dette Kongeriges Luft, Grund, Fielde, Vande, Växter, Metaller, Steenarter, Dyr...1e deel. Kjöbenhavn...4°. German transl. 1753; English transl. 1755.

1753 Hanow, M. C. ...Von etlichen biesigen Teichschwämmen und den darinn befindlichen wespenartigen Wasserthierchen. In: Hanows Seltenh. der Natur. & Oekon. i, p. 615–635.

1753 Linné, C. von. Musaeum Tessinianum...Holmiae...Fol.

1753 (α) Linné, C. von. ...Species Plantarum...Holmiae.

1755 Argenville, A. J. D. d'. L'Histoire Naturelle éclaircie dans une de ses parties principales, l'Oryctologie, qui traite des terres, des pierres, des métaux, des minéraux, et autres fossiles, ouvrage dans lequel on trouve une nouvelle méthode Latine et Françoise de les diviser, et une notice critique des principaux Ouvrages qui ont paru sur ces matières...par M...A Paris...4°.

1755 Ellis, J. An Essay towards a Natural History of the Corallines and other Marine Products of the like Kind Commonly found On the Coasts of Great Britain and Ireland...London, Millar. 4°. 103 pp. and 39 Plates. Dutch transl. 1756; French transl. 1756; German transl. 1767.

1755 Ginanni, G. Opere postume del Conte...Tomo i...Venezia...Fol.

1755 Guettard, J. E. Mémoire sur quelques corps fossiles peu connus. Par...In: Hist. Acad. Paris, 1751, hist. p. 29–35, mém. p. 239–267, Pl. ix–xvii. Amsterdam Ed. 1758.

1755 Knorr, G. W. Lapides ex celeberrimorum virorum sententia Diluvii universalis testes...i. Theil...Fol.

1756 Browne, P. The civil and natural history of Jamaica...London. Fol.

1757 Baier, J. J. ...Monumenta Rerum Petrificatorum...Norimbergae ...Fol.

1757 Hasselquist, F. ...Iter palaestinum...Stockholm.

1758 Linné, C. von. ...Systema Naturae...Tom. i, Animalia. Editio decima, reformata. Holmiae...

1758 Peyssonel, J. A. New Observations upon the Worms that form Spongos. By...Translated from the Fronch. In: Phil. Trano. l, p. 590–594. On the title-page of vol. l, Pl. 2, is printed 1759, The paper of Peyssonel seems to have appeared in 1758.

1758 Seba, A. Locupletissimi rerum naturalium thesauri accurata descriptio et iconibus artificiosissimis expressio per universam physices historiam...Tomus iii. Fol. There seems to have been published another edition of which vol. iii bears the year 1761 on the title-page.

1758 (**Turgot, E. F.**) Mémoire instructif sur la manière de rassembler, de préparer, de conserver, et d'envoyer les diverses curiosités d'histoire naturelle...Paris...German transl. 1761.

1759 **Guettard, J. E.** Mémoire où on examine en général le terrain, les pierres et les différens fossiles de la Champagne, et de quelques endroits des provinces qui l'avoisinent. In: Hist. Acad. Paris, 1754, mém. p. 435–494.

1759 **Linné, C. von.** ...Systema Naturae...Tom. ii, Vegetabilia...

1760 (**Bianchi, G.**) Jani Planci ariminensis de conchis minus notis liber. Cui accessit specimen aestus reciproci maris superi ad Littus Portumque Arimini. Editio altera...Romae. In aedibus Palladis...4°.

1760 **Meese, D.** Flora Frisica of Lyst der Planten welke in de Provintie Friesland in het Wilde gevonden worden...Door...Franeker.

1760 **Schultze, C. F.** Betrachtung der versteinerten Seesterne und ihrer Theile, Abgefasst von...Warschau und Dresden...4°. iii and 58 pp., 3 Pl.

1761 **Ledermüller, M. F.** ...Mikroskopische Gemüths- und Augen-Ergötzung: Bestehend in Ein Hundert nach der Natur gezeichneten und mit Farben erleuchteten Kupfertafeln, sammt deren Erklärung...Nürnberg. 4°. French transl. 1764; Dutch transl. 1771.

1762 (**Ginanni, F.**) Produzioni naturali che si ritrovano nel museo Ginanni in Ravenna, methodicamente disposte e con annotazioni illustrate. Lucca...4°.

1762 **Ström, H.** Physisk og oeconomisk beskrivelse over fogderiet Söndmör, beliggende i Bergens stift i Norge. 1 Deel. Sörve...4°.

1762 **Walch, J. E. I.** Das Steinreich systematisch entworfen von... i. Theil...Halle...147 pp., 24 Pl.

1762 **White, C.** An account of the topical application of the spunge in the stoppage of haemorrhages. London.

1763 **Bianchi, G.** De duplici Tethyi genere et de manu marina. In: Atti accad. Siena, ii, p. 217–219, Pl. viii.

1764 **Ellis, J.** An Account of the Sea Pen, or Pennatula Phosphorea of Linnaeus; likewise a Description of a new Species of Sea Pen, found on the Coast of South-Carolina, with Observations on Sea-Pens in general. In a Letter to the Honourable Coote Molesworth,...from...In: Phil. Trans. liii, p. 419–435, tab. xix–xxi.

1764 **Gleditsch, J. G.** ...Systema Plantarum a Staminum Situ...Berolini.

1764 **Kirkland, T.** On the Use of Sponge after Amputations. In: Medic. observ. and inq. ii, p. 278.

1764 **Müller, O. F.** Fauna Insectorum Fridrichsdalina...Hafniae et Lipsiae.

1766 **Ellis, J.** On the Nature and Formation of Sponges: In a Letter from...to Dr Solander...In: Phil. Trans. lv, p. 280–289, tab. x–xi*. Transl. Neues Hamb. Mag. xvi (1767), p. 324–337.

1766 **Gunner, J. E.** Flora norvegica. Pars i. Nidrosiae...Fol.

1766 **Pallas, P. S.** Elenchus Zoophytorum...Hagae-comitum apud Petrum van Cleef. Dutch transl. by Boddaert (Lijst der Plant-dieren), 1768, Utrecht, ed. ii Amsterdam, 1798; German transl. Wilkens (Charakteristik der Thierpflanzen), 1787, Nürnberg, 4º.

1767 **Linné, C. von.** ...:Systema Naturae...Editio duodecima, reformata. Tom. i, Pars 2, Insecta, Vermes. Holmiae...

1767 **(Meuschen, F. C.)** Catalogue systématique d'un magnifique cabinet de coquillages et crustacés, rassemblé par feu M. Arnoud Leers...Amsterdam.

1767 **(Romé de L'isle, J. B. L. de & Dugnat).** Catalogue systématique et raisonné des curiosités de la nature et de l'art, qui composent le cabinet de Mr Davila,...Tom. i and Tom. iii. Briasson.

1768 **Boddaert, P.** (Transl. of Pallas) vide 1766 Pallas.

1768 **Gunner, J. E.** Om adskillige Söeswampe (Spongiae). In: Norske vidensk. selsk. skrifter, iv, p. 74– .

1768 **Linné, C. von.** Systema Naturae...Editio duodecima reformata. Tom. iii. Holmiae.

1769 **Dutour, E. F.** Observation sur une éponge de mer dont le noyau était une coquille du genre des vis. In: Hist. Acad. Paris, 1766, mém. p. 39.

1769 **(Jonston, J.,** cura **Ruysch** rendered German Anon.) Naturgeschichte ...Heidelberg. 7 vols. fol. Heilbronn 1772–4 fol. Vide 1650.

1769 **Walch, J. E. I.** Lapides ex celeberrimorum virorum sententia diluvii universalis testes...ii, 2. Nürnberg. Fol. Cont. of Knorr, 1755.

1770 **Guettard, J. E.** Mémoires sur différentes parties des sciences et arts. Par...Paris. 4º. (ii and iii.)

1770 **Strange, J.** An Account of some very perfect and uncommon Speci-mens of Spongiae from the Coast of Italy: In a Letter to James West,...from...In: Phil. Trans. lx, p. 179–183, Pl. vi. Abstr. in: Phil. Trans. abridged xiii (1809), p. 32–34, Pl. i. Ital. transl. in Olivi 1792.

1771 **Pallas, P. S.** ...Reise durch verschiedene Provinzen des Russischen Reichs. Erster Theil. St Petersburg...4º.

1772 **Houttuyn, M.** Natuurlijke historie of uitvoerige beschrijving der Dieren, Planten en Mineralen, volgens het samenstel van den Heer Linnaeus. Door...1 Deel. 17 Stuk...Amsterdam.

1772 **Scopoli, J. A.** ...Flora carniolica, exhibens plantas Carniolae indi-genas. Editio secunda aucta et reformata. Tom. ii. Viennae.

---

* The paper was read Dec. 19, 1765. It is, therefore, hardly possible that the (printed) *publication* can have been before 1766.

9

1776 **Forskål, P.** Icones rerum naturalium, quas in itinere orientali depingi curavit...Post mortem auctoris ad regis mandatum aeri incisas edidit Carsten Niebuhr. Havniae...4º.

1776 **(Nicolson.)** Essai sur l'histoire naturelle de l'Isle de Saint-Domingue ...Paris.

1776 **Maratti, G. F.** De Plantis zoophytis et lithophytis in mari mediterraneo viventibus commentatio. Romae.

1776 **Müller, O. F.** Zoologiae Danicae prodromus, seu animalium Daniae et Norvegiae indigenarum characteres, nomina et synonyma imprimis popularium. Auctore...Havniae. 4º.

1776 **Pallas, P. S.** Reise durch verschiedene Provinzen des Russischen Reichs. Dritter Theil. St Petersburg...4º.

1776 **Walch, J. E. I.** Beyträge zur Naturgeschichte der Saugschwämme. In: Naturforscher, 8, p. 179–213.

1776 **Wy, G. J. van.** Proeve met de spongia in verouderde ulcera. In: Handelingen Servandis civibus, i, p. 339–353.

1777 **Cartheuser, F. A.** Examen chymicum plantae cujusdam aquaticae Badiaga dictae. In: Acta acad. mogunt. 1776, p. 58–60.

1777 **Dicquemare, J. F.** Le reclus marin. In: Observat. phys. ix, p. 356–357.

1777 **Müller, O. F.** Zoologiae Danicae...Fasc. i, continens tab. i–xl. Hafniae...Fol.

1777 **Scopoli, J. A.** Introductio ad historiam naturalem sistens genera lapidum, plantarum, et animalium hactenus detecta...Pragae. 506 pp.

1778 **M(euschen), F. C.** Museum Gronovianum...Lugduni Batavorum... 251 pp.

1778 **Schröter, J. S.** Vollständige Einleitung in die Kenntniss und Geschichte der Steine und Versteinerungen. iii. Theil...Altenburg...4º.

1779 **Dicquemare, J. F.** Sur le Floriforme;—La Coralline spongieuse;—La fausse Éponge. In: Observat. phys. xiii, p. 416–419.

1779 **Fabricius, J. C.** Reise nach Norwegen mit Bemerkungen aus der Naturhistorie und Oekonomie. Hamburg. French transl. 1802.

1779 **Leske, N. G.** Anfangsgründe der Naturgeschichte. i. Theil. Allgemeine Natur- und Thiergeschichte...Leipzig.

1779 **Müller, O. F.** Zoologia Danica...Vol. i. Hafniae et Lipsiae. Ed. ii, 1788.

1780 **Fabricius, O.** Fauna groenlandica...Hafniae et Lipsiae.

1780 **Gallandat, D. H.** Aanmerkingen en Waarnemingen over het nuttig gebruik van de spons in de uitzakkingen der lijfmoeder en schede. In: Verhandel. maatsch. Haarlem, xix, 2, p. 203.

1780 **Schmidel, C. C.** Vorstellung einiger merkwürdigen Versteinerungen mit kurzen Anmerkungen versehen. Nürnberg...4º.

1781 **Gorter, D. de.** Flora septem provinciarum Belgii foederati indigena. Harlemi. x and 378 pp.

1781 **Gronovius, L. F.** Zoophylacium Gronovianum. Fasc. iii...Lugduni Batavorum...Fol.

1781 **Picot de Lapeyrouse, P.** De novis quibusdam Orthoceratitum et Ostracitum speciebus dissertatiuncula...Erlange...Fol.

1782 **Molina, J. I.** Saggio sulla storia naturale del Chili. Bologna. German transl. 1786; French transl. 1789; English transl. 1809.

1782 **Roques de Maumont, J. E.** Mémoires sur les Polypiers de mer... Zelle...4º. German transl. 1783.

1783 **Gosse, H. A.** Mémoire sur cette question: Déterminer les maladies auxquelles sont exposées les doreurs sur métaux et la meilleure manière de les en préserver. Paris.

1783 **Guettard, J. E.** Mémoires sur différentes parties de la Physique, de l'Histoire naturelle, des Sciences et Arts. Tom. iv. Paris...4º.

1784 **Jacob, E.** Testacea minuta rariora, nuperrime detecta in arena littoris Sandvicensis...London...4º.

1784 **Spallanzani, L.** Lettere del Ab....al Sig. Bonnet relativa a diverse produzioni marini. In: Mem. soc. Italian. ii, 2, p. 603–661.

1785 **Cavolini, F.** Memorie per servire alla storia de' polipi marini di... Napoli. 4º. 279 pp., 9 Pl.

1785 **Gosse, H. A.** Mémoire sur cette question: Déterminer la nature et les causes des maladies des ouvriers employés dans la fabrique des chapeaux. Paris.

1785 **Wartmann, B.** Von dem Fisch-brod, von...In: Naturforscher, xxi, p. 113–128.

1786 **Ellis, J.** The Natural History of many curious and uncommon Zoophytes, collected from various parts of the Globe. By the late...Systematically arranged and described by the late Daniel Solander...London. 4º. 206 pp. and 63 Pl.

1786 **(Marsigli, L. F.)** Dutch translation of Marsigli, 1725.

1787 **Cestoni, G.** Memorie concernenti la storia naturale, e la medicina, tratte dalle lettere inedite di...In: Opuscoli scelti, x.

1787 **(Meuschen, F. C.)** Museum Geversianum, sive index rerum naturalium, continens instructissimam copiam pretiosissimorum omnis ex tribus regnis naturae objectorum, quam, dum in vivis erat, comparavit Abrah. Gevers...A la Haye.

1787 **Wartmann, B.** Von dem Fisch-brod. In: Naturforscher, xxii, p. 113–122.

1787 **Wilkens** (transl. of Pallas). Vide 1766, Pallas.

1788 **Herrenschwand, J. F.** Traité des principales et des plus fréquentes maladies externes et internes.

11

1788 **Lane, F.** An Account of two Cases of Bronchocele, wherein the use of Burnt Sponge appeared to have a considerable Effect. In: Med. com. xiii, p. 80-...

1788 **Linnaeus, C.** Systema naturae... Tomus i. See 1791.

1788 **Müller, O. F.** Zoologia Danica... Volumen primum... Ad formam tabularum denuo edidit frater auctoris... Havniae... Fol.

1788 **Schröter, J. S.** Beschreibung einer neuen Spongie des süssen Wassers, Spongia canalium. In: Naturforscher, xxiii, p. 149–158, Pl. II.

1789 **Müller, O. F.** Zoologia danica... Vol. iii. Descripsit et tabulas addidit **Abildgaard, P. C.** Havniae. Fol.

1789 **Poiret, J. L. M.** Voyage en Barbarie, ou Lettres écrites de l'ancienne Numidie, pendant les années 1785 et 1786...; avec un essai sur l'histoire naturelle de ce pays. 2me Partie. Paris...315 pp.

1790 **Esper, E. J. C.** Die Pflanzenthiere in Abbildungen nach der Natur mit Farben erleuchtet nebst Beschreibungen von... Erster Theil... Nürnberg, 1791. 4º. xii and 320 pp. This part appeared 1788–1790 and contains Liefr. 1–6; something about sponges, pp. 15–16, 137, 213, 242–245, 278. Herewith 68 Pl. (Isis, Madrepora, Millepora, Cellepora and Tubipora.)

1790 **Nozeman, C.** Verhandeling over de inlandsche Zoetwater-spongie, Eene Huisvesting der Maskers van Puistenbijteren. Door... In: Verh. Bataafsch Gen. Rotterdam, ix, p. 242–257, 1 Pl. According to Weltner, 1893 (a), p. 212, no sponge but Plumatella.

1791 **Esper, E. J. C.** Oryctographiae Erlangensis specimina quaedam, imprimis spongiarum petrificatarum. In: Nova acta acad. nat. curios. viii, p. 194–204, Pl. VIII.

1791 **Linnaeus, C.** Systema naturae... Tomus i, Pars 6. Editio decima tertia... (Gmelin)... Lipsiae*.

1791 **Poli, I. X.** Testacea Utriusque Siciliae eorumque Historia et Anatome tabulis aeneis illustra a... Tom. i. Parmae... Fol.

1792 **Bruguière, J. G.** Encyclopédie méthodique. Histoire naturelle des vers. Tome premier. Par...4º.

1792 **Olivi, G.** Zoologia Adriatica ossia Catalogo ragionato degli Animali del Golfo e delle Lagune di Venezia; preceduto da una Dissertazione sulla Storia fisica e naturale del Golfo; e accompagnato da Memorie, ed Osservazioni di Fisica Storia naturale ed Economia dell' Abate... Bassano, MDCCXCII. 4º. 334 and xxxi pp., 9 Tavv. Herein transl. of Strange, 1770, and a letter from Vio, entitled: Della natura delle Spongie di Mare, etc.

1792 **Vio, G.** Della natura delle Spongie di Mare, e particolarmente delle più rare, che allignano nel Golfo di Smirne. Lettera del P. Lettore D... In: Olivi, Zoologia Adriatica (1792), p. ix–xxxi.

* According to Apstein, Zool. Ann. xlvii (1916), p. 32, Pars 6 (in which the sponges are mentioned) appeared in 1791. On the title-page of the first volume is printed 1788.

1793 **Linné, C. von.** Systema naturae...Tomus iii. (Gmelin.) Lipsiae.

1793 **Renier, S. A.** Lettera al Sign. Ab. Gius. Olivi sopra il Botrillo, piantanimale marino. In: Opuscoli scleti, xvi, p. 256–267, 1 Pl.

1793 **Vahl, M.** Beskrivelse af en nye Söe-svamp. In: Skrivter naturhist. selskab. ii, p. 51–55.

1794 **Esper, E. J. C.** Die Pflanzenthiere in Abbildungen nach der Natur mit Farben erleuchtet nebst Beschreibungen von...Zweyter Theil...Nürnberg. 4°. 303 pp. This second part appeared 1791–1794 and contains Lief. 7–12. About Sponges, pp. 102, 165–282, 289–294. Spong, Tab. I–VII, VII A, VII B, VIII–IX, IX A, X–XXI, XXI A, XXII–XXIII, XXIII A, XXIV–XLVII, XLVII A, XLVIII–XLIX.

1795 **Berkenhout, J.** Synopsis of the natural history of Great Britain and Ireland, containing a systematic arrangement and concise description of all the animals, vegetables, and fossiles, which have hitherto been discovered in these kingdoms. Vol. i. London.

1795 **Lichtenstein, A. A. H.** Ueber die Badeschwämme. In: Intellig. Bl. zur Alg. Lit. Zeit., no. 140, p...

1796 **Beckmann, J.** Vorbereitung zur Waarenkunde, oder zur Kenntniss der vornehmsten ausländischen Waaren. ii, 1. Göttingen.

1797 **Esper, E. J. C.** Fortsetzungen der Pflanzenthiere in Abbildungen nach der Natur mit Farben erleuchtet nebst Beschreibungen von...Erster Theil...Nürnberg. 4°. 230 pp. The "Fortsetzungen" appeared *before* the third part of the "Pflanzenthiere." Sponges: Spong Tab. L–LXI.

1797 **(Humphrey, G.)** Museum Calonnianum...London.

1797 **Lichtenstein, A. A. H.** Ny Opløsning af det gamle mørke Spørgsmaal: Hvad ere Suesvampene og hvorledes fremkomme de? af Mag... In: Skrivt. Naturhist. selskab. iv, p. 104–120.

1797 **Zeller, S.** Praktische Bemerkungen ueber den Nutzen der Spongien und des kalten Wassers bei chirurgischen Operationen, Wunden und Hemorrhagien. Wien.

1798 **Boddaert, P.** (transl. of Pallas, ed. ii). Vide 1766, Pallas.

1798 **Cuvier, G. L. C. F. D.** Tableau élémentaire de l'histoire naturelle des animaux...Paris.

1798 **Esper, E. J. C.** Fortsetzungen der Pflanzenthiere...von...Zweyter Theil...Nürnberg. 4°. p. 1–24. Sponges: Spong. Tab. LXII–LXIV.

1798 **Fabricius, J. C.** Entomologia systematica...Supplementum. Hafniae.

1798 **Welter, J. J.** Notice sur quelques matières particulières, trouvées dans les substances animales, traités par l'acide nitrique; Par...In: Ann. de chimie, xxix, p. 301–305.

1799 **Faujas de Saint-Fond, B.** Histoire naturelle de la montagne de Saint-Pierre de Maestricht...Paris...4°. Dutch transl. 1803–1804.

1800 **Hatchett, C.** Chemical Experiments on Zoophytes; with some Observations on the component Parts of Membrane. By...In: Phil. Trans. R. Soc. Edinburgh, p. 327–402.

1801 **Draparnaud, J. P. R.** Sur l'Alcyonium domuncula. In: Bull. Soc. philom. ii, p. 169–170.

1801 **Lamarck, J. B. P. A. de Monet.** Système des animaux sans vertèbres ...Paris.

1801 **Ramond de Carbonnière, L. F. E.** Voyage au Mont Perdu et dans la partie adjacente des Hautes-Pyrénées. Paris.

1801 (a) **Ramond de Carbonnière, L. F. E.** Nouveau genre de Polypiers fossiles, Ocellaria. In: Bull. Soc. phylom. ii, p. 177.

1802 **Bosc, L. A. G.** Histoire naturelle des vers...Vol. iii. Paris...18ᵛᵒ.

1802 **Georgi, J. G.** Geographisch-physikalische und naturhistorische Beschreibung des Russischen Reiches...Königsberg, 1797–1802.

1802 **Stewart, C.** Elements of the natural history of the animal Kingdom...Vol. ii. Edinburgh.

1803 **Bory de St Vincent, J. B. G. M.** Essais sur les isles Fortunées et l'antique Atlantide...Paris...4ᵒ.

1803 **Martinet, J. F.** Traité des maladies chroniques. Paris.

1803 **Sage, B. G.** Sur l'organisation intérieure des Alcyonites. In: Journ. de phys. lvii, p. 281–282.

1803 **Tilesius von Tilenau, W. G.** Zwei verschiedene Species in einem Röhrenschwamm vereinigt. In: Mag. neuesten Zust. Naturk. vi, p. 277–296, 1 Pl.

1803 **Treviranus, G. R.** Biologie...ii. Göttingen.

1804 **Duvernoy, G. L.** (Article Alcyon.) In: Dictionn. Sc. natur. i, p. 457–460.

1804 (a) **Duvernoy, G. L.** (Article Balanes.) In: Dictionn. Sc. natur. iii, p. 408–414.

1805 **Tromsdorff, J. B.** Einige Versuche mit dem Meerschwamm. In: Journ. Pharm. xiii, p. 205–216.

1806 **Defrance, J. L. M.** (Article Cappadox.) In: Dictionn. Sc. nat. vi, p. 507.

1806 **Duméril, A. J. C.** Zoologie analytique...Paris.

1806 **Esper, E. J. C.** Fortsetzungen der Pflanzenthiere...von...Zweyter Theil...Nürnberg. 4ᵒ. p. 25–48. Sponges: Spong. Tab. LXV, LXV A, LXVI–LXVIII, LXXIX (read LXIX), LXX.

1806 **Müller, O. F.** Zoologia danica...Vol. iv. (written and added to by **Abildgaard, P. C., Vahe, M., Holten, J. S., Rathke, J.**) Havniae ...Fol.

1806 **Sowerby, J.** British Miscellany, or coloured figures of new rare or little known animal subjects, not before ascertained to be inhabitants of the British Isles...2 vols. London, 1804–1806.

1806 **Turton, W.** A general system of nature, through the three grand Kingdoms of animals, vegetables and minerals. Translated, amended and enlarged from Linné by...Vol. iv. London.

1807 **Renier, S. A.** Tavole per servire alla classificazione e connoscenza degli animali del dottor...Padova...Fol.

1807 **Turton, W.** British fauna...Swansea...12⁰.

1808 **Montagu, G.** Testacea britannica...Supplement...4⁰.

1808 **Parkinson, J.** Organic Remains of a Former World...Vol. i–iii. London. Cf. 1833.

1808 **Walker, J.** Essays on Natural History and Rural Oeconomy. Edinburgh.

1810 **Bertolini, A.** Rariorum Italiae plantarum decas tertia. Accedit specimen Zoophytorum portus lunae. Pisae.

1810 **Péron, F.** Voyage de découvertes aux terres Australes...Paris. 4⁰. and fol. Ed. ii, 1824–25.

1810 **Savigny, M. J. C. L. de.** Description de l'Égypte...Tome ii...Fol.

1811 **Fourcroy & Vauquelin.** Analyse d'une espèce de Madrépore pêché à la sonde à 35 brasses de profondeur aux environs du Cap l'Ewin, et rapporté par M. Péron. Par...In: Ann. Mus. H. N. xviii¹ p. 354–356.

1811 **Jameson (R.).** Catalogue of Animals, of the class Vermes, found in the Frith of Forth, and other Parts of Scotland. By...In: Mem. Werner. Soc. i, p. 556–565*.

1811 **Sowerby, J.** Account of a pit about two miles from Farringdon, commonly called the Farringdon Gravel Pit. In: Trans. Linn. Soc. London, x, p. 405–406.

1811 **Spix, J.** Geschichte und Beurtheilung aller Systeme in der Zoologie ...Nürnberg.

1812 **Bertolini, A.** Specimen Zoophytorum portus lunae. In: Giorn. di fisica, v, p. 462–468.

1812 **Lamarck, J. B. P. A. de Monet.** Extrait du cours de zoologie du Muséum d'histoire naturelle sur les animaux invertébrés... Paris.

1812 **Lamouroux, J. V. F.** Mémoire sur la classification des Polypiers coralligènes non entièrement pierreux...In: Nouveau bull. scienc. soc. philom. iii (v), p. 181–188.

1813 **Bertolini, A.** Specimen Zoophytorum portus lunae. In: Giorn. di fisica, vi, p. 434–449.

1813 **Lamarck, J. B. P. A. de Monet.** Sur les Polypiers empâtés. Par... Suite du mémoire intitulé: sur les Polypiers empâtés. Par... Suite des Éponges. Par...In: Ann. du Muséum, xx, pp. 294–312, 370–386, 432–458.

1813 **Prichard, J. C.** Researches into the Physical History of Mankind. London.

* The article was read 9th Dec. 1809. The first volume of the Mem. Werner. Soc. bears, however, the year 1811 on the title.

1813 **Townsend, J.** The Character of Moses established for Veracity as an Historian, recording Events from the Creation to the Deluge. Vol. i. London and Bath. 4°.

1814 **Bertolini, A.** Specimen Zoophytorum portus lunae. In: Giorn. di fisica, vii, p. 40–66.

1814 **Conybeare, W. D.** On the Origin of a remarkable class of organic impressions occurring in nodules of flint. In: Trans. Geol. Soc. London, ii, p. 328–335, Pl. xiv.

1814 **Oken, L.** Lehrbuch der Naturgeschichte. 3 Band. Lehrbuch der Zoologie. 1e Abtheilung...Jena...4°.

1814 **Rafinesque Schmaltz, C. S.** Delle spugne di Sicilia. In: Specchio delle scienze, i (ii?), p...vide 1818.

1814 **Webster, F.** On some new varieties of fossil Alcyonia. In: Trans. Geol. Soc. London, ii, p. 377–387, Pl. xxvii–xxviii.

1815 **Lamarck (J. B. P. de Monet de).** Suite des polypiers empâtés (dont l'exposition commence au 20e vol. des Annales, p. 294) par...In: Mém. du Muséum, i, p. 69–80, 162–168, 331–340.

1815 **Mantell, G. A.** Description of a fossil Alcyonium, from the Chalk strata near Lewes...In: Trans. Linn. Soc. London, xi, p. 401–407, Pl. xxvii–xxx.

1816 **Blainville, M. H. D. de.** Prodrome d'une nouvelle distribution systématique du règne animal. In: Bull. soc. philom. Paris, p. 105–124. Also in: Journ. Physique, lxxxiii, p. 244–267.

1816 **Blumenbach, J. F.** Specimen Archaeologiae telluris terrarumque imprimis Hannoveranarum alterum. In: Comment. Soc. scient. Götting. iii, p. 3–25.

1816 **Lamarck (J. B. P. A. de Monet).** Histoire naturelle des animaux sans vertèbres, présentant les caractères généraux et particuliers de ces animaux, leur distribution, leurs classes, leurs familles, leurs genres, et la citation des principales espèces qui d'y rapportent ...par...Tome second. Paris. p. 568.

1816 **Lamouroux, J. V. F.** Histoire des Polypiers coralligènes flexibles, vulgairement nommés Zoophytes. Par...A Caen, de l'Imprimerie de F. Poisson...8°.

1816 **Savigny, M. J. C. L. de.** Mémoires sur les animaux sans vertèbres; par...Paris.

1816 **Smith, W.** Strata identified by organised fossils...London...4°.

1817 **Buckland, W.** Description of the Paramoudra...In: Trans. Geol. Soc. London, iv, p. 413–423, Pl. xxiv.

1817 **Cuvier, G. L. C. F. D.** Le règne animal...Tome iv...Paris.

1817 **Goldfusz, G. A.** Ueber die Entwickelungsstufen des Thieres...Nürnberg.

1817 **(Oken, L.)** Cuviers und Okens Zoologien neben einander gestellt. In: Isis, i, coll. 1145–1185. Also separately, Jena, 1821, 4°.

1817 **Sowerby, J.** British Mineralogy. Vol. i–v. 1802–1817.

1818 **Goldfusz, G. A.** Ueber die Classification der Zoophyten, von...In: Isis, i, coll. 1008–1013.

1818 (a) **Goldfusz, G. A.** Probe aus Goldfusz Handbuch der Zoologie... In: Isis, ii, coll. 1670–1676.

1818 **Lamarck, J. B. P. A. de Monet.** Histoire naturelle des animaux sans vertèbres...Tome cinquième. Paris.

1818 **Montagu, G.** An Essay on Sponges, with Descriptions of all the Species that have been discovered on the Coast of Great Britain. By...In: Mem. Werner. Soc. ii, p. 67–122, Pl. iii–xvi*.

1818 **(Oken, L.)** Spongia fluviatilis neue Pflanze. In: Isis, ii, col. 1273.

1818 **Parkinson, J.** Remarks on the fossils collected by Mr Will. Phillips near Dover and Folkstone. In: Trans. Geol. Soc. London, v, 1, p. 52–59.

1818 **Rafinesque, C. S.**† Descriptions of species of Sponges observed on the shores of Long Island. In: Amer. Journ. Sc. & Arts, i, p. 149–151.

1818 **Stewart, C.** Elements of the Natural History of the Animal Kingdom, being an Introduction to the "Systema Naturae" of Linnaeus... ii. Edinburgh-London.

1819 **Bertolini, A.** Amoenitales italicae sistentes opuscula ad rem herbariam et zoologiam Italiae spectantia...Bononiae Typis Annesii de Nobilibus...4º.

1819 **Blainville, M. H. D. de.** (Article Éponge.) In: Dict. Sc. Nat. xv, p. 93–133.

1819 (a) **Blainville, M. H. D. de.** (Article Éphydatie.) In: Dict. Sc. Nat. xv, p. 55–56.

1819 **Fyfe, A.** Account of some experiments, made with the view of ascertaining the different substances from which Iodine can be procured. In: Edinburgh Phil. Journ. i, p. 254–258.

1819 **Lyngbye, H. C.** Tentamen Hydrophytologiae Danicae Hafniae 1819. pp. 248+xxxii cum tabulis aeneis lxx (Spongilla as Echinella acuta.)

1819 **(Oken, L.)** Okens Pflanzensystem. In: Isis, i, coll. 445–474.

1819 **Schweigger, A. F.** Beobachtungen auf naturhistorischen Reisen von...Anatomisch-physiologische Untersuchungen über Corallen; nebst einem Anhange, Bemerkungen über den Bernstein enthaltend...Berlin. 4º.

1819 **Uhlo, A. P.** Dissertatio pharmacologico-medica de Spongia marina ...auct...Lipsiae. 4º. 27 and xii pp. Acad. dissert.

1820 **Blainville, M. H. D. de.** (Article Géodie, Geodia.) In: Dict. Sc. Nat. xviii, p. 351–352.

---

* Some authors (Haeckel, Pick) place this work on 1814, I don't know why; in taking 1818 I follow Carus and Engelmann's Bibl. Zool.
† So in Amer. Journ. Sc. & Arts, 2nd ed. 1819, but Roy. Soc. Cat. gives Rafinesque-Schmaltz and Vosmaer's MS. gives Rafinesque Schmaltz for this title. Compare 1814.
—G.P.B.

1820 **Choulant, L.** Wird salzsaures Natrum durch kohlensaures Ammoniak zersetzt? In: Ann. Phys. lxviii.

1820 **Coindet, C.** Découverte d'un nouveau remède contre le goître. In: Biblioth. Univers. xiv, p. 120–...Also in: Ann. Chimie & Phys. xv, p. 49–59, and in: Journ. Pharm. vi, p. 485–493.

1820 **Fleming, J.** Observations on the natural history of the Sertularia gelatinosa of Pallas. In: Edinburgh Phil. Journ. ii, p. 82–89.

1820 **Goldfuss, G. A.** Handbuch der Naturgeschichte...3 Theil. Handbuch der Zoologie, ii...Nürnberg.

1820 **Schlotheim, E. F. von.** Die Petrefactenkunde auf ihrem jetzigen Standpunkte durch die Beschreibung seiner Sammlung versteinerter und fossiler Ueberreste des Thier- und Pflanzenreichs der Vorwelt erläutert...Gotha. Ed. ii, 1832.

1820 **Schweigger, A. F.** Handbuch der Naturgeschichte der skelet-losen ungegliederten Thiere von...Leipzig.

1821 **(Defrance, J. L. M.** Art. Hippalime.) In: Dict. Sc. Nat. xxi, p. 171.

1821 **Goebel, F.** Jodine in den Schwämmen. In: Repert. Pharm. xi, p. 44–48.

1821 **Gray, S. F.** A Natural Arrangement of British Plants. Vol. i. London, Baldwin. xxviii and 824 pp., 21 Pl. (The description of the sponges is by J. E. Gray.)

1821 **Lamouroux, J. V. F.** Exposition méthodique des genres de l'ordre des polypiers, avec leur description et celle des principales espèces, figurées dans 84 planches...Par...Paris. 4°.

1821 **Miller, J. S.** A natural history of the Crinoidea...Bristol...4°.

1822 **Conybeare, W. D. & W. Phillips.** Outlines of the Geology of England and Wales...Part i. London.

1822 **(Defrance, J. L. M.** Art. Ierée.) In: Dict. Sc. Nat. xxiii, p. 1–3.

1822 **Fleming, J.** The Philosophy of Zoology; or a general View of the Structure, Functions, and Classification of Animals. By... Edinburgh.

1822 **Hardwicke, Th.** Description of a Zoophyte, commonly found about the Coasts of Singapore Island, with a Plate. By...In: Asiatick Researches, xiv, p. 180–181.

1822 **Mantell, G. A.** The Fossils of the South Downs, or illustrations of the geology of Sussex. With 42 Pl. London. 4°.

1822 **Parkinson, J.** Outlines of Oryctology...London. Ed. ii, 1831.

1823 **Conybeare, W. D.** Memoir illustrative of a general Geological Map of the principal mountain chains of Europe. In: Ann. of Phil. (2), v, pp. 1–16, 135–149, 210–218, 278–289, 356–359; vi, p. 214–219.

1823 **Fleming, J.** Gleanings of Natural History, gathered on the Coast of Scotland during a voyage in 1821. By...In: Edinburgh Phil. Journ. ix, p. 248–254.

1823 **Krüger, J. F.** Geschichte der Urwelt, in Umrissen entworfen von...
2 Theil. Quedlinburg und Leipzig.

1824 **Beche, H. F. de la.** On the geology of the coast of France and of the inland country adjoining, from Fécamp...to St Vaast...In: Trans. Geol. Soc. London (2), i, p. 73–89, Pl. ix.

1824 **Bell, Th.** Remarks on the Animal Nature of Sponges. By...In: Zool. Journ. London, i, p. 202-204*.

1824 **B(ertrand?) G(esclin?).** Ardennes.—Rosiers près Grandpré.—Géognosie.—Polypiers fossiles. In: Revue Encyclop. xxi, p. 482–483. Also in: Bull. Sc. Nat. (Férussac), iii, p. 323–324.

1824 **Bory de Saint-Vincent, J. B. G. M.** (Art. Polypes.) In: Encyclop. Méthod., Zoophytes, ii, p. 636–646.

1824 (a) **Bory de Saint-Vincent, J. B. G. M.** (Art. Psychodiaire.) In: Encyclop. Méthod., Zoophytes, ii, p. 657–663.

1824 **Caumont, A. de.** Rapport sur les travaux de la Société linnéenne du Calvados...In: Mém. Soc. Linn. Calvados, p. 1–48.

1824 (a) **Caumont, A. de.** Extrait d'un mémoire sur la Géologie de l'arrondissement de Bayeux. In: Mém. Soc. Linn. Calvados, pp. 67–72, 179–209.

1824 **Defrance, J. L. M.** Tableau des corps organisés fossiles, précédé de remarques sur leur pétrification, par...Paris et Strasbourg.

1824 (a) **Defrance, J. L. M.** [Art. Millépore (Foss.).] In: Dict. Sc. Nat. xxxi, p. 82–85.

1824 (β) **Defrance, J. L. M.** [Art. Alcyon (Foss.).] In: Dict. Sc. Nat. Supplém. i, p. 106–109.

1824 **Deslongchamps, J. A. E.** (Art. Lymnorée.) In: Encyclop. Méthod., Zoophytes, ii, p. 502–503.

1824 **Gray, J. E.** On the situation and Rank of Sponges in the scale of Nature, and on their internal structure. In: Zool. Journ. i, p. 46-52.

1824 **Hall, H. C. van.** Over de polypen en polypenhuizen (Polyparia) welke men tot nu toe in Nederland gevonden heeft, door...In: Recensent. xvii, 2, p. 215–235.

1824 **Keferstein, C.** Ueber das Vorkommen der Formation des Calcaire grossier im nördlichen Teutschland...In: Teutschland geogn.-geol. dargestellt, mit charten...iii, p. 1–23. Weimar.

1824 (**König, C.**) Icones fossilium sectiles. Centuria i...Fol. London.

1824 **Lamouroux, J. V. F.** (Art. Alcyon, Alcyonium.) In: Encyclop. Méthod., Zoophytes, ii, p. 20–38.

1824 (a) **Lamouroux, J. V. F.** (Art. Éponge, Spongia.)...In: Encyclop. Méthod., Zoophytes, ii, p. 326–369.

---

* On the title-page of the first volume of Zool. Journ. is printed 1825. The above article of Thomas Bell, however, appeared in June 1824.

1824 (β) **Lamouroux, J. V. F.** (Art. Éphydatie.) In: Encyclop. Méthod., Zoophytes, ii, p. 322–324.

1824 **Martens, G. M. v.** Reise nach Venedig von...Erster Theil. Von Stuttgart über Ulm, Wien und Triest nach Venedig. Mit einem Kupfer und einer Charte. Ulm...1824. 8°. xiv and 472 pp Zweiter Theil. Venedig. Euganeen. Alpen von Bellano. Tirol. Baiern. Naturgeschichtlicher Anhang. Mit zwei Kupfern und sieben lithographirten Abbildungen. vi and 664 pp.

1824 **Quoy, J. R. C. & P. Gaimard.** Voyage autour du monde...Zoologie par...Paris...4°. Atlas, Fol.

1824 **Taylor, R. C.** On the Alluvial Strata and on the Chalk of Norfolk, and on the fossils by which they are accompanied. In: Trans. Geol. Soc. London (2), i, p. 374–378.

1824 **Vogel, H. A. von.** Bleichung der Badeschwämme. In: Arch. gesammte Naturlehre, i, p. 243–245.

1825 **Lamouroux, J. V. F.** (Art. Hallirhoé.) In: Encyclop. Méthod., Zoophytes, ii, p. 453–454.*

1825 (α) **Lamouroux, J. V. F.** (Art. Hippalime.) In: Encyclop. Méthod., Zoophytes, ii, p. 455.

1825 (β) **Lamouroux, J. V. F.** (Art. Ierée.) In: Encyclop. Méthod., Zoophytes, ii, p. 462–463.

1825 **Boué, A.** Geological distribution of the Fossil Organic Remains enumerated by Baron von Schlotheim. In: Edinburgh Phil. Journ. xii, pp. 142–146, 270–288.

1825 **Bronn, H. G.** System der urweltlichen Pflanzenthiere...Heidelberg ...Fol.

1825 **Defrance, J. L. M.** [Art. Ocellaire (Foss.).] In: Dict. Sc. Nat. xxxv, p. 327–328.

1825 **Desmarest, A. G.** Considérations générales sur la classe des Crustacés et description des espèces de ces animaux, qui vivent dans la mer, sur les côtes, ou dans les eaux douces de la France, par...Paris.

1825 **Desnoyers, J.** Mémoire sur la Craie et sur les Terrains Tertiaires du Cotentin. In: Mém. Soc. d'Hist. Nat. Paris, ii, pp. 176–248, 397–408, Pl. ix.

1825 **Grant, R. E.** Observations and Experiments on the Structure and Functions of the Sponge. By...In: Edinburgh Phil. Journ. xiii, pp. 94–107, 333–346.

1825 (α) **(Grant,** on the Ova of the Sponge.) In: Edinburgh Phil. Journ. xiii, p. 381–383.

1825 **Gray, J. E.** On the chemical composition of sponges. In: Ann. Phil. (2), ix, p. 431–432.

1825 (α) **Gray, J. E.** A Synopsis of the genera of Cirripedes arranged in natural families, with the description of some new species. In: Ann. Phil. (2), x, p. 97–107.

* Following, for the date, Sherborn & Woodward, P.Z.S., 1893, p. 584.

1825 **Lanfossi, P.** Saggio di una storia naturale dei contorni di Mantova. In: Giorn. di Fisica (2), viii, p.

1825 **Latreille, P. A.** Familles naturelles du règne animal,...2e édition. Paris.

1826 **Blainville, M. H. D. de.** (Art. Polypiers empatés.) In: Dict. Sc. Nat. xlii, p. 397.

1826 (α) **Blainville, M. H. D. de.** (Art. Polypiers fluviatiles.) In: Dict. Sc. Nat. xlii, p. 397.

1826 (β) **Blainville, M. H. D. de.** (Art. Psychodiaires.) In: Dict. Sc. Nat. xliii, p. 516–536.

1826 **Defrance, J. L. M.** [Art. Polypiers (Foss.).] In: Dict. Sc. Nat. xlii, p. 377–397.

1826 **Donovan, E.** The naturalist's repository...Vol. v. London*...4º.

1826 **Dutrochet, R. H. J.** L'agent immédiat du mouvement vital devoilé dans sa nature et dans son mode d'action chez les végétaux et chez les animaux; par...Paris-Londres. viii and 228 pp. Vide p. 179–180.

1826 **Goldfuss, G. A.** Petrefacta Musei universitatis regiae Borussicae Rhenanae Bonnensis nec non Hoeninghusiani Crefeldiensis, iconibus et descriptionibus illustrata, oder Abbildungen und Beschreibungen der Petrefakten der Königl. Rhein-Universität zu Bonn...von...1es Heft. Zoophytorum reliquiae Düsseldorf. Fol.

1826 **Grant, R. E.** Notice of a New Zoophyte (Cliona celata, Gr.) from the Frith of Forth...In: Edinburgh New Phil. Journ. i, p. 78–81. Transl. Ann. Sc. Nat. x (1827), p. 162–168; Froriep's Notizen, xvi (1826), Coll. 52–56.

1826 (α) **Grant, R. E.** Observations and Experiments on the Structure and Functions of the Sponge. By...continued from Vol. xiii, p. 346. In: Edinburgh Phil. Journ. xiv, pp. 113–124, 336–341†.

1826 (β) **Grant, R. E.** On the Structure and Nature of the Spongilla friabilis. By...In: Edinburgh Phil. Journ. xiv, p. 270–284.

1826 (γ) **Grant, R. E.** Remarks on the Structure of some Calcareous Sponges. By...In: Edinburgh New Phil. Journ. i, p. 166–170. Transl. Froriep's Notizen, xvi, Coll. 85–88.

1826 (δ) **Grant, R. E.** Observations on the Spontaneous Motions of the Ova of the Campanularia dichotoma, Gorgonia verrucosa, Caryophyllea calycularis, Spongia panicea, Sp. papillaris, cristata, tomentosa, and Plumularia falcata. By...In: Edinburgh New Phil. Journ. i, p. 150–156. Transl. Froriep's Notizen, xv, Coll. 321–327; Ann. Sc. Nat. xiii (1828), p. 52–61.

1826 (ε) **(Grant, R. E.)** On the Siliceous Spicula of two Zoophytes from Shetland. By...In: Edinburgh New Phil. Journ. i, p. 195–196.

* On the title 1827; it appeared in 1826.
† The part of the Journal in which Grant's article was published appeared in 1826; on the title-page of the volume is printed 1827.

1826 (ζ) **Grant, R. E.** Observations on the Structure of some Siliceous Sponges. By...In: Edinburgh New Phil. Journ. i, p. 341–351 Transl. Froriep's Notizen, xvi, Coll. 113–121.

1826 (η) **Grant, R. E.** Observations on the Structure and Functions of the Sponge. By...In: Edinburgh New Phil. Journ. ii, p. 121–141, Pl. ii*. Transl. Ann. Sc. Nat.; Froriep's Notizen, xviii (1827), Coll. 3–10, 19–26.

1826 (θ) **(Grant, R. E.)** Notice of two new species of British Sponges. In: Edinburgh New Phil. Journ. ii, p. 203–204†.

1826 (ι) **Grant, R. E.** On the Spicula of the Spongilla friabilis, Lamarck. In: Edinburgh Phil. Journ. xiv, p. 183–185.

1826 **Link, H. F.** Entwurf eines phytologischen Pflanzensystems nebst einer Anordnung der Cryptophyten. In: Abhandl. Akad. Berlin, 1824, p. 145–149.

1826 **Osler, E.** On burrowing and boring marine animals. By...In: Phil Trans. R. Soc. London, p. 342–371.

1826 **Pylaye, B. de la.** Examen de la question de savoir si les Cristatelles ou Éponges d'eau douce sont des végétaux; par...In: Ann. Soc. Linn. Paris (Mém. vi), iv, p. 407–

1826 **Risso, A.** Histoire naturelle des principales productions de l'Europe méridionale et particulièrement de celles des environs de Nice et des Alpes maritimes; par...Tome cinquième. Paris et Strasbourg. F. G. Levrault...viii and 403 pp., 10 Pl.

1826 **Tilesius, A. von.** Naturhistorische Abhandlungen und Erläuterungen besonders die Petrefacten Kunde betreffend von...Cassel. Fol. xiv and 154 pp., 8 Pl.

1827 **Bosc, L. A. G.** Histoire naturelle des Vers...2 édit. Vol. iii. Paris.

1827 **Defrance, J. L. M.** (Art. Réceptaculite.) In: Dict. Sc. Nat. xlv, p. 5–7.

1827 **Guilding, L.** Observations on the Zoology of the Caribean Islands. In: Zool. Journ. iii, pp. 403–408, 527–544.

1827 **Hogg, J.** On the Natural History of the Vicinity of Stockton-on-Tees. Stockton...

1827 **Nardo, G. D.** Ueber die Nadeln im Innern des Alcyonium lynceum und Cydonium von...In: Zeitschr. f. organ. Physik (Heusinger), i, p. 67–68.

1827 **Raspail, F. V.** Existence dans les végétaux de cristaux d'oxalate de chaux...In: Bull. Sc. Nat. & Géol. xi, p. 376–377. Also in: Revue Medicale, p. 161, and in: Zeitschr. organ. Physik (Heusinger), i, p. 437–438. They are abstracts from a lecture in the "Institut," published in extenso 1828.

---

* On the title-page is printed 1827; the above article of Grant, however, appeared in 1826.
† On the title-page printed 1827; Grant's article is from 1826.

1828 **Audouin & Milne Edwards.** Résumé des Recherches sur les Animaux sans vertèbres, faites aux îles Chausey; par...In: Ann. Sc. Nat. xv, p. 5–19.

1828 **Blainville, M. H. D. de.** (Art. Téthie.) In: Dict. Sc. Nat. liii, p. 285–288.

1828 **Caumont, A. de.** Essai sur la Topographie Géognostique du département du Calvados. In: Mém. Soc. Linn. Normandie, p. 58–242.

1828 **Chiaje, S. delle.** Memorie sulla storia a notomia degli animale senze vertebre del regno di Napoli, scritte da...Corredate di vignetta e di figure incise in rame...Napoli...4º*. xx and 232 pp.

1828 **Dujardin, F.** Notice sur la Constitution géognostique de la Touraine; par...In: Ann. Sc. Nat. xiii, p. 122–134.

1828 **Dutrochet, R. J. H.** Observations sur la Spongille rameuse (Spongilla ramosa, Lamarck, Ephydatia lacustris, Lamouroux); par... In: Ann. Sc. Nat. xv, p. 205–217.

1828 **Ende, W. P. van den.** Lijst van nederlandsche ongewervelde dieren, welke niet in de Nederlandsche fauna of derselver supplementen gevonden worden,...door...In: Natuurk. verhand. maatsch. Haarlem, xvi, 2, p. 301–303.

1828 **Fée, A. L. A.** Cours d'histoire naturelle pharmaceutique...Vol. i. Paris.

1828 **Fleming, J.** A History of British Animals, exhibiting the Descriptive Characters and Systematical Arrangement of the Genera and species of Quadrupeds, Birds, Reptiles, Fishes, Mollusca and Radiata of the United Kingdom...By...Edinburgh-London.

1828 **Forchhammer, J. G.** On the Chalk Formation of Denmark. In: Edinburgh Journ. Sc. ix, p. 56–68.

1828 **Hornemann, G. L.** (Chemical analysis of sponge.) In: Berlin. Jahrbuch. Pharm. xxx, p. 199.

1828 **Jonas, [L. E.?].** (Analysis of sponge.) In: Journ. Chimie Medic. iv, p. 383.

1828 **Lesson, R. P.** Note sur un nouveau zoophyte, nommé Cliona celata, des côtes du golfe de Forth, par R. E. Grant. (Edinburgh New Phil. Journ. avril 1826, p. 78.) In: Bull. Sc. Nat. et de Géol. (Férussac), xiii, p. 263–264.

1828 **Meyen, F. J. F.** Naturgeschichte der Polypen, von...In: Isis, Coll. 1225–1233.

1828 **Münster, G. zu.** Nachtrag zu dem Aufsatze des Professor Germar im Theil 4, Heft 2, dieser Zeitschrift über die Versteinerungen von Solnhofen...In: Teutschland geogn.-geol. Dargest. v, p. 578–581.

* This is the title on the volume, which obviously is the third. Vol. i appeared in 1823; vol. ii in 1825.

23

1828 **Raspail (F. V.).** Expériences de chimie microscopique, ayant pour but de démontrer l'analogie qui existe entre la disposition qu'affecte la silice dans les spongilles et dans certaines éponges, et celle qu'affecte l'oxalate de chaux dans les végétaux; accompagnées de l'anatomie microscopique des spongilles; Par M...(Mémoire lu à la Société d'Histoire naturelle le 22 juin, et à l'Institut le 25 juin 1827*.) In: Mém. Soc. Hist. Nat. Paris, iv, p. 204–237, Pl. 21–22.

1828 (a) **Raspail, F. V.** Analyse physiologique du Spongilla friabilis, par ...In: Bull. Sc. Nat. et Géol. xiii, p. 170–173.

1828 (β) **Raspail, F. V.** Histoire naturelle de l'Alcyonelle fluviatile...et de tous les genres voisins...In: Mém. Soc. Hist. Nat. Paris, iv, p. 75–165, Pl. xii–xvi.

1828 (γ) **Raspail, F. V.** Notes additionelles relatives aux mémoires sur l'Alcyonelle et sur les Spongilles; par...In: Mém. Soc. Hist. Nat. Paris, iv, p. 246.

1828 **Stark, J.** Elements of Natural History...Vol. ii...Edinburgh.

1829 **Beche, H. T. de la.** On the Geology of Tor and Babbacombe Bays, Devon. In: Trans. Geol. Soc. London (2), iii, 1, p. 160–170, Pl. xx.

1829 **Blainville, M. H. D. de.** (Art. Vespetum.)...In: Dict. Sc. Nat. lviii, p. 55.

1829 **Chiaje, S. delle.** Memorie sulla storia e notomia degli animale senza vertebre del regno di Napoli scritte da...Corredate di vignetta e di figure incise in rame. Napoli, stamperia della Societa tipografica†...4º. Atlas in fol.‡

1829 **Defrance, J. L. M.** (Art. Verticillite.) In: Dict. Sc. Nat. lviii, p. 5–6.

1829 **Dujardin, F.** Sur les Poudingues siliceux qui surmontent la craie grossière en Tourraine; Par...In: Ann. Sc. Nat. xvi, p. 110–112.

1829 **Eichwald, (K.) E. von.** Zoologia specialis, quam expositis animalibus, tum vivis, tum fossilibus potissimum Rossiae in universum, et Poloniae in specie, in usum lectionum publicarum in universitate Caesarea Vilnensi habendarum edidit. Paris, i...Vilnae.

1829 **Goldfuss, G. A.** Petrefacta Musei universitatis regiae Borussicae Rhenanae Bonnensis nec non Hoeninghusiani Crefeldiensis, iconibus et descriptionibus illustrata...von Dr...2es Heft... Düsseldorf...Fol. p. 77–164, Pl. xxvi–l.

1829 **(Hisinger, W.)** Esquisse d'un Tableau des Pétrifications de la Suède ...Stockholm...27 pp.

1829 **Hornemann, G. L.** Nachtrag zur chemischen Untersuchung des Meerschwammes. In: Berlin. Jahrb. Pharm. xxxi, p. 65–71.

* Published Sept. 1828.
† Title of Part iv.
‡ On the title of Atlas is printed 1822. The plates appeared probably between 1822 and 1829.

1829 **Kuntzmann, J. H. L.** Ueber die Gehäuse der Blutegel, besonders über den schwammartigen Ueberzug derselben. In: Verhandl. Ges. naturf. Freunde. Berlin, i, pp. 374–381, 413.

1829 **Mantell, G. A.** A tabular Arrangement of the Organic Remains of the County of Sussex. In: Trans. Geol. Soc. London (2), iii. p. 201–216.

1829 **Phillips, J.** Illustrations of the Geology of Yorkshire...By...York ...4⁰.

1829 **Rose, C. B.** On the Anatomy of the Ventriculites of Mantel*. By... In: Mag. Nat. Hist. (London), ii, p. 332–341, fig. 93–101.

1829 (a) **Rose, C. B.** On the Organic Remains of the Diluvium in Norfolk. In: Quart. Journ. Sc. iv, p. 308–314.

1829 **Taylor, R. C.** On the Blackdown Fossils, by...In: Mag. Nat. Hist. ii, p. 294–295.

1830 **Beche, H. T. de la.** Notes on the Geographical Distribution of Organic Remains contained in the Oolitic Series of the Great London and Paris Basin, and in the same Series of the South of France. By...In: Phil. Mag. or Ann. vii, pp. 250–268, 334–351.

1830 **Blainville, M. H. D. de.** (Art. Zoophytes, Zoophyta.) In: Dict. Sc. Nat. lx, p. 1–546.

1830 (a) **Blainville, M. H. D. de.** Dictionnaire des sciences naturelles. Planches. 2ᵉ partie: règne organisé. Zoologie. Vers et Zoophytes, par...Paris.

1830 **[Chevreul, M. E.** Art. Zoophytes (Chim.).] In: Dict. Sc. Nat. lx, p. 547–548.

1830 **Coldstream, J.** Additions to the Natural History of British Animals. By...In: Edinburgh New Phil. Journ. ix, p. 234–241, Pl. ii†, fig. 3–5.

1830 **Eichwald, K. E. von.** Naturhistorische Skizze von Lithauen, Volhynien und Podolien, in geognostisch-mineralogischer, botanischer und zoologischer Hinsicht entworfen, von...Wilna... 4⁰. 256 pp. and 3 Pl.

1830 **Esper, E. J. C.** Die Pflanzenthiere...von...Dritter Theil. Nürnberg. 4⁰. (pp.   .) This part appeared 1805–1830, and contains Lief. 13–17. See for sponges: Alcyon. Tab. viii–xxv. It is possible that some of these plates were published before 1800.

1830 **Grant, R. E.** (Art. Zoophytes.) In: Edinburgh Encyclop. xviii, p. 844.

1830 **Greville, R. K.** Algae britannicae...Edinburgh. lxxxviii and 218 pp., 19 color. Pl.

1830 **Hartmann, F.** Systematische Uebersicht der Versteinerungen Würtembergs...Tübingen...55 pp.

---

* Read Mantell.
† On the Plate erroneously iv.

1830 Hennah, R. On the animal remains found in the Transition Limestone of Plymouth. In: Proc. Geol. Soc. London, i, p. 169–170. Also in: Phil. Mag. or Ann. vii, p. 209–210.

1830 Lesson, R. P. Voyage autour du Monde...Zoologie par...Tome 2, 1ere Partie. Paris.

1830 Link, H. F. Ueber Pflanzenthiere überhaupt und die dazu gerechneten Gewächse besonders...In: Abhandl. Akad. Berlin, p. 109–123, 1 Pl.

1830 Meyen, F. J. F. Nachträgliche Bemerkungen zur Naturgeschichte der Polypen des süssen Wassers, von...In: Isis, Coll. 185–188.

1830 Pidgeon, E. The fossil remains of the Animal Kingdom.

1830 Taylor, R. C. Illustrations of Antediluvian Zoology and Botany. By...In: Mag. Nat. Hist. iii, p. 262–287.

1830 Woodward, S. A synoptical table of British organic remains... London...50 pp. and 1 Pl.

1831 Beche, H. F. de la. A geological Manual; by...London...12º. Ed. ii, 1832.

1831 Benett, E. A catalogue of the Organic Remains of the County of Wilts...Warminster...4º.

1831 Blumenbach, J. F. Handbuch der Naturgeschichte; von...Göttingen. Ed. xii.

1831 Bronn, H. G. Italiens Tertiär-Gebilde und deren organische Einschlüsse...Heidelberg...192 pp.

1831 Deshayes, G. P. Description des coquilles caractéristiques des terrains...Strasbourg.

1831 Hermbstädt, S. F. Ueber das Brom, sein Vorkommen in verschiedenen Substanzen, und die Darstellungen desselben. In: Abhandl. Akad. Berlin, 1828, p. 85–95.

1831 Hisinger, W. Anteckningar i Physik och Geognosi under Resor i Sverige och Norrige. Af...Feinte Häftet...Stockholm. 174 and 7 pp. and 8 Pl.

1831 Lesson, R. P. Illustrations de zoologie...par...7e Livraison. Paris...

1831 Murchison, R. I. Notes on the secondary formations of Germany, as compared with those of England. In: Proc. Geol. Soc. London, i, p. 325–331. Also in: Phil. Mag. or Ann. x, p. 45–51.

1831 Ritgen, F. A. von. Andeutungen zu einer natürlichen Gruppirung der Pflanzenwelt. In: Schriften Ges. Beförd. Naturwiss. Marburg. ii, p. 79–138.

1831 Steiniger, J. Bemerkungen über die Versteinerungen, welche in dem Uebergangskalkgebirge der Eifel gefunden werden. Trier... 44 pp.

1832 Audouin, J. V. & H. Milne Edwards. Recherches pour servir à l'histoire naturelle du Littoral de la France...Tome i...Paris.

1832 **Bertoloni, A.** Memoria del...sopra alcune produzioni naturali nel Golfo della Spezia...In: Mem. Soc. Ital. Sc. Modena, xx, p. 434–...Pl. xxv. Sep. copy, p. 1–15.

1832 **Bley, L. F.** Chemische Untersuchung der Schwammsteine. In: N. Journ. Pharm. xxvi, 2, p. 287–291.

1832 **Johnston, G.** A descriptive Catalogue of the recent Zoophytes found on the Coast of North Durham...In: Trans. Nat. Hist. Soc. Northumberland, ii, 2, p. 239–272, with 6 Pl.

1832 **Kressler, K.** Bleichen der Badeschwämme. In: Journ. Chim. und Physik. lxiv (N. J. iv), p. 371–372.

1832 **Morren, C. F. A.** Responsio ad quaestionem...: "Queritur descriptio coralliorum fossilium in Belgio repertorum," quae praemium reportavit. In: Ann. Acad. Groningan, 1827–28.

1832 **Passy, A.** Description géologique du Département de la Seine-Inférieure...Rouen...4º. 2 vols.

1832 **Raspail, F. V.** Beylage I zur Isis, 1832, Heft i. Réponse aux remarques additionnelles de M. F. I. F. Meyen sur les Polypes d'eau douce ...par...In: Isis, col. 1–6.

1833 **Goldfuss, G. A.** Petrefacta Germaniae...von...Erster Theil. Düsseldorf...Fol.*

1833 **Klöden, K. F. von.** Die Versteinerungen der Mark Brandenburg... Berlin, 1834†.

1833 **Mantell, G. A.** The Geology of the South-East of England. London.

1833 **(Münster, G. zu.)** Verzeichniss der Versteinerungen, welche in der Kreis-Naturalien-Sammlung zu Baireuth vorhanden sind. Baireuth.

1833 **Nardo, G. D.** Auszug aus einem neuen System der Spongiarien, wornach bereits die Aufstellung in der Universitäts-Sammlung zu Padua gemacht ist, von...In: Isis (Oken), Coll. 519–523.

1833 **Parkinson, J.** Organic Remains of a Former World. An examination of the mineralized remains of the vegetables and animals of the antediluvian world; generally termed extraneous fossils. By... Second edition. Vol. i–iii. London. 4º.

1833 **Quoy, J. R. C. & P. Gaimard.** Voyage des découvertes de l'Astrolabe ...Zoologie...Tome iv. Paris. With atlas in folio.

1833 **Sternberg, K. M. von.** Versuch einer geognostisch-botanischen Darstellung der Flora der Vorwelt. v und vi Heft. Leipzig und Prag...Fol.

1833 **Vallon, J. N.** Sur l'Alcyonelle des étangs et sur l'éponge fluviatile. In: Mém. Acad. Dyon. Part. Sc. p. 25–44.

1833 **Woodward, S.** An Outline of the Geology of Norfolk...Norwich, London...64 pp. and 6 Pl.

---

* The Petrefacta must have been published between 1826 and 1833. Blainville, e.g., mentions several species in 1830.

† Appeared in 1833.

1834 **Agassiz, L. J. R.** Abgerissene Bemerkungen über fossile Fische. In:
N. Jahrb. Mineral. p. 379–390.

1834 **Blainville, M. H. D. de.** Manuel d'actinologie ou de zoophytologie,
contenant: 1º Une histoire abrégée de cette partie de la zoologie,
avec des considérations générales sur l'anatomie, la physiologie,
les mœurs, les habitudes et les usages des Actinozoaires; 2º Un
système général d'Actinologie, tiré à la fois des animaux et de
leurs parties solides ou polypiers; 3º Un catalogue des principaux
auteurs qui ont écrit sur ce sujet. Avec un atlas de 100 planches
...Paris... Strasbourg. 8º. viii and 695 pp. It seems that in
1834 it was complete, but with 50 Pl. only.

1834 **Ehrenberg, C. G.** Beiträge zur physiologischen Kenntniss der Corallenthiere im Allgemeinen, und besonders des rothen Meeres,
nebst einem Versuche zur physiologischen Systematik derselben.
In: Abh. K. Akad. Wiss. Berlin. Aus dem Jahre 1832, p. 225–
380*. Also sep. sub titulo: "Die Corallenthiere des rothen Meeres
physiologisch untersucht und systematisch verzeichnet..."
Berlin. 4º. 1834.

1834 **Fischer de Waldheim, G.** Bibliographia palaeontologica Animalium
systematica. Auctore...Ed. ii...Mesquare.

1834 **Johnston, G.** Illustrations in British Zoology. By...In: Mag. Nat.
Hist. London, vii, p. 490–492, woodcut.

1834 **Mérat, F. V. & A. J. de Lens.** Dictionnaire universel de matière
médicale et de thérapeutique générale...par...Tome sixième
...Paris.

1834 **Münster, G. zu.** Ueber das Kalkmergel-Lager von St Cassian in Tyrol
und die darin vorkommenden Ceratiten, Von...In: N. Jahrb.
Mineral. p. 1–15, Pl. i–ii.

1834 (**Nardo, G. D.**) De Spongiis. In: Isis (Oken), Coll. 714–716.

1834 (α) **Nardo, G. D.** Possibile applicazione alle arti degli aghi silicei.
costituenti il tessuto solido di alcuni Spongiali del Mare Adriatico.
In: Giorn. Tecnol. e Belle Arti, p. 83.

1834 **Roget, P. M.** The Bridgewater Treatises...No. v. Animal and
vegetable physiology, considered with reference to natural
theology. Vol. i...London.

1834 **Sommer, P. J.** Ueber den Meerschwamm als Heilmittel gegen Kropf,
von...In: Arch. Chemie und Meteor. viii, p. 82–90.

1835 **Brown, J.** A Synoptical Table of some of the Mineral Substances,
and of the Organic Remains, found in the Gravel at Stanway,
Essex...By...In: Mag. Nat. Hist. viii, p. 349–353.

1835 (**Gervais**, Lettre de M. Paul...sur les éponges d'eau douce.) In:
C. R....p. 260–262, and in: Ann. Sc. Nat. iv, p. 254–255.

---

* I place this article in the year 1834 as this is printed on the title of the volume.
It was read in 1831 and printed Dec. 1833. It is, therefore, not probable that it should
have been *published* in the same month. At any rate I think Topsent is wrong (1888,
p. 4) in mentioning Ehrenberg's paper for the year 1832.

1835 **Grant, R. E.** Outlines of Comparative Anatomy...By...London. p. 1–?*.

1835 (a) **Grant, R. E.** (Art. Animal Kingdom.) In: Cyclop. Anat. and Phys. (Todd), i, p. 107–118, fig. 29.

1835 **Gray, J. E.** (On the Coral known as the Glass Plant.) In: Proc. Zool. Soc. London, iii, p. 63–65. Transl. in: Institut, iii (1835), p. 426; Isis (Oken), 1837, Coll. 128–129.

1835 **Kirby, W.** The Bridgewater Treatises. No. vii. On the power, wisdom, and goodness of God, as manifested in the creation of Animals, and in their History, Habits and Instincts. Vol. i. London.

1835 **Klöden, K. F. von.** Das älteste Naturdenkmal Pommerns. In: Baltische Studien, iii, p. 1-27.

1835 **Kutorga, S.** Beitrag zur Geognosie und Palaeontologie Dorpat's und seiner nächsten Umgebungen...Petersburg.

1835 **Owen, R.** Remarks on the Entozoa, and on the structural Differences existing among them: including Suggestions for their Distribution into other Classes. By...In: Trans. Zool. Soc. London, i, p. 387–394.

1835 **Sars, M.** Beskrivelser og Jagttagelser over nogle maerkelige eller nye i Havet ved den Bergenske Kyst levende Dyr af Polypernes... af...Bergen. 4º.

1836 **Ehrenberg, C. G.** Das Leuchten des Meeres...In: Abh. Akad. Berlin, 1834, p. 411–575, Pl. i–ii.

1836 (a) **Ehrenberg, C. G.** Bestätigung und nähere Bestimmung thierloser lebender Polypenstöcke. In: Ber. Akad. Berlin, i, p. 33.

1836 (β) **Ehrenberg, C. G.** (Ueber fossile Infusionsthiere.) In: Ber. Akad. Berlin, i, p. 50–54.

1836 (γ) **Ehrenberg, C. G.** (Weitere Nachrichten über das Vorkommen fossiler Infusorien.) In: Ber. Akad. Berlin, i, p. 83–84.

1836 (δ) **Ehrenberg, C. G.** Mittheilungen über die in den Feuersteinen bei Delitzsch vorkommenden mikroskopischen Algen und Bryozoen als Begleiter der fossilen Infusorien. In: Ber. Akad. Berlin, i, p. 114–115.

1836 **Fitton, W. H.** On some of the Strata between the Chalk and the Oxford Oolite in the South-east of England...In: Trans. Geol. Soc. London (2), iv, p. 103–388.

1836 **Johnston, G.** The Natural History of British Zoophytes. By... 1. History of Zoophytology. In: Mag. Zool. Bot. i, p. 64–81.

1836 **Lamarck, J. B. P. A. de Monet.** Histoire naturelle des Animaux sans vertèbres...par...Deuxième édition. Revue et augmentée de notes présentant les faits nouveaux dont la science s'est enrichie jusqu'à ce jour; Par MM. G. P. Deshayes et H. Milne Edwards. Tome deuxième. Histoire des Polypes. Paris, J. B. Baillière.

* Cf. 1841 Grant.

1836 **Phillips, J.** Illustrations of the Geology of Yorkshire; or a Description of the Strata and Organic Remains of the Yorkshire Coast. Part i and ii...(Ed. ii.) London. 4º.

1836 **Pusch, G. G.** Polens Palaeontologie...1ste Lieferung. Stuttgart...4º.

1836 **Quenstedt, F. A.** Ueber den Rautenberg bei Schöppenstedt. In: Arch. Naturgesch. ii, 1, p. 254–256.

1836 **Sharpey, W.** (Art. Cilia.) In: Cyclop. Anat. Phys. (Todd), i, p. 606–638.

1836 **Templeton, J.** A Catalogue of the Species of Rayed Animals found in Ireland...In: Mag. Nat. Hist. ix, p. 466–472.

1836 **Zeuker, J. K.** Historisch-topographisches Taschenbuch von Jena und seiner Umgebung...Jena. x and 388 pp.

1837 **Bronn, H. G.** Lethaea geognostica...1. Band. Stuttgart. 1834–1837.

1837 **Ehrenberg, C. G.** Die fossilen Infusorien und die lebendige Dammerde ...Berlin...Fol.

1837 (α) **Ehrenberg, C. G.** Ueber ein aus fossilen Infusorien bestehendes, 1832 zu Brod verbacknes Bergmehl von den Grenzen Lapplands in Schweden. In: Ber. Akad. Berlin, ii, p. 43–45. Also in: Ann. Phys. und Chemie, x, p. 148–150.

1837 (β) **Ehrenberg, C. G.** Ueber den ebenfalls aus mikroskopischen Kiesel-Organismen gebildeten Polirschiefer von Oran in Afrika. In: Ber. Akad. Berlin, ii, p. 59–61. Also in: Ann. Phys. und Chemie, x, p. 635–637.

1837 (γ) **Ehrenberg, C. G.** Ueber das Massenverhältniss der jetzt lebenden Kiesel-Infusorien und über ein neues Infusorien-Conglomerat als Polirschiefer von Jastraba in Ungarn. In: Ber. Akad. Berlin, ii, p. 105–107. Also in: Ann. Phys. und Chemie, xi, p. 555–558. Prelim. Account of 1838.

1837 (δ) **Ehrenberg, C. G.** Ueber ein am südlichen Rande der Lüneburger Haide entdecktes mehr als 28 Fuss mächtiges Infusorien-Lager. In: Ber. Akad. Berlin, ii, p. 165–167.

1837 **Fischer de Waldheim, G.** Oryctographie du gouvernement de Moscou ...Moscou, 1830–1837...Fol. xvii and 202 pp., 62 Pl.

1837 **Hasbach, J. D.** Verzeichniss der im Uebergangskalk in der Gegend von Bensberg vorkommenden Petraefakten; von...In: Arch. Pharm. ix, p. 174–179.

1837 **Hisinger, W.** Lethaea Suecica...Holmiae...4º.

1837 **Johnston, G.** The Natural History of British Zoophytes. By...In: Mag. Zool. and Bot. i, pp. 225–247, 440–448.

1837 **Kröyer, H. N.** De danske Ostersbanker, et Bidrag til Kundskap om Danmarks Fiskerier...Kjöbenhavn.

1837 **Phillips, J.** Cabinet of natural history. Vol. xiii. A Treatise on Geology. By...Vol. i. London.

1837 **Preuss,** . Ueber Schwämme und Schwammkohle; von...In: Arch. Pharm. ix, p. 134–143.

1837 **Turpin, P. J. F.** Analyse ou étude microscopique des différents corps organisés et autres corps de nature diverse qui peuvent, accidentellement, se trouver enveloppés dans la pâte translucide des silex; par...In: C. R. Paris, iv, p. 304–314, 351–362, figs. A–C, C′, D–E. Also in: Ann. Sc. Nat. vii, p. 129–156, Pl. VI–VII.

1838 **Bronn, H. G.** Lethaea geognostica...2. Band. Stuttgart. p. 769–1346, Pl....

1838 **Buch, C. L. von.** Ueber die Muscheln im Granaten-Lager von Trziblitz. In: Arch. Mineral. xi, p. 315–318.

1838 **Costa, O. G.** Cf. 1843.

1838 (α) **Costa, O. G.** Lezioni di zoologia...Napoli...4º.

1838 **Dujardin, F.** Observations sur les Éponges et en particulier sur la Spongille ou Éponge d'eau douce; par...In: Ann. Sc. Nat. (2), x, p. 5–13, 34, Pl. I, fig. 1–5.

1838 (α) **Dujardin, F.** Observations sur les éponges; par...In: C. R. vi (1838), p. 676.

1838 (β) **Dujardin, F.** Observations sur les Éponges. In: Institut, vi, p. 202. Also in: Soc. Philom. Extraits procès-verbaux, p. 81–82.

1838 **Ehrenberg, C. G.** Ueber das Massenverhältniss der jetzt lebenden Kiesel-Infusorien und über ein neues Infusorien-Conglomerat als Polirschiefer von Jastraba in Ungarn. In: Abh. Akad. Berlin, 1836, Physik. Klasse, p. 109–135, Pl. I–II.

1838 (α) **Ehrenberg, C. G.** On the Organic Origin of the Potstones or Paramoudras of Whitlingham, near Norwich. By...In: Ann. or Mag. Nat. Hist. ii, p. 161–162.

1838 (β) **Ehrenberg, C. G.** Nachrichten über drei neue Lager fossiler Infusorien-Schalen und über die schon ältere Gewohnheit des Essens von Infusorien-Erden in Schweden und Finnland. In: Ber. Akad. Berlin, p. 5–8.

1838 (γ) **Ehrenberg, C. G.** Beobachtungen über neue Lager fossiler Infusorien und das Vorkommen von Fichten-Blüthenstaub neben deutlichen Fichtenholz, Hayfischzähnen, Echiniten und Infusorien in volhynischen Feuersteinen der Kreide. In: Ber. Akad. Berlin, p. 102–104.

1838 (δ) **Ehrenberg, C. G.** Ueber dem blossem Auge unsichtbare Kalkthierchen und Kieselthierchen als Hauptbestandtheile der Kreidegebirge. In: Ber. Akad. Berlin, p. 192–200.

1838 **Flourens, M. J. P.** Analyse d'un ouvrage manuscrit intitulé: Traité du corail...par le sieur de Peysonnel...Par...In: Ann. Sc. Nat. ix, p. 334–351.

1838 **Hausmann, J. L. F.** Nachträge zu den Bemerkungen über das im Amte Ebstorf entdeckte Lager einer aus Infusorien-Schaalen bestehenden Kieselerde. In: Götting. gelehrte Anzeigen, p. 1065–1077.

1838 **(Hogg, J.** On the action of light upon the colour of the River Sponge.) In: Abstracts Phil. Trans, iv, p. 72. Also in: Phil. Mag. and Journ. Sc. xiii, p. 457, and in: Ann. or Mag. ii, p. 370.

1838 (a) **Hogg, J.** Observations on the Spongilla fluviatilis. In: Proc. Linn. Soc. London, i, p. 8.

1838 **Johnston, G.** A History of the British Zoophytes. By...Edinburgh ...pp., 44 Pl., 80 woodcuts.

1838 **Jones, F. Rymer.** A General Outline of the Animal Kingdom. By... Part i. London.

1838 **Laurent, J. L. M.** Recherches sur la Spongille fluviatile; par...In: C. R. vii, p. 325–335, 617–619. Also in: Ann. Franc. Étrang. Anat. et Phys. ii, p. 316–318.

1838 **Mantell, G. A.** The Wonders of Geology...2 vols...London...12º.

1838 **Michelotti, G.** Specimen zoophytologiae diluvianae.

1838 **Quenstedt, F. A.** Ueber die Geschiebe der Umgegend Berlins. In: N. Jahrb. Mineral. p. 136–157.

1838 **Troost, G.** Description d'un nouveau genre de fossiles. In: Mém. Soc. Géol. France, iii, p. 87–96, Pl....

1838 **Turpin, P. J. F.** Rapport sur une note de M. Dujardin, relative à l'animalité des Spongilles (Commissaires, MM. Audouin, Turpin rapporteur). In: C. R. vii, p. 556–567, Pl...(no number), fig. 1–6.

1838 **Westwood, J. O.** Anomalous Insect found in Spongilla fluviatilis. (Entomol. Soc. London, 3 Dec. 1838.) In: Mag. Nat. Hist. iii, 1839, p. 200.

1839 **Archiac, E. J. A. d'.** Observations sur le groupe moyen de la formation crétacée. In: Mém. Soc. Géol. France, iii, p. 261–311.

1839 **Bean, W.** A Catalogue of the Fossils found in the Cornbrash Limestone of Scarborough: with Figures and Descriptions of some of the undescribed Species. By...In: Mag. Nat. Hist. iii, p. 57–62.

1839 **Bellamy, J. C.** The Natural History of South Devon. By...Plymouth.

1839 **Buch, C. L. von.** Ueber den Jura in Deutschland. In: Abh. Akad. Berlin, 1837, p. 49–135, Pl. I–III.

1839 **Carpenter, W. B.** Principles of General and Comparative Physiology ...London. Ed. ii, 1841; ed. iii, 1851.

1839 **Charlesworth, E.** Illustrated Zoological Notices. By...In: Mag. Nat. Hist. iii, p. 347–353.

1839 **Costa, O. G.** Intorno alle uova dalle Spugne. In: Corrispond. Zool. i, p. 34–36.

1839 **Ehrenberg, C. G.** Ueber die Bildung der Kreidefelsen und des Kreidemergels durch unsichtbare Organismen. Von...In: Abh. Akad. Berlin, A. d. J. 1838, p. 59–147, Pl. ɪ–ɪv, and 3 tables. Also separately: Die Bildung der europ. lyb. und arab. Kreidefelsen, etc. Leipzig, 1839. Fol.

1839 (α) **Ehrenberg, C. G.** Ueber wieder 2 neue Lager fossiler Kiesel-Infusorien in Frankreich und New-York. In: Ber. Akad. Berlin, p. 30–31.

1839 (β) **Ehrenberg, C. G.** Ueber fossile Infusorien in Süd-amerika. In: Ber. Akad. Berlin, p. 126–127.

1839 **Gould, A. A.** On Spongia called Neptune's Goblet. In: Amer. Journ. Sc. and Arts, xxxvi, p. 386.

1839 **Hagenow, F. von.** Monographie der Rügen'schen Kreide-Versteinerungen, 1. Abtheilung. Phytolithen und Polyparien, von...In: N. Jahrb. mineral. p. 253–296, Pl. ɪv–v.

1839 **Hogg, J.** Additional Observations on the Spongilla fluviatilis. By... In: Proc. Linn. Soc. London, i, p. 36–39.

1839 (α) **(Hogg, J.** Further Observations on the Spongilla fluviatilis, with some remarks on the nature of the Spongiae marinae...by...). In: Proc. Linn. Soc. London, i, p. 226–227. Also in: Ann. or Mag. iii, p. 458–461.

1839 **Laurent, J. L. M.** Règne animal d'après M. de Blainville...par... Tableau supérieurement gravé. Paris...Pl.

1839 (α) **(Laurent, J. L. M.** Sur les mouvements de spongilles très-jeunes et non encore fixées.) In: C. R. Paris, ix, p. 302.

1839 **Lee, J. E.** Notice of Undescribed Zoophytes from the Yorkshire Chalk. By...In: Mag. Nat. Hist. iii, p. 10–17, fig. 1–14.

1839 **Mantell, G. A.** Note on the Chalk Ventriculite figured in page 352. In: Mag. Nat. Hist. iii, p. 604–605, fig. 75.

1839 **Meyen, F. J. F.** Beiträge zur näheren Kenntniss unseres Süsswasser-Schwammes (Spongilla lacustris). Von...In: Arch. Anat. Phys. p. 83–86.

1839 **Murchison, R. I.** The Silurian System...By...London...4º. In two parts.

1839 **Nardo, G. D.** Sopra un nuovo genere di Spugne, le quali perforano le pietri ed i gusci marini. Letta al veneto Ateneo nel giorno 29 Aprile 1839, ed all' Assemblea de' Naturalisti tenutaoi in Pisa il giorno 7 Ottobre dell' anno stesso. In: Ann. delle Scienze del Regno Lombardo-Veneto, ix, p. 221–226. Cf. 1840, Nardo.

1839 **Roemer, F. A.** Die Versteinerungen des norddeutschen Oolithen-Gebirges...Hannover...4º.

1839 (α) **Roemer, F. A.** Bestimmung Wiener Foraminiferen und Entomostraceen und über Kreide-Versteinerungen von Ilseberg am Harze. In: N. Jahrb. Mineral. p. 430–431.

33

1839 **Wetherell, N. T.** A Notice of some undescribed Organic Remains which have recently been discovered in the London Clay Formation. By...In: Mag. Nat. Hist. iii, p. 496–499, Pl. VIII–IX.

1840 **Bowerbank, J. S.** On the Origin and Structure of Chalk-flints and Greensand cherts. In: Proc. Geol. Soc. London, iii, p. 278–281.

1840 **Braun, C. F. W. J.** Verzeichniss der in der Kreis-Naturalien-Sammlung zu Bayreuth befindlichen Petrefacten...Leipzig...4º.

1840 **Duvernoy, G. L.** Note sur une espèce d'Éponge qui loge dans la coquille de l'huître à pied de cheval (Ostrea hippopus Lamarck), en creusant des canaux dans l'épaisseur des valves de cette coquille, par...In: C. R. xi, p. 683–686*.

1840 (a) **Duvernoy, G. L.** Note additionnelle sur les Éponges perforantes; par...In: C. R. Paris, xi, p. 1021–1023.

1840 **Ehrenberg, C. G.** Ueber 274 Arten in dem 1838 erschienenen grösseren Infusorienwerke noch nicht abgebildeter Infusorien. In: Ber. Akad. Berlin, p. 197–219.

1840 **Eichwald, K. E. von.** Geognostische Uebersicht von Esthland und den Nachbar-Gegenden, von...In: N. Jahrb. Mineral. p. 421–430.

1840 (a) **Eichwald, K. E. von.** Ueber das Silurische Schichten-system in Esthland. In: Zeitschr. f. Nat. und Heilk. Akad. St. Petersburg, i, p. 208–210. Also separately?.

1840 **Forbes, E. & J. Goodsir.** Notice of zoological researches in Orkney and Shetland during the month of June 1839. In: Rep. Brit. Assoc. 9 meeting, pt. 2, p. 79–82.

1840 **Geinitz, H. B.** Charakteristik der Schichten und Petrefakten des Sächsischen Kreide-Gebirges. Zweites Heft...Dresden und Leipzig...4º. p. 30–66, Pl. IX–XVI.

1840 **Grube, A. E.** Actinien, Echinodermen und Würmer des Adriatischen und Mittelmeers, nach eigenen Sammlungen beschrieben... Königsberg...4º.

1840 **Gunn, J.** On Paramoudras, and the Drift of Norfolk and Suffolk. In: Proc. Geol. Soc. London, iii, p. 180.

1840 **Hagenow, F. von.** Monographie der Rügen'schen Kreide-Versteinerungen, 2. Abtheilung: Radiarien und Annulaten, von...In: N. Jahrb. Mineral. p. 631–672.

1840 **Hassall, A. H.** Catalogue of Irish Zoophytes. By...In: Ann. and Mag. vi, p. 166–175.

1840 **Hogg, J.** Observations on the Spongilla fluviatilis. In a Letter to the Secretary. By...In: Trans. Linn. Soc. London, xviii, p. 363–368.

* Number 18, in which D.'s article is printed, appeared in 1840.

B. S. 3

1840 (α) **Hogg, J.** Further Observations on the Spongilla fluviatilis; with some Remarks on the Nature of the Spongiae Marinae. In a Letter to the Secretary. By...In: Trans. Linn. Soc. London, xviii, p. 363–407.

1840 (β) **Hogg, J.** On the Action of Light upon the Colour of the River Sponge. By...In: Mag. Nat. Hist. iv, p. 259–268.

1840 (γ) **Hogg, J.** On the Tentacular Classification of Zoophytes. By... In: Ann. or Mag. iv, p. 364–367.

1840 (δ) **Hogg, J.** River-sponge Insect. Correction of a mistake relating to the River-sponge Insect, and to the Freshwater Sponge. By ...In: Ann. and Mag. vi, p. 315–316*.

1840 **Laurent, J. L. M.** Deuxième Mémoire sur la Spongille; par... (Extrait par l'auteur.) In: C. R. Acad. Paris, xi, p. 693–696.

1840 (α) **Laurent, J. L. M.** Recherches sur les Spongilles et spécialement sur leur mode de reproduction; par...In: C. R. Paris, xi, p. 478–479.

1840 (β) **Laurent, J. L. M.** Études des masses spongillaires; par...(Extrait par l'auteur.)...In: C. R. Paris, xi, p. 1048–1050.

1840 (γ) **Laurent, J. L. M.** Résultats d'observations microscopiques faites sur les corps reproducteurs libres et vaguants de l'Ectosperma clavata, et sur les embryons ciliés libres de la Spongille; par... In: C. R. Paris, xi, p. 1050–1051.

1840 (δ) **Laurent, J. L. M.** Nature de la Spongille fluviatile. In: Institut, viii, p. 223. Also in: Soc. Philom. Extraits procès-verbaux, p. 69–70.

1840 (ε) **Laurent, J. L. M.** Spongilles. In: Institut, viii, p. 231–232. Also in: Soc. Philom. Extraits procès-verbaux, p. 71–73.

1840 (ζ) **Laurent, J. L. M.** Reproductions des Spongilles. In: Institut, viii, p. 240–241. Also in: Soc. Philom. Extraits procès-verbaux, p. 73–74.

1840 **Lee, J. E.** New Species of Siphonia from the Yorkshire Chalk. In: Mag. Nat. Hist. iv, p. 46–47, fig. 2–3.

1840 **Murchison, R. I.** Sur les Roches Dévoniennes du Boulonnais. In: Bull. Soc. Géol. France (2), xi, p. 229–256, Pl. II.

1840 **Nardo, G. D.** Sopra un nuovo genere di spongiali silicei intitolato Vioa il quale vive nell' intorno delle pietre e de' guscj marini perforandoli in mille guise. Memoria letta al Veneto Ateneo il giorno 29 Aprile 1829 (sic) ed all' assemblea dei medici e naturalisti tenutasi in Pisa il giorno 7 Ottobre 1839 dal dottor... Venezia...4°. Reprint of Nardo, 1839.

1840 **Roemer, F. A.** Die Versteinerungen des Norddeutschen Kreidegebirges. 1ᵉ Lieferung...Hannover...4°.

1840 (α) **Roemer, F. A.** Ueber das Norddeutsche Kreidegebirge, von...In: N. Jahrb. Mineral, p. 192–195.

* "Mag. Nat. Hist." was united Sept. 1840 with "Ann. or Mag." to form No. xxxiv (beginning Vol. vi) with new title "Ann. and Mag."

1840 **Sedgwick, A. & R. I. Murchison.** On the Distribution and Classification of the older of Palaeozoic Deposits of the North of Germany and Belgium, and their comparison with Formations of the same age in the British Isles. In: Proc. Geol. Soc. London, iii, p. 300–311. Also in: Phil. Mag. and Journ. Sc. xviii (1841), p. 398–409.

1840 (a) **Sedgwick, A. & R. I. Murchison.** On the Physical Structure of Devonshire, and on the Subdivisions and Geological Relations of its older stratified Deposits. In: Trans. Geol. Soc. London, v, 3, p. 633–704, Pl. L–LVIII.

1840 **Steiniger, J.** Geognostische Beschreibung des Landes zwischen der untern Saar und dem Rheine...Trier...4°.

1840 **Thompson, W.** Additions to the Fauna of Ireland. By...In: Ann. and Mag. N. H. v, p. 245–257.

1840 **Troost, G.** Organic remains discovered in the strata of Tennessee. In: 5 Geol. Rep. Tennessee, p. 45–76.

1841 **B(ianconi), G. G.** Sopra alcuni Zoofiti descritti sotto i nomi di Cliona celata (Grant), Vioa (Nardo) i Spongia terebrans (Duvernoy). In: Nuovi Annali Sc. Nat. Anno iii, tomo vi, p. 455–469.

1841 **Bowerbank, J. S.** Description of three species of Sponge, containing some new forms of organization*. In: Microsc. Journ. London, i, p. 161–162. Also in: Ann. and Mag. N. H. viii (1842), p. 393–394.

1841 (a) **Bowerbank, J. S.** (On the Keratose or Horny Sponges of Commerce.) In: Ann. and Mag. N. H. vii, p. 72–74. Reprinted p. 1–4.

1841 (β) **Bowerbank, J. S.** Observations on a Keratose Sponge from Australia. By...In: Ann. and Mag. N. H. vii, p. 129–132, Pl. III.

1841 (γ) **Bowerbank, J. S.** On the Keratose or Horny Sponges of Commerce. By...In: Trans. Micr. Soc. London, i, p. 32–39, Pl. 3.

1841 (δ) **Bowerbank, J. S.** On Moss Agates and other Siliceous Bodies, by...In: Proc. Geol. Soc. London, iii, p. 431–436. Also in: Phil. Mag. and Journ. Sc. xix, p. 542–547.

1841 (ε) **Bowerbank, J. S.** On the Siliceous Bodies of the Chalk, Greensands, and Oolites. In: Trans. Geol. Soc. London, vi, 1, p. 181–194, Pl. XVIII–XIX. Also in: Microsc. Journ. i, pp. 99–103, 113–115, 131–135; and in: Phil. Mag. and Journ. Sc. xviii, p. 220–...

1841 **Chiaje, S. delle.** Descrizione e notomia degli animali invertebrati della Sicilia Citeriore osservati vivi negli anni 1822–1830 da... Tomo quinto Polipi, Amorfi, Descrizione tecniche. Napoli...4°. With atlas in Fol. 165 pp., Pl. 87–181†.

* In a footnote is added: "Abstract of a paper read at the Microscopical Society of London, Nov. 24th, 1841. Communicated by the Author."

† Tomo i–iv appeared in the same year. The plates form Tomo v–vii; there seems to be published also Tomo viii (as a Supplement) in 1844. The book is to be considered as the second enlarged edition of delle Chiaje's Memorie sulla storia e notomia, etc. Several of the copperplates are the same. This 2nd edition appears to be rather rare and very seldom to be had complete. I have seen specimens of the first edition with coloured plates.

1841 **Dujardin, F.** Histoire naturelle des Zoophytes. Infusoires, comprenant la physiologie et la classification de ces animaux, et la manière de les étudier à l'aide du microscope. Par...Ouvrage accompagné de planches. Paris, Roret. 8º. Vide: p. 305–306, Pl. III, fig. 19.

1841 **Duvernoy, G. L.** Note additionnelle sur les Éponges perforantes; par...In: C. R. xi, p. 1021–1023*.

1841 **Ehrenberg, C. G.** Ueber Verbreitung und Einfluss des mikroskopischen Lebens in Süd- und Nordamerika. In: Ber. Akad. Berlin, p. 139–144.

1841 (a) **Ehrenberg, C. G.** Nachtrag zu dem Vortrage über Verbreitung und Einfluss des mikroskopischen Lebens in Süd- und Nord-Amerika. In: Ber. Akad. Berlin, p. 202–209.

1841 (β) **Ehrenberg, C. G.** Ueber ein Lager fossiler mikroskopischer Organismen in Berlin. In: Ber. Akad. Berlin, p. 231–235. Also in: Ann. Physik und Chemie, xxiv, p. 436–444.

1841 (γ) **Ehrenberg, C. G.** Weitere Resultate seiner Untersuchungen über die in Berlin lebenden mikroskopischen unterirdischen Organismen. In: Ber. Akad. Berlin, p. 362–364.

1841 **Glocker, E. F. von.** Ueber die Kalkführende Sandsteinformation auf beiden Seiten der mittleren March, in der Gegend zwischen Kwassitz und Kremsier...In: Nov. Act. Acad. nat. Curios. xix, Suppl. ii, p. 309–334, Pl. IV.

1841 **Gould, A. A.** Report on the Invertebrate Animals of Massachusetts, comprising the Mollusca, Crustacea, Annelidae and Radiata... By...Cambridge, Mass.

1841 **Grant, R. E.** Outlines of Comparative Anatomy...By...London. p. –656†. Porifera: pp. 5-9, 310–313, figs. II, III, IV.

1841 **Hogg, J.** Remarks on the Horny Sponges, with proposed divisions of the Order Spongiae. By...In: Ann. and Mag. viii, p. 3–6.

1841 **Laurent, J. L. M.** Sur les embryons ciliés et libres des Spongilles. In: Institut, ix, p. 242.

1841 **Lereboullet, A.** Sur une espèce d'Éponge perforante qui occupe l'épaisseur des valves de l'Huître commestible (Ostrea edulis)... In: Institut, ix, p. 131–132.

1841 **Leymerie, A.** Mémoire sur le terrain crétacé du département de l'Aube, contenant des considérations générales sur le terrain néocomien. In: Mém. Soc. Géol. France, iv, p. 291–364, Pl. XVI–XVII.

* This paper was read before the Academy on Dec. 28, 1840. It is, therefore, hardly possible that it could be printed and published in that same year. An abstract of the session appeared in the Revue Zoolog. for the year 1840. Though this year appears also on the title as the year of publication, the greater probability is that this is a mistake, for the same reasons as given above.
† Cf. 1835, Grant. The book appeared in seven parts from 1835–1841.

1841 **Münster, G. zu.** Beiträge zur Petrefacten-Kunde. 4tes Heft von Wissmann und Graf...Bayreuth...4°. 152 pp. and 16 Pl.

1841 **Owen, R.** (Description of a New Genus and Species of Sponge, Euplectella Aspergillum.) In: Proc. Zool. Soc. London, ix, p. 3–5. Also in: Ann. and Mag. N. H. viii, p. 222–224.

1841 **Perty, J. A. M.** Allgemeine Naturgeschichte...Band iii. Bern.

1841 **Phillips, J.** Figures and descriptions of the palaeozoic fossils of Cornwall, Devon and West Somerset...London.

1841 **Roemer, F. A.** Die Versteinerungen des norddeutschen Kreidegebirges. 2^e Lieferung...Hannover...4°. p. 49–145, Pl. VIII–XVI.

1841 **Sandberger, G.** Einige Beobachtungen über mehre ältere...Gebirgs-Schichten der Gegend von Weilburg...In: N. Jahrb. Mineral. p. 238–241.

1841 **Stutchbury, S.** Description of a new Sponge from Barbadoes. In: Proc. Zool. Soc. London, ix, p. 86–87. Also in: Ann. and Mag. ix (1842), p. 504–506.

1841 **Thompson, W.** Additions to the Fauna of Ireland. By...In: Ann. and Mag. vii, p. 477–482.

1841 **Weaver, T.** On the Composition of Chalk Rocks and Chalk Marl by invisible Organic Bodies: from the Observations of Dr Ehrenberg. By...In: Ann. and Mag. vii, p. 296–315. Also in: Phil. Mag. and Journ. of Sc. xviii, pp. 375–397, 443–464.

1841 **Zimmermann, K. G.** Ueber die Geschiebe der norddeutschen Ebene ...In: N. Jahrb. Mineral., p. 642–661.

1842 **Archiac, E. J. A. d'.** Description géologique du département de l'Aisne. In: Mém. Soc. Géol. France, v, p. 129–420, Pl. XXI–XXXI.

1842 **Archiac, E. J. A. d' & P. E. P. de Verneuil.** Memoir on the Fossils of the older deposits in the Rhenish provinces:...In: Trans. Geol. Soc. London, vi, 2, p. 303–410, Pl. XXV–XXVII.

1842 **Bowerbank, J. S.** On the spongeous origin of Moss Agates and other siliceous bodies. By...In: Ann. and Mag. x, pp. 9–18, 84–91, Pl. I–III.

1842 **Buckland, W.** Anniversary Address to the Geological Society of London, 1841. In: Proc. Geol. Soc. London, iii, 2, p. 469–. Also in: Phil. Mag. and Journ. Sc. xx, pp. 418–434, 512–528.

1842 **Conrad, F. A.** Descriptions of new species of organic remains belonging to the Silurian, Devonian and Carboniferous systems of the United States. In: Journ. Acad. Sc. Philadelphia, viii, 2, p. 235–280.

1842 **Couch, R. Q.** Observations on the Sponges of Cornwall. In: 10 Ann. Rep. Cornwall Polytech. Soc. p. 41–62.

1842 **Ehrenberg (C. G.).** Ueber einen plastischen Kreidemergel von Aegina aus mikroskopischen Organismen und über die Möglichkeit, durch mikroskopische Untersuchung des Materials den Ursprung gewisser alter ächt-griechischer Kunstdenkmäler aus gebrannten Erde (Terracotten) mit bisher unbekannter Sicherkeit zu bestimmen. In: Ber. Akad. Berlin, p. 263–268.

1842 (a) **Ehrenberg, C. G.** Ueber 3 neue Lager fossiler Infusorien in Frankreich. In: Ber. Akad. Berlin, p. 270–273.

1842 (β) **Ehrenberg, C. G.** Ueber zwei durch die Countess of Caledon in Irland beobachtete Lager fossiler Infusorien. In: Ber. Akad. Berlin, pp. 321, 335–339.

1842 **Fleming, J.** History of British Animals, exhibiting the descriptive Characters and systematical Arrangement of the genera and species of Quadrupeds, Birds, Reptiles, Fishes, Mollusca, and Radiata of the United Kingdom;...by...Second edition. 565 pp. London, Duncan and Malcolm.

1842 **Geinitz, H. B.** Charakteristik der Schichten und Petrefakte des Sächsisch-Böhmischen Kreidegebirges. Drittes Heft...Dresden und Leipzig...4º. pp. i–xxvi, 63–166, Pl. xvii–xxiv.

1842 **Hehl, J. C. L.** Résumé über die Petrefacten Würtembergs in Hinsicht ihrer geognostischen Verhältnisse. Von...In: Schriften St Petersburg Ges. Mineral. i, 2, p. 269–342.

1842 **Johnston, G.** A History of British Sponges and Lithophytes by... Edinburgh, London, Dublin. 8º. xii and 264 pp., 25 Pl.

1842 **Koninck, L. G. de.** Description des animaux fossiles qui se trouvent dans le terrain carbonifère de Belgique, par...Liège, Paris, Bonn. 1842–1844...4º*.

1842 **Kutorga, S.** Beitrag zur Paläontologie Russlands. In: Verh. Mineral. Ges. St. Petersburg. p. 1–34, Pl. i–vi.

1842 **Leymerie, A. & G. P. Deshayes.** Mémoire sur le terrain cretacé du département de l'Aube, contenant des considérations générales sur le terrain néocomien. 2ᵐᵉ partie. In: Mém. Soc. Géol. France, v, 1, p. 1–34, Pl. i–xviii.

1842 **Lyell, C.** On the Silurian Strata in the Neighbourhood of Christiania, in Norway. In: Proc. Geol. Soc. London, iii, p. 465–...

1842 **Macgillavray, J.** Catalogue of the Marine Zoophytes of the Neighbourhood of Aberdeen. By...In: Ann. and Mag. ix, p. 402–409.

1842 **M(ulder, G. J.).** Zijde en spons. In: Scheik. Onderzoek. Laborat. Utrecht, i, p. 486.

1842 **Quenstedt, F. A.** Geognostisches Verhalten Schwäbischer Formationen und deren bemerkenswerthesten Versteinerungen...In: N. Jahrb. Mineral. p. 304–309.

* The part in which sponges are mentioned is from 1842.

1842 **Sandberger, G.** Vorläufige Übersicht über die eigenthümlichen bei Villmar an der Lahn auftretenden jüngeren Kalk-Schichten des älteren (sog. Übergangs-) Formation...von...In: N. Jahrb. Mineral. p. 379–402.

1842 **Sauvage, S. & A. Buvignier.** Statistique minéralogique et géologique du département des Ardennes...

1842 **Sedgwick, A. & R. I. Murchison.** On the Distribution and Classification of the older or Palaeozoic Deposits of the North of Germany and Belgium, and their comparison with Formations of the same age in the British Isles. In: Trans. Geol. Soc. London, vi, 2, p. 221–302, Pl. xxiii–xxiv, xxxviii.

1842 **Vanuxem, L.** Natural History of New York. Part iv. Geology. Vol. iii...by...4°.

1842 **Westwood, J. O.** Description of some Insects which inhabit the tissue of Spongilla fluviatilis. In: Trans. Entomol. Soc. London, iii, 2, p. 105–108.

1843 **Costa, O. G.** Fauna del Regno di Napoli ossia Enumerazione di tutti gli animali che abitano le diverse regioni di questo Regno e le acque che le bagnano, contenente la descrizione de' nuovi o poco esattamente conosciuti con figure ricavate da originali viventi e dipinti al naturale di...(Polipi a polipario flessibile...Tethya. Napoli. 4°. p. 1–24, Pl. i, figs. 1–4, 6, Pl. ii*.)

1843 **Croockewit, J. H.** Over de zamenstelling van spons, door...In: Scheik. Onderzoek. Laborat. Utrecht, ii, p. 1–18.

1843 **Ehrenberg, C. G.** Verbreitung und Einfluss des mikrokopischen Lebens in Süd- und Nord-Amerika. In: Abh. Akad. Berlin, 1841, p. 291–446, Pl. i–iv. Also separately intituled: Amerika. Berlin und Leipzig. Fol.

1843 (α) **Ehrenberg, C. G.** Über 2 neue asiatische Lager fossiler Infusorien-Erden aus russischen Trans-Kaukasien (Grusien) und Siberien. In: Ber. Akad. Berlin, p. 43–49.

1843 (β) **Ehrenberg, C. G.** Die dritte Abtheilung seiner Beobachtungen über die Verbreitung des jetzt wirkenden kleinsten Organischen Lebens in Asien, Australien und Afrika...In: Ber. Akad. Berlin, p. 137–143.

1843 (γ) **Ehrenberg, C. G.** Neue Beobachtungen über den sichtlichen Einfluss der mikroskopischen Meeres-Organismen auf den Boden des Elbbettes bis oberhalb Hamburg. In: Ber. Akad. Berlin, p. 161–167.

1843 (δ) **Ehrenberg, C. G.** Über das Gehalt an unsichtbar kleinen Lebensformen aus einigen von Hrn. Prof. Koch aus Constantinopel eingesandten Proben der Meeres-Ablagerungen im Marmara-Meer und im Bosporus. In: Ber. Akad. Berlin, p. 253–257.

* Costa's Fauna del Regno di Napoli appeared in several parts, between the years 1829 and 1858; each part is paged separately. I have not been able to get certainty about the year in which the part is published wherein Tethya is described; 1834 is generally accepted.

1843 (ε) **Ehrenberg, C. G.** Mittheilungen über seine fortgesetzten Beobachtungen des bedeutenden Einflusses unsichtbar kleiner Organismen auf die unteren Stromgebiethe, besonders der Elbe, Jahde, Ems und Schilde. In: Ber. Akad. Berlin, p. 259–272.

1843 **Fischer de Waldheim, G.** Sur quelques Polypiers fossiles du Gouvernement de Moscou. In: Bull. Soc. Nat. Moscou, xvi, p. 663–670, Pl. xiv–xvi.

1843 (α) **Fischer de Waldheim, G.** Notice sur deux fossiles de Sibérie. In: Bull. Soc. Nat. Moscou, xvi, p. 792–795, Pl. xvii–xviii.

1843 **Geinitz, H. B.** Die Versteinerungen von Kieslingswalda, und Nachtrag zur Charakteristik des sächsisch-böhmischen Kreidegebirges... Dresden und Leipzig. 4º. 23 pp. and 6 Pl.

1843 **Gray, J. E.** Catalogue of Radiated Animals, Annelids and Sponges found in New-Zealand, with description of some new Genera and species lately discovered there. In: Travels New-Zealand (Dieffenbach), ii, p. 264...

1843 **Grube, A. E.** Beschreibung einer auffallenden an Süsswasserschwämmen lebenden Larve. Von... In: Arch. f. Naturgesch. ix, 1, p. 331–337, Pl. x...

1843 **Leuchtenberg, M. E. J. N. von.** Beschreibung einiger neuer Thierreste der Urwelt aus den silurischen Kalkschichten von Zarskoje Selo ...St Petersburg...4º.

1843 **Morris, J.** A catalogue of British fossils...By...London. xii and 222 pp. Ed. ii, 1854.

1843 **Owen, R.** Description of a new Genus and Species of Sponge Euplectella Aspergillum, O. By...In: Trans. Zool. Soc. London, iii, 2, p. 203–205, Pl. xiii.

1843 **Pasquier, A. du.** Sur l'existence du Brôme et de l'Iode dans le Fucus crispus, et sur un procédé propre à y faire reconnaître par une seule expérience, ainsi que dans les Éponges...la présence de ces deux principes. In: Journ. Pharm. et Chimie (3), iii, p. 112–114.

1843 **Peach, C. W.** Synopsis of the land and freshwater Shells, Star Fishes, Sea Urchins, Corallines, Alcyonites, Sponges, and marine Algae of the parish of Gorran, Cornwall. In: 11 Ann. Rep. Cornwall Polytech. Soc. p. 44–46.

1843 **Philippi, R. A.** Beiträge zur Kenntniss der Tertiärversteinerungen des nordwestlichen Deutschlands...Kassel...4º. 88 pp. and 4 Pl.

1843 **Portlock, J. E.** Report on the Geology of the County of Londonderry and of parts of Tyrone and Fermanagh. London...xxxii and 784 pp., 39 Tab.

1843 **Posselt, L.** Ueber die Zusammensetzung der Badeschwämme; von ...In: Ann. Chemie und Pharm. xlv, p. 192–198.

1843 **Quenstedt, F. A.** Das Flözgebirge Würtembergs;...Tübingen...

1843 **Roemer, F. A.** Die Versteinerungen des Harzgebirges...Hannover... 4°.

1843 **Ross, L.** Reisen und Länderbeschreibungen der ältern und neuesten Zeit...25 Lief. Reisen auf den griechischen Inseln des ägäischen Meeres. Von...2er Bd...Stuttgart...212 pp.

1843 **Sandberger, G. & K. L. F. Sandberger.** Ueber das Vorkommen von Versteinerungen im Rotheisenstein von Weilburg an der Lahn von...In: N. Jahrb. Mineral. p. 775-782.

1844 **Anstedt, D. F.** On the Zoological condition of Chalk Flints...By ...In: Ann. and Mag. xiii, p. 241-249.

1844 **Beyrich, H. E.** Ueber die Entwicklung des Flötzgebirges in Schlesien. In: Arch. Mineral. xviii, p. 3-86.

1844 **Bowerbank, J. S.** On three species of Sponge containing some new forms of organization. In: Trans. Microsc. Soc. London, i, p. 63-76, Pl. vi-.

1844 (a) **Bowerbank, J. S.** On the Keratose or Horny Sponges of Commerce. In: Trans. Microsc. Soc. London, i, 1, p. 32-39, Pl. iii.

1844 **Deshayes, G. P.** Sur les fossiles des Pyrénées. In: Bull. Soc. Géol. France, (2), i, p. 576-579.

1844 **Ehrenberg, C. G.** Ueber 2 neue Lager von Gebirgsmassen aus Infusorien als Meeres-Absatz in Nord-Amerika und eine Vergleichung derselben mit den organischen Kreide-Gebilden in Europa und Afrika. In: Ber. Akad. Berlin, p. 57-97.

1844 (a) **Ehrenberg, C. G.** Einige vorläufige Resultate seiner Untersuchungen...über das Verhalten des kleinsten Lebens in den Oceanen und den grössten bisher zugänglichen Tiefen des Weltmeers. In: Ber. Akad. Berlin, p. 182-207.

1844 (β) **Ehrenberg, C. G.** Untersuchungen über die kleinsten Lebensformen im Quellenlande des Eufrats und Araxes, so wie über eine an neuen Formen sehr reiche marine Tripelbildung von den Bermuda-Inseln. In: Ber. Akad. Berlin, p. 253-275.

1844 (γ) **Ehrenberg, C. G.** Über einen deutlichen Einfluss des unsichtbar kleinen organischen Lebens als vulkanisch gefrittete Kieselmasse auf die Massenbildung von Bimstein, Tuff, Trass, vulkanischem Conglomerat und auch auf das Muttergestein des nord-asiatischen Marekanits. In: Ber. Akad. Berlin, p. 324-344.

1844 **Eichwald, K. E. von.** Über silurisch-devonische Schichten in Petersburgischen Gouvernement und auf den Inseln der Ostsee, von... In: N. Jahrb. Mineral. p. 41-48.

1844 **Fischer de Waldheim, G.** Observations sur le genre de polypiers Coeloptychium de Goldfuss. In: Bull. Soc. Nat. Moscou, xvii, p. 276-284, Pl. vii-ix.

1844 **Forbes, E.** Report on the Mollusca and Radiata of the Aegean Sea and on their distribution considered as bearing on Geology. In: Rep. 13 Meet. Brit. Assoc. p. 130-193.

1844 Gervais, P. (Art. Éponges.) In: Dict. Univers. Hist. Nat. v, p. 374.

1844 Laurent, J. L. M. Recherches sur l'hydre et l'éponge d'eau douce...
par...Paris...

1844 MacCoy, F. A Synopsis of the characters of the carboniferous limestone Fossils of Ireland. Dublin. 4º.

1844 Mantell, G. A. The medals of creation...Vol.i...London. Ed.ii,1854.

1844 Murchison, R. I. & P. E. P. de Verneuil. Note sur les équivalents du système permien en Europe...In: Bull. Soc. Géol. France, (2), i, p. 475–517.

1844 Nardo, G. D. Rischiarimenti e rettificazioni ai generi ed a qualche specie della famiglia de' Zoofitari sarcinoidi od Alcionari stabilita dal Sig. de Blainville...In: Nuovi Ann. d. Sc. Nat. (2), ii, p. 460–474. Cf. 1845, Nardo.

1844 (a) Nardo, G. D. Nuova classificazione dei Zoofiti. In: Atti 5 Riun Sc. Ital. p. 430–436; and in: Nuovi Ann. Sc. Ital. iii (1845), p. 104–109. Abstr. and Rev. in: Isis (Oken), 1845, coll. 635–637.

1844 Owen, D. D. Report of a geological exploration of part of Iowa, Wisconsin and Illinois...

1844 Reuss, A. E. Geognostische Skizzen aus Böhmen. Zweiter Band. Das Kreide-Gebirge des westlichen Böhmens...Prag...4º. 304 pp. and 3 Pl.

1844 Roemer, C. F. Das rheinische Uebergangsgebirge...Hannover...4º.

1844 Roemer, F. A. Gänge im Harz; Hilsthon; Rutsch-Flächen im Zechstein; Gebirgsarten vom Swanriver in Neuholland. In: N. Jahrb. Mineral. p. 56–58.

1844 Scouler, [J.]. [Not] in: Proc. Dublin Nat. Hist. Soc. i. p... [Query, 1841 Dub. J. Med. Chem. Sci. xviii, p. 398? or 1838 D. Geol. Soc. J. i, p. 382?]

1844 Thompson, W. Additions to the Fauna of Ireland. By...In: Ann. and Mag. N. H. xiii, p. 430–440.

1844 Wood, S. V. Descriptive Catalogue of the Zoophytes from the Crag. By...In: Ann. and Mag. N. H. xiii, p. 10–21.

1845 Anderson, J. (Art. Nervous System.) In: Cyclop. Anat. and Phys. (Todd), iii, p. 601–626.

1845 Austin, F. Note on Mr Bowerbank's Paper on the genus Dunstervillia (Bowerbank), with Remarks on the Ischadites Königii, the Tentaculites and the Conularia. By...In: Ann. and Mag. N. H. xv, p. 406–407.

1845 Bailey, J. W. Notice of some new localities of Infusoria, fossil and recent. In: Amer. Journ. Sc. and Arts, xlviii, p. 321–343, Pl. iv.

1845 Bowerbank, J. S. Observations on the Spongiadae, with descriptions of some new genera. By...In: Ann. and Mag. N. H. xvi, p. 400–410, Pl. xiii–xiv. Reprinted 1850, p. 1—12, with 1845 (a) and the plates.

1845 (a) **Bowerbank, J. S.** Description of a new genus of Calcareous Sponge. By...In: Ann. and Mag. N. H. xv, p. 297–300, Pl. xvii.

1845 **Brandt, J. F.** Second rapport sur l'expédition zoologique et paléontologique dirigée par lui. In: Bull. Acad. St Pétersbourg, iii, p. 74–84.

1845 **Ehrenberg, C. G.** Neue Untersuchungen über das kleinste Leben als geologisches Moment. In: Ber. Akad. Berlin, p. 35–87.

1845 (a) **Ehrenberg, C. G.** Vorläufige zweite Mittheilung über die weitere Erkenntniss der Beziehungen des kleinsten organischen Lebens zu den vulkanischen Massen der Erde. In: Ber. Akad. Berlin, p. 133–157.

1845 (β) **Ehrenberg, C. G.** Nachrichten über das kleinste organische Leben an mehreren bisher nicht untersuchten Erdpunkten. In: Ber. Akad. Berlin, p. 304–321.

1845 (γ) **Ehrenberg, C. G.** Einige Zusätze zu seinen letzten Mittheilungen über die mikroskopischen Lebensformen von Portugall...In: Ber. Akad. Berlin, p. 357–377.

1845 (δ) **Ehrenberg, C. G.** Über einen am 15. Mai 1830 in Malta gefallenen atmosphärischen Staub...In: Ber. Akad. Berlin, p. 377–381.

1845 (ε) **Ehrenberg, C. G.** Untersuchung des am 2 Sept. d. J. auf und bei den Orkney-Inseln gefallenen Meteorstaubes...In: Ber. Akad. Berlin, p. 398–405.

1845 **Gressly, A.** Übersicht der Geologie des nordwestlichen Aargau's von...In: N. Jahrb. Mineral. p. 153–163.

1845 **Hogg, J.** Additional Remarks on the Spongilla fluviatilis. In: Proc. Linn. Soc. London, i, p.

1845 **Klipstein, A. von.** Mittheilungen aus dem Gebiete der Geologie und Palaeontologie. 1. Band...Giessen...4°. 311 pp. and 20 Pl.

1845 **Lonsdale, W.** Report on the Corals from the Tertiary Formations of North America...by...In: Quart. Journ. Geol. Soc. London, i, p. 495–533.

1845 **Lyell, C.** Observations on the White Limestone and other Eocene or Older Tertiary Formations of Virginia, South Carolina, and Georgia. In: Quart. Journ. Geol. Soc. London, i, p. 429–442.

1845 **Mantell, G. A.** Notes of a Microscopical Examination of the Chalk and Flint of the South-East of England;...By...In: Ann. and Mag. N. H. xvi, p. 73–88.

1845 **Murchison, R. I., P. E. P. de Verneuil, & A. von Keyserling.** The Geology of Russia and the Ural Mountains. Vol. ii. London... 4°. 512 pp. and 50 Pl.

1845 **Nardo, G. D.** Rischiarimenti e rettificazioni ai generi ed a qualche specie della famiglia de' Zoofitari sarcinoidi od alcionari stabilita dal Sig. de Blainville...In: Ann. Sc. Regno Lombardo-Veneto, xiv, p. 3–12. Reprint of 1844, Nardo.

1845 (a) **Nardo, G. D.** Illustrazioni relative al genere Vioa Nardo. In: Atti 6 Ruin. Sc. Ital. pp. 372, 428.

1845 **Reuss, A. E.** Die Versteinerungen der böhmischen Kreideformation beschrieben von...Stuttgart...4º. 58 pp. and 13 Pl.

1845 **Sandberger, G.** Schalstein mit Versteinerungen und Porphyr-Geschieben bei Weilburg. In: N. Jahrb. Mineral. p. 457–458.

1845 **Zeuschner, L.** Die Glieder des Jura an der Weichsel. In: Arch. Mineral. xix, p. 605–626.

1846 **Archiac, E. J. A. d'.** Description des fossiles recueillis par M. Thorent dans les couches à Nummulines des environs de Bayonne. In: Mém. Soc. Géol. France, (2), ii, p. 189–217, Pl. v–ix.

1846 **Bruckmann,** . Cidaris vesiculosa u. a. Versteinerungen aus Neocomien bis Kreide in der Sentis-Kette...In: N. Jahrb. Mineral. p. 716–717.

1846 **Costa, O. G.**

1846 **Dana, J. D.** United States exploring Expedition,...Vol. vii. Zoophytes by...Philadelphia,...4º.

1846 **Duchassaing de Fonbressin, P. & H. Michelin.** Note sur deux polypiers de la famille des Coraux appartenant aux genres Solanderia et Pterogorgia. In: Revue Zool. p. 218-220.

1846 **Ehrenberg, C. G.** Über die geformten unkrystallinischen Kieseltheile von Pflanzen, besonders über Spongilla Erinaceus in Schlesien und ihre Beziehung zu den Infusorienerde-Ablagerungen des Berliner Grundes. In: Ber. Akad. Berlin, p. 96–101.

1846 (a) **Ehrenberg, C. G.** Weitere Untersuchungen des mikroskopischen organischen Verhältnisses zu den vulkanischen Ablagerungen beim Laacher-See am Rheine,...In: Ber. Akad. Berlin, p. 158–173, Pl. IA, IB, and II.

1846 (β) **Ehrenberg, C. G.** Über die Beziehungen des kleinsten organischen Lebens zu den Auswurfsstoffen des Imbaburu Vulkans in Quito ...In: Ber. Akad. Berlin, p. 189-207.

1846 (γ) **Ehrenberg, C. G.** Mittheilung seiner mikroskopischen Untersuchungen des Scirocco-Staubes...In: Ber. Akad. Berlin, p. 319-328.

1846 (δ) **Ehrenberg, C. G.** Weitere Mittheilungen über die mikroskopisch-organischen Beimischungen der vulkanischen Auswurfsmassen in Island...In: Ber. Akad. Berlin, p. 376-381.

1846 **Geinitz, H. B.** Grundriss der Versteinerungskunde...Dresden und Leipzig...813 pp. and 28 Pl.

1846 **Giebel, C. G. A.** Formationen und Versteinerungen um Quedlinburg; Sickleria. In: N. Jahrb. Mineral. p. 712–716.

1846 **Johnston, G.** Description of a new British Sponge (Halichondria macularis). In: Hist. Berwickshire Natural. Club, ii, p. 196. Also in: Ann. and Mag. N. H. xviii (1847), p. 475.

1846 **Keyserling, A. von.** Geognostiche Beobachtungen, von...In: Wiss. Beobacht. Reise Petschora-Land.

1846 **King, W.** An Account of some Shells and other Invertebrate Forms found on the coast of Northumberland and of Durham. By... In: Ann. and Mag. xviii, p. 233–251.

1846 **MacCoy, F.** A Synopsis of the Silurian Fossils of Ireland...by... Dublin. 4°.

1846 **Michelin, H.** Note sur différentes espèces du genre Vioa (famille des Spongiaires), par...In: Revue Zool. Année ix (1846), p. 56–61, Pl. I.

1846 **Michelotti, (G.) A. J.** Übersicht miocener Organismen in Ober-Italien. In: N. Jahrb. Mineral., p. 52–56.

1846 **Reuss, A. E.** Die Versteinerungen der Böhmischen Kreideformation, beschrieben von...2 Abtheilung...Stuttgart...4°.

1846 **Rominger, C.** Vergleichung des Schweitzer Jura's mit der Württembergischen Alp, von...In: N. Jahrb. Mineral. p. 293–306.

1846 **Schafhäutl, K. F. E.** Beiträge zur nähern Kenntniss der Bayerischen Voralpen, von...In: N. Jahrb. Mineral. p. 641–695, Pl. VIII, fig. 7–31, Pl. IX.

1846 **Schmid, E. E. & M. J. Schleiden.** Die geognostischen Verhältnisse des Saalthales bei Jena...Leipzig...Fol. 76 pp., 4 Pl. and 1 map.

1846 **Scouler, J.** Notes on some rare species of Animals found on the Coasts of Ireland. By...In: Ann. and Mag. xvii, p. 176–177.

1846 (a) **Scouler, J.** In: Proc. Dublin Nat. Hist. Soc. i, p....

1846 **Thompson, W.** Additions to the Fauna of Ireland, including a few species unrecorded in that of Britain; with the description of an apparently new Glossiphonia. By...In: Ann. and Mag. xviii, p. 310–315, 383–397.

1846 **Williamson, W. C.** On the real nature of the Minute Bodies in Flints, supposed to be Sponge Spiculae. By...In: Ann. and Mag. N. H. xvii, p. 467–468.

1847 **Bensbach, A.** Über die Feuersteine im Kreide-Gebirge, von...In: N. Jahrb. Mineral. p. 769–777.

1847 **Blanchet**...Sur deux préparations d'éponge destinées aux usages de la chirurgie. In: C. R. Paris, xxv, p. 343.

1847 **Bowerbank, J. S.** On the Siliceous Bodies of the Chalk and other Formations, in reply to Mr J. Toulmin Smith. By...In: Ann. and Mag. N. H. xix, p. 249–262.

1847 **Carter, H. J.** Notes on the Species, Structure, and Animality of the Freshwater Sponges in the Tanks of Bombay (Genus Spongilla). By...In: Trans. Bombay Med. and Phys. Soc. viii, p. 101–107*. Also in: Ann. and Mag. (2), i (1848), p. 303–311.

---

* I did not see this paper in the Bombay Transactions. In the Annals (2), i, it is, however, stated that the article is "reprinted." Where the paper is quoted, this is done after the reprint in the Annals.

1847 **Delbos, J.** Notice géologique sur les terrains du bassin de l'Adour. In: Bull. Soc. Géol. France, (2), iv, p. 712–725.

1847 **Ehrenberg, C. G.** Über die mikroskopischen kieselschaligen Polycystinen als mächtige Gebirgsmasse von Barbados...In: Ber. Akad. Berlin, p. 40–60.

1847 (*a*) **Ehrenberg, C. G.** Über durch das Mikroskop erkennbare organische Beimischungen der am 1. Mai 1812 gefallenen meteorischen Asche...In: Ber. Akad. Berlin, p. 152–159.

1847 (*β*) **Ehrenberg, C. G.** Über den rothen Schneefall mit Föhn im Pusterthale in Tyrol...In: Ber. Akad. Berlin, p. 285–304.

1847 (*γ*) **Ehrenberg, C. G.** Über die zimmt- und ziegelfarbenen...Staub-Meteore...In: Ber. Akad. Berlin, p. 319–360.

1847 (*δ*) **Ehrenberg, C. G.** Nachtrag vom 11. November...In: Ber. Akad. Berlin, p. 360–366, 427.

1847 (*ε*) **Ehrenberg, C. G.** Über vor Kurzem von dem Preuss. Seehandlungs Schiffe, der Adler, aus Canton mitgebrachte verkäufliche chinesische Blumen-Cultur-Erde...In: Ber. Akad. Berlin, p. 476–485.

1847 **Haliday, A. H.** On the Branchiotoma Spongillae and on Coniopteryx. In: Trans. Entom. Soc. London, v, p. 31–32.

1847 **Hall, J.** Natural History of New York. Part vi. Palaeontology of New York by...Vol. i...Albany...4º. xxiii and 338 pp., 100 Pl.

1847 **Heyl, A.** Analyse der officinellen Schwammkohle; von...In: Ann. Chemie und Pharm. lxii, p. 87–89.

1847 **Johnston, G.** A History of the British Zoophytes. By...2nd edition ...London. 1st ed. 1838.

1847 **Jones, T. Rymer.** (Art. Porifera.) In: Cyclop. Anat. and Phys. (Todd), iv, p. 64–70, fig. 67–73.

1847 **Michelin, H.** Iconographie zoophytologique. Description par localités et terrains des Polypiers fossiles de France et pays environnants...Paris. 4º. 1840–1847. The chief Parts in which Sponges are mentioned appeared in 1847.

1847 **Michelotti, G.** Description des Fossiles des terrains miocènes de l'Italie septentrionale, par...In: Natuurk. Verh. Mij. Haarlem, iii, 2, p. 1–408, Pl. i–xvii. Also separately.

1847 **Nardo, D.** Prospetto della Fauna marina volgare del Veneto estuario con cenni sulle principali specie commestibili dell' Adriatico, ecc. Del....Estratto dall' opera: Venezia e le sue lagune. Sep. p. 1–45.

1847 (*a*) (**Nardo, G. D.**) Osservazioni anatomiche sopra l' animale marino detto volgarmente Rognone di mare. In: Atti Instituto Veneto, vi, p. 267–268.

1847 (*β*) **Nardo, G. D.** Osservazioni sui generi di Spongiali Suburites e Litumena. In: Osservaz. postume (Renier), p.

1847 **Oswald, F.** Ueber die Petrefacten von Sadewitz. In: Uebersicht Arb und Veränd. Schles. Ges. 1846, p. 56–65.

1847 **Rominger, C.** Beiträge zur Kenntniss der Böhmischen Kreide, von... In: N. Jahrb. Mineral. p. 641–664.

1847 (α) **Rominger, C.** Beobachtungen über das Alter des Karpathen-Sandsteins und des Wiener-Sandsteins, von...In: N. Jahrb. Mineral. p. 778–784.

1847 **Sandberger, K. L. F.** Geologische Verhältnisse in Nassau. In: N. Jahrb. Mineral. p. 816–818.

1847 **Schomburgk, R. H.** The history of Barbados...London.

1847 **Smith, J. Toulmin.** On the Formation of the Flints of the Upper Chalk. By...In: Ann. and Mag. N. H. xix, p. 1–16, Pl. I.

1847 (α) **Smith, J. Toulmin.** Further Observations on the Formation of the Flints of the Upper Chalk, with Remarks on the Sponge Theory of Mr Bowerbank. By...In: Ann. and Mag. N. H. xix, p. 289–309.

1847 (β) **Smith, J. Toulmin.** On the Ventriculidae of the Chalk...By... In: Ann. and Mag. N. H. xx, pp. 73–97, 176–191, Pl. VII–VIII.

1847 **Verneuil, E. P. de.** Note sur le parallélisme des roches des dépôts paléozoiques de l'Amérique Septentrionale avec ceux de l'Europe ...In: Bull. Soc. Géol. France, (2), iv, p. 646–710.

1847 **Williamson, W. C.** On some of the microscopical objects found in the mud of the Levant and other deposits...In: Mem. Liter. and Phil. Soc. Manchester, viii, p. 1–128, Pl. I–IV.

1847 **Zeuschner, L.** (Über die Entwickelung der Jura und der Pläner-Schichten der Umgebung von Krakau.) In: Ber. Mittheil. Freunden Naturwiss. Wien, ii, p. 479–485.

1847 (α) **Zeuschner, L.** Über die Entwickelung der Jura-Formation bei Ciechocinek unweit Thoren, von...In: N. Jahrb. Mineral. p. 156–160.

1847 (β) **Zeuschner, L.** Gesteine und Fossil-Reste im Jura-Kalke von Krakau. In: N. Jahrb. Mineral. p. 331–332.

1848 **Carter.** Vide 1847.

1848 **Charlesworth, E.** On the mineral condition and general affinities of the Zoophytes of the Chalk at Flamborough and Bridlington. In: Proc. Yorkshire Phil. Soc. i, p. 73–77, Pl. I.

1848 **(Desor, E.** Description of two new sponges: Spongia urceolata and Spongia sulphurea.) In: Proc. Boston Soc. Nat. Hist. iii (1848–51) (1851), p. 67–68. The paper was read Oct. 18, 1848.

1848 **Ehrenberg, C. G.** i, über in den Magen eines peruanischen Fluss-fisches als speise gefundene mikroskopische Organismen; ii, über zwei neue Genera Kieselschaliger Polygastern aus dem Guano...; iii, über 3 neue Infusorien-Biolithe der Braunkohle...; iv, über ...in Königsberg aufgefundene Polygastern in Bernstein; v, Bericht über...Befruchtung der Farnkräuter. In: Ber. Akad. Berlin, pp. 3–24.

1848 (α) **Ehrenberg, C. G.** Eine auf den am 31. März 1847...im Pusterthale
...vorgekommenen rothen Staubregen bezügliche Nachricht...
In: Ber. Akad. Berlin, p. 65–69.

1848 (β) **Ehrenberg, C. G.** Über den Meteorstaubfall in Schlesien...In:
Ber. Akad. Berlin, p. 107–120.

1848 (γ) **Ehrenberg, C. G.** Über den Meteorstaub von Muhrau in Schlesien
...In: Ber. Akad. Berlin, p. 195–200.

1848 (δ) **Ehrenberg, C. G.** Über eigenthümliche auf den Bäumen des
Urwaldes in Süd-Amerika zahlreich lebende mikroskopische...
Organismen. In: Ber. Akad. Berlin, p. 213–220.

1848 (ε) **Ehrenberg, C. G.** Uber die Ampo oder Tanah-ampo...genannte
essbare Erde von Samarang auf Java...In: Ber. Akad. Berlin,
p. 220- 225.

1848 (ζ) **Ehrenberg, C. G.** Über die in der heissen Quelle des Rio Taenta-
Flusses in Afrika...vorkommenden mikroskopischen Organ-
ismen. In: Ber. Akad. Berlin, p. 225–227.

1848 (η) **Ehrenberg, C. G.** Über eine neue einflussreiche Anwendung des
polarisirten Lichtes...In: Ber. Akad. Berlin, p. 238–247.

1848 (θ) **Ehrenberg, C. G.** Eine Mittheilung neuer Beobachtungen über das
gewöhnlich in der Atmosphäre unsichtbar getragene formen-
reiche Leben...In: Ber. Akad. Berlin, p. 325–345.

1848 (ι) **Ehrenberg, C. G.** Fortgesetzte Beobachtungen über jetzt herr-
schende atmosphärische mikroskopische Verhältnisse. In: Ber.
Akad. Berlin, p. 370–381

1848 **Fischer de Waldheim, G.** Notice sur quelques fossiles du gouvernement d'Orel. In: Bull. Soc. Nat. Moscou, xxi, (2), p. 455–469, Pl. xi.

1848 **Girard, H.** Über die metamorphischen Schiefer und Porphyre der
Gegend von Rübeland von...In: N. Jahrb. Mineral. p. 260–268.

1848 **Goeppert, H. R.** Zur Flora des Quader-Sandsteins in Schlesien...
In: N. Jahrb. Mineral. p. 269–278.

1848 (α) **Goeppert, H. R.** Über Pflanzenähnliche Einschlüsse in den Chal-
cedonen, besonders über die Dendriten. In: Uebers. Arb. und
Veränd. Schles. Ges. 1847, p. 135–147.

1848 (**Gray, J. E.**) List of the Specimens of British Sponges in the collection of the British Museum. London: printed by order of
the trustees...12°. pp. viii and 24.

1848 **Hancock, A.** On the Boring of the Mollusca into Rocks, etc.; and on
the Removal of portions of their Shells. By...In: Ann. and
Mag. (2), ii, p. 225–248, Pl. viii.

1848 **Hauer, F. von.** (Fossilien aus den Venetianischen Alpen.) In: Ber.
Mitth. Freunden Naturw. Wien, iv, p. 373–377.

1848 **King, W.** Catalogue.

1848 **Koerte, F. F. E.** Utrum Spongiae officinalis tela fibrosa animali an
vegetabili sit natura. Diss. Berlini. 44 pp.

1848 **Laurent, J. L. M.** Sur les corps reproducteurs des Éponges. In: Institut, xvi, p. 60. Also in: Soc. Philom. Extraits Procès-verbaux, p. 40–41.

1848 **MacCoy, F.** On some new Mesozoic Radiata. By...In: Ann. and Mag. N. H. (2), ii, p. 397–420.

1848 **Mantell, G. A.** Reply to Mr Smith's Remarks on Dr Mantell's Account of the Ventriculites. In: Ann. and Mag. N. H. (2), i, p. 435–436.

1848 (a) **Mantell, G. A.** On Mr Smith's "Observations"...In: Ann. and Mag. N. H. (2), ii, p. 133–134.

1848 **Marcou, J.** Recherches géologiques sur le Jura salinois. In: Mém. Soc. Géol. France (2), iii, p. 1–151.

1848 **Phillips, J. & J. W. Salter.** Palaeontological appendix to Prof. John Phillips's memoir on the Malvern Hills...by...In: Mem. Geol. Survey Great Britain, ii, 1, p. 331–386.

1848 **Quekett, J. E.** A practical treatise on the use of the Microscope...

1848 **Quenstedt, F. A.** Petrefactenkunde Deutschlands...Erste Abtheilung. 1. Band...Tübingen, 1846 (1845)–1849*. 580 pp., with atlas in Fol., Pl. i–xxx.

1848 **Roemer, C. F.** Über eine neue Art der Gattung Blumenbachium (König) und mehrere unzweifelhafte Spongien in obersilurischen Kalkschichten der Grafschaft Decatur im Staate Tennessee in Nord-Amerika, von...In: N. Jahrb. Mineral. p. 680–686, Pl. ix.

1848 **Smith, J. Toulmin.** On the Ventriculidae of the Chalk; their classification. By...In: Ann. and Mag. N. H. (2), i, pp. 36–48, 203–220, 279–295, 352–372, Pl. xiii–xvi.

1848 (a) **Smith, J. T.** Observations on Dr Mantell's "Reply" to Mr Toulmin Smith's Account of the Ventriculidae. In: Ann. and Mag. N. H. (2), ii, p. 48–51.

1848 **Westwood, J. O.** The Spongilla Insect. In: Gardener's Chron. p. 557.

1849 **Bowerbank, J. S.** On a siliceous Zoophyte, Alcyonites parasiticum. In: Quart. Jour. Geol. Soc. London, v, p. 319–328, Pl. viii.

1849 **Bronn, H. G.** Naturgeschichte der drei Reiche...Stuttgart†.

1849 **Carter, H. J.** A descriptive Account of the Freshwater Sponges (genus Spongilla) in the Island of Bombay, with Observations on their Structure and Development. By...In: Ann. and Mag. (2), iv, p. 81–100, Pl. iii–v (a). Also in: Journ. Bombay Branch Asiat. Soc. iii, p. 29–50, Pl....

1849 **Cunnington, W.** On a Peculiarity in the Structure of one of the Fossil Sponges of the Chalk, Choanites Königi, Mantell. By... In: Rep. 18 Meet. Brit. Assoc. p. 67.

---

* This part appeared at the following dates: 1845, Heft i (p. 1–104), Pl. i–vi; 1846, Heft ii (p. 105–184), Pl. vii–xii; 1847, Heft iii (p. 185–264), Pl. xiii–xviii; 1848, Heft iv and v (p. 265–472), Pl. xix–xxx; 1849, Heft vi (p. 473–580).
† 1848 on the title.

1849 **Debey, M. H.** Entwurf zu einer geognostisch-geogenetischen Darstellung der Gegend von Aachen...Aachen...4°. 67 pp. and 1 Pl.

1849 **Ehrenberg, C. G.** Passatstaub und Blutregen...In: Abh. Akad. Berlin, 1847, p. 269–460, Pl. i–vi. Also separately.

1849 (a) **Ehrenberg, C. G.** Weitere Mittheilungen über Resultate bei Anwendung des chromatisch-polarisirten Lichtes für mikroskopische Verhältnisse. In: Ber. Akad. Berlin, p. 55–76.

1849 (β) **Ehrenberg, C. G.** Über das mächtigste bis jetzt bekannt gewordene ...Lager von mikroskopischen reinen Kieselschaligen Süsswasser-Formen am Wasserfall-Flusse im Oregon. In: Ber. Akad. Berlin, p. 76–87.

1849 (γ) **Ehrenberg, C. G.** Über das mikroskopische Leben in Texas. In: Ber. Akad. Berlin, p. 87–91.

1849 (δ) **Ehrenberg, C. G.** Über weitere atmosphärische mikroskopische Verhältnisse...In: Ber. Akad. Berlin, p. 91–97, 1 Pl.

1849 (ε) **Ehrenberg, C. G.** Über mikroskopische Untersuchungen des Jordan-Wassers...In: Ber. Akad. Berlin, p. 187–193.

1849 (ζ) **Ehrenberg, C. G.** Über ein infusorienhaltiges Gypslager in Klein-Asien. In: Ber. Akad. Berlin, p. 193–195.

1849 (η) **Ehrenberg, C. G.** Eine erste Mittheilung über das mikroskopische Leben der Alpen und Gletscher der Schweiz. In: Ber. Akad. Berlin, p. 287–301.

1849 **Geinitz, H. B.** Über das Quader-Gebirge von Regensburg. In: Jahresber. Ges. Natur- und Heilk. Dresden, p. 13–15.

1849 **Gray, J. E.** On the Velvet-like Periostraca of Trigona. By...In: Ann. and Mag. N. H. (2), iv, p. 296–297.

1849 **Hancock, A.** On the Excavating Powers of certain Sponges belonging to the Genus Cliona; with Descriptions of several new Species, and an allied generic Form. By...In: Ann. and Mag. (2), iii, p. 321–348, Pl. xii–xv. Reviews: Froriep's Tagsberichte, i (1850), Zool., pp. 170–176, 177–180.

1849 (a) **Hancock, A.** Observations on Mr Morris's paper on the Excavating Sponges. By...In: Ann. and Mag. (2), iv, pp. 355–357.

1849 **Harting, P.** De Magt van het kleine...Utrecht. 218 pp. with woodcuts.

1849 **Kützing, F. T.** Species Algarum. Lipsiae. vi and 922 pp. Cf. p. 508.

1849 **Lonsdale, W.** Notes on fossil Zoophytes found in the deposits described by Dr Titton in his Memoir on a Section from Atherfield to Rocken End. In: Quart. Journ. Geol. Soc. London, v, p. 55–103, Pl. iv–v.

1849 **MacCoy, F.** On some new genera and species of Palaeozoic Corals and Foraminifera. By...In: Ann. and Mag. N. H. (2), iii, pp. 1–20, 119–136.

1849 **Mayor,**...Recherches sur l'organisation des Éponges. In: Comptes Rendus Soc. Biol. i, p. 153–156.

1849 **Morris, J. A.** Observations on Mr Hancock's paper on the Excavating Sponges. By...In: Ann. and Mag. (2), iv, p. 239–242.

1849 **Orbigny, A. D. d'.** Cours élémentaire de paléontologie...Vol. i. 1re–3me partie...12º. 299 pp.

1849 (α) **Orbigny, A. D. d'.** Prodrome de paléontologie stratigraphique... Vol. i. Paris...12º. 394 pp.

1849 (β) **Orbigny, A. D. d'.** Note sur la classe des Amorphozoaires. In: Revue et Mag. Zool. i, p. 545–550.

1849 **Schafhäutl, K. T. E.** Chemische Analyse des sogenannten Trasses aus dem Riese (Riesgau) bei Nördlingen in Bayern...von...In: N. Jahrb. Mineral. p. 641–670, Pl. ix.

1849 **Strombeck, A. von.** Über die Neocomien-Bildung in der Umgegend von Braunschweig. In: Zeitschr. d. Geol. Ges. i, p. 462–465.

1850 **Alth, A. von.** Geognostisch-paläontologische Beschreibung der nächsten Umgebung von Lemberg. 1. Abtheilung. In: Naturw. Abhandl. iii, 2, p. 171–284, Pl. vii–xiii.

1850 **Damour, A. A.** Sur la composition chimique des Millépores. In: Bull. Soc. Géol. France (2), vii, p. 675–678.

1850 **Dixon, F.** The Geology and Fossils of the Tertiary and Cretaceous Formations of Sussex. By...London...4º. 456 pp. and 41 (44?) Pl.

1850 **Duchassaing, P.** Animaux radiaires des Antilles, par...Paris, typographie Plon Frères. 8º. 35 pp. 2 planches. Spongia isidis figured on Pl. ii; the name is erroneously printed on Pl. i.

1850 **Edwards, H. Milne & J. Haime.** A Monograph of the British fossil Corals. Part i...London...4º. lxxxv and 71 pp., 11 Pl.

1850 **Ehrenberg, C. G.** Über einen die Sonne zwei Tage lang trübenden Staub-Nebel in Russland...In: Ber. Akad. Berlin, p. 9–12.

1850 (α) **Ehrenberg, C. G.** Über einen...bei Detmold gefallenen russartigen Staub...In: Ber. Akad. Berlin, p. 123–128.

1850 (β) **Ehrenberg, C. G.** Über den sehr merkwürdigen Passatstaub- oder rothen Schnee-Fall...auf den höchsten Gotthard-Alpen...In: Ber. Akad. Berlin, p. 169–189.

1850 (γ) **Ehrenberg, C. G.** Vorläufige Bemerkungen über die mikroskopische Bestandtheile der Schwarzerde, Tscherno-Sem, in Russland. In: Ber. Akad. Berlin, p. 268–272.

1850 (δ) **Ehrenberg, C. G.** Weitere Erläuterungen über die für Russland sehr wichtige Schwarz-Erde...In: Ber. Akad. Berlin, p. 364–370.

1850 **Fleming, J.**

1850 **Fraas, O. F.** Versuch einer Vergleichung des deutschen Jura's mit dem Französischen und Englischen, von...In: N. Jahrb. Mineral. p. 139–185.

1850 **Geinitz, H. B.** Das Quadersandsteingebirge oder Kreidegebirge in Deutschland. 2. Hälfte. Freiberg.

1850 **Giebel, C.°G. A.** Scyphia uvaeformis n. sp. In: Jahresber. naturw. Ver. Halle, ii, p. 57–60.

1850 **Gray, J. E.** On the genus Hyalonema. By...In: Ann. and Mag. N. H. (2), vi, p. 306–307.

1850 **Hauer, F. von.** Über die Gliederung der geschichteten Gebirgs-Bildungen in den östlichen Alpen und Karpathen. In: Sitzungsber. Akad. Wien.

1850 **King, W.** A Monograph of the Permian Fossils of England. London ...4°. xxxvii and 258 pp., 29 Pl.

1850 **MacCoy, F.** Descriptions of three new Devonian Zoophytes. By... In: Ann. and Mag. N. H. (2), vi, p. 377–378.

1850 **Mantell, G. A.** Pictorial atlas of fossil remains...London...4°.

1850 (a) **Mantell, G. A.** Notice of the Remains of the Dinornis and other Birds, and of Fossils and Rock-specimens, recently collected by Mr Walter Mantell in the Middle Island of New Zealand...In: Quart. Journ. Geol. Soc. London, vi, p. 319–344, Pl. xxviii–xxix.

1850 **Orbigny, A. D. d'.** Prodrome de Paléontologie stratigraphique universelle des Animaux Mollusques et rayonnés faisant suite au Cours élémentaire de Paléontologie et de Géologie stratigraphiques...Vol. ii. Paris. 8°. 1850.

1850 (a) **Orbigny, A. D. d'.** Recherches zoologiques sur la marche successive de l'animalisation à la surface du globe...In: Ann. Sc. Nat. (3), xiii, p. 218–228.

1850 (β) **Orbigny, A. D. d'.** Note sur les fossiles de l'étage danien. In: Bull. Soc. Géol. France, vii, p. 126–135.

1850 (γ) **Orbigny, A. D. d'.** Catalogue des espèces fossiles de Mollusques Bryozoaires, de Polypiers et d'Amorphozoaires de l'étage néocomien. In: Revue et Mag. Zool. ii, p. 170–181.

1850 **Oswald, F.** Über silurische Schwämme (Aulocopium) von Sadewitz. In: Zeitschr. d. geol. Ges. ii, p. 83–86.

1850 **Roemer, F. A.** Beiträge zur geologischen Kenntniss des nordwestlichen Harz-gebirges, von...(i) In: Palaeontogr. iii, p. 1–67, Pl. i–x.

1850 **Zeuschner, L.** Über die Entwickelung der oberen Glieder der Kreideformation nordlich von Krakau. In: Jahrb. geol. Reichsanst. i, p. 242–255.

1850 (a) **Zeuschner, L.** Geognostische Beschreibung des Nerineen-Kalkes von Inwald und Roczyny. Von...In: Naturwiss. Abhandl. iii, 1, p. 133–146, Pl. xvi–xvii.

1851 **Abich, W. H.** Verzeichniss einer Sammlung von Versteinerungen von Dhaghestan, mit Erläuterungen. Von...In: Zeitschr. d. geol. Ges. iii, p. 15–48, Pl. i–ii.

1851 **Bailey, J. W.** Microscopical Examination of Soundings made by the U.S. Coast Survey off the Atlantic Coast of the U.S. In: Smithsonian Contrib. Knowl. ii, art. iii (16 pp.).

1851 (a) **Bailey, J. W.** Microscopical Observations made in South Carolina, Georgia and Florida. In: Smithsonian Contrib. Knowl. ii, art. iv (48 pp.).

1851 **Beaudouin, J.** Sur le terrain Kelloway-Oxfordien du Châtillonnais. In: Bull. Soc. Géol. France (2), viii, p. 582–599.

1851 **Bischof, K. G.** Lehrbuch der chemischen und physikalischen Geologie. 2. Band. 5. Abth. Bonn. xx and 363 pp.

1851 **Dechen, H. von.** Über die Schichten im Liegenden des Steinkohlen-Gebirges an der Ruhr. In: Verh. naturw. Ver. Rheinland und Westfalens, vii, p. 186–208.

1851 **Edwards, H. Milne & J. Haime.** Recherches sur les Polypiers. 6e mémoire. Monographie des Fongides. In: Ann. Sc. Nat. (3) xv, p. 73–144.

1851 (a) **Edwards, H. Milne & J. Haime.** Monographie des Polypiers fossiles des terrains palaeozoiques...In: Arch. Muséum, v, p. 1–502, Pl. i–xx. Also separately.

1851 **Ehrenberg, C. G.** Mittheilung über einen zu Ende März 1850 in Ningpo in China gefallenen Staub...In: Ber. Akad. Berlin, p. 26–33.

1851 (a) **Ehrenberg, C. G.** Einige Bemerkungen über den...in Graubünden gefallenen rothen Schnee...In: Ber. Akad. Berlin, p. 158–166.

1851 (β) **Ehrenberg, C. G.** Nachricht über die unsichtbar kleinen Lebensformen im central-afrikanischen neu entdeckten Reiche Ukamba ...In: Ber. Akad. Berlin, p. 224–231.

1851 (γ) **Ehrenberg, C. G.** Nachricht des Hrn. Dr. Weisse über dessen mikroskopische Analyse eines...am Argun-Flusse gefallenen Meteorstaubes...In: Ber. Akad. Berlin, p. 309–323.

1851 (δ) **Ehrenberg, C. G.** Über den Gehalt an festen Theilen und an mikroskopischen Lebensformen in der Wassertrübung des Mississipi. In: Ber. Akad. Berlin, p. 324–333.

1851 (ε) **Ehrenberg, C. G.** Über die neuesten...Bewegungen in den organischen Naturwissenschaften. In: Ber. Akad. Berlin, p. 761–795.

1851 **Eichwald, K. E. von.** Geognostischer Ausflug nach Tyrol. In: Nouv. Mém. Soc. Nat. Moscou, ix, p. 75–204.

1851 **Fehling, H. von & J. G. Kurr.** Untersuchung verschiedener Württembergischer Kalksteine. In: Jahresh. Ver. vaterl. Naturk. Württemberg, vii, p. 95–126.

1851 **Giebel, C. G. A.** Über einige Versteinerungen aus den Plänerkalk bei Quedlimburg. In: Jahresber. naturw. Ver. Halle, iii, p. 49–57, Pl. ii.

1851 **Hogg, J.** On Dr Nardo's Classification of the Spongiae, and further notices of the Spongilla fluviatilis. By...In: Ann. and Mag. (2), vii, p. 190–193.

1851 **Huxley, T. H.** Zoological Notes and Observations made on board H.M.S. Rattlesnake during the years 1846–50. By...II. On the Anatomy of the genus Tethya. In: Ann. and Mag. (2), vii, p. 370–373, Pl. XIV, fig. 4–10.

1851 **Leidy, J.** On Spongilla. In: Proc. Acad. Nat. Sc. Philadelphia, v, p. 278.

1851 **MacCoy, F.** A Synopsis of the Classification of the British Palaeozoic rocks, by Adam Sedgwick. Part 2...by...London...4°. p. 1–184, Pl. I A–G.

1851 (a) **MacCoy, F.** On some new Devonian Fossils. By...In: Ann. and Mag. N. H. (2), viii, p. 481–489.

1851 **Morris, J.** Palaeontological Notes. By...In: Ann. and Mag. (2), viii, p. 85–90, Pl. IV; Rev. Neues Jahrb. Mineral. für 1853, p. 758.

1851 **Orbigny, A. D. d'.** Cours élémentaire de paléontologie...Vol. ii, 1er fasc...12°.

1851 **Schafhäutl, K. F. E.** Über die Gliederung des Süd-Bayern'schen Alpen-Kalkes, von...In: N. Jahrb. Mineral. p. 129–149.

1851 **Siebold, C. F. E. von.** Ueber die organisirten Kalkablagerungen der Haut der Strahlthiere (Radiata). In: Jahresber. Schles. Ges. xxviii, p. 35–36.

1851 **Thurmann, J.** Abraham Gagnebin de la Ferrière...Porrentruy. x and 143 pp. with 2 Pl.

1851 **Weisse, J. F.** Nachricht über einen Staubfall...im Gouvernement Irkutzk...In: Bull. Acad. St Pétersbourg, ix, coll. 313–318.

1851 **Wineberger, L.** Versuch einer geognostischen Beschreibung des Bayerischen Waldgebirges...Passau. iv and 141 pp., 3 Pl.

1852 **Bauer,...** (Verzeichniss der Petrefakten...von Gummersbach...) In: N. Jahrb. Mineral. p. 192–193.

1852 **Beyrich, H. E.** Korallen und Schwämme im Muschelkalk. In: Zeitschr. d. geol. Ges. iv, p. 216–219.

1852 **Bornemann, J. G.** Über die geognostischen Verhältnisse des Ohm-Gebirges bei Worbis, von...In: N. Jahrb. Mineral. p. 1–34.

1852 **Bowerbank, J. S.** On Ciliary Action in the Spongiadeae. By...(Read November 13, 1850.) In: Trans. Micro. Soc. London, iii, p. 137–142, Pl. XIX.

1852 **Bronn, H. G. & C. F. Roemer.** Prof. Dr. H. G. Bronn's Lethaea geognostica...3 stark vermehrte Auflage bearbeitet von...Band v. Stuttgart. 412 pp.

1852 **Dobie, W. M.** Note on the Observation of Cilia in Grantia. By...(Proc. R. Med. Soc. Edinburgh)...In: Ann. and Mag. (2), x, p. 317–318, and in: Goodsir, Ann. Anat. and Phys. ii, p. 127–128.

1852 (a) **Dobie, W. M.** Second notice on cilia in two species of Grantia. In: Ann. Anat. and Phys. (Goodsir), ii, p. 129–137, Pl. v, fig. 1.

1852 **Edwards, H. Milne & J. Haime.** A Monograph of the British fossil Corals. Part iii...London...4º. p. 147–210, Pl. XXXI–XLVI.

1852 **Ehrenberg, C. G.** Über die nun gewonnene Übersicht des mikroskopischen Lebens in Californien. In: Ber. Akad. Berlin, p. 528–535.

1852 (a) **Ehrenberg, C. G.** Neue Untersuchungen der Nilerden. In: Ber. Akad. Berlin, p. 617–620.

1852 **Eyton, T. C.** Some Account of a Dredging Expedition on the coast of the Isle of Man during the months of May, June, July and August 1852. By...In: Ann. and Mag. N. H. (2), x, pp. 282–285, 434–436.

1852 **Gaudry, A.** Sur la formation des silex de la craie et des meulières des terrains tertiaires. Thèse. Paris...4º.

1852 **Giebel, C. G. A.** Deutschlands Petrefacten...Leipzig. xvi and 706 pp.

1852 (a) **Giebel, C. G. A.** Allgemeine Palaeontologie...1. Abth. Palaeozoologie. 2. Aufl. 2. Abth. Palaeophytologie. Leipzig. viii and 414 pp.

1852 **Göppert, H. R.** Über den Cylindrites spongioides. In: Jahresb. Schles. Ges. xxix, p. 46–48.

1852 **Gras, A.** Catalogue des corps organisés fossiles qui se recontrent dans le département de l'Isère...In: Bull. Soc. Statist. Isère (2), ii, p. 1–54, Pl. I–IV.

1852 **Hall, J.** The Natural History of New York. Part vi. Palaeontology of New York by...Vol. ii...Albany...4º. viii and 362 pp., 103 Pl.

1852 **MacCoy, F.** A synopsis of the classification of the British palaeozoic rocks by Adam Sedgwick. Part 2...by...Fasciculus ii... London–Cambridge...4º. pp. i–viii, 185–406, Pl. I H–L, II A–B.

1852 **Merian, P.** Über den Aargauischen Jura. In: Ber. Verh. naturf. Ges. Basel, x, p. 137–144.

1852 **Orbigny, A. D. d'.** Cours élémentaire de Paléontologie et de Géologie stratigraphiques. Vol. ii...Paris. 12º, and Atlas 4º.

1852 (a) **Orbigny, A. D. d'.** Prodrome de Paléontologie.

1852 **Otto, E. von.** Additamente zur Flora des Quadergebirges in der Gegend um Dresden...Dippoldiswalde-Meissen...Fol. iv and 29 pp.

1852 **Owen, D. D.** Report of a geological Survey of Wisconsin...Philadelphia...4º. 2 vols. with atlas.

1852 **Perty, J. A. M.** Zur Kenntniss kleinster Lebensformen nach Bau, Funktionen, Systematik...Bern...4º. viii and 228 pp.

1852 **Quenstedt, F. A.** Handbuch der Petrefaktenkunde...Tübingen. vii and 792 pp., 61 Pl.

1852 **Roemer, F. A.** Beiträge zur geologischen Kenntniss des nordwestlichen Harzgebirges von...2. Abtheilung. In: Palaeontogr. iii,
pp. i–iii, 69–111, Pl. xi–xv.

1852 **Ross, L.** Reisen auf den griechischen Inseln des Aegeischen Meeres.
Band iv. Reisen nach Kos, Halikarnassos, Rhodos und der Insel
Cypern...Halle. viii and 216 pp., 3 Pl.

1852 **Wetherell, N. T.** Note on a new species of Clionites. By...In: Ann.
and Mag. N. H. (2), x, p. 354–355, Pl. v(c), fig. 1–2.

1853 **Baranowsky,...** Die Heilkräfte der Spongia fluviatilis; von...In:
Medic. Zeitung Russlands, x, p....

1853 **Barrande, J.** (Über Bestätigungen seiner Ansichten.) In: N. Jahrb.
Mineral. p. 335–347.

1853 **Brown, J.** Note on the Artesian Well at Colchester...By...In: Ann.
and Mag. N. H. (2), xii, p. 240–242, Pl. viii–ix.

1853 **Cavolini, F.** Memorie postume sceverate dalle schede autografe di...
per cura ed a spese di S. D(elle) Chiaie...Benevento...4°.
xxxiv and 344 pp., 23 Pl.*

1853 **Ehrenberg, C. G.** Über das mikroskopische Leben der Galapagos-
Inseln...In: Ber. Akad. Berlin, p. 178–182.

1853 (α) **Ehrenberg, C. G.** Über das vorweltliche kleinste Süsswasserleben
in Ägypten. In: Ber. Akad. Berlin, p. 200–203.

1853 (β) **Ehrenberg, C. G.** Über das jetzige mikroskopische Leben als
Flusstrübung und Humusland in Florida. In: Ber. Akad. Berlin,
p. 252–271.

1853 (γ) **Ehrenberg, C. G.** Über die auf den höchsten Gipflen der europäischen Central-Alpen...lebenden mikroskopischen Organismen...In: Ber. Akad. Berlin, p. 315–333.

1853 (δ) **Ehrenberg, C. G.** Über einige neue Materialien zur Übersicht des
kleinsten Lebens. In: Ber. Akad. Berlin, p. 505–533, Pl. i–ii.

1853 **Eichwald, K. E. von.** Einige paläontologische Bemerkungen über den
Eisensand von Kursk. In: Bull. Soc. Natur. Moscou, xxvi, 1,
p. 209–232.

1853 **Giebel, C. G. A.** (Über Kreide-Versteinerungen aus Texas.) In: N.
Jahrb. Mineral. p. 165.

1853 **Goodsir, J.** On the structure and economy of Tethea, and on an
undescribed species from the Spitzbergen seas. In: Proc. Roy.
Soc. Edinburgh, iii, p. 181 182. (Abstr. of 1868 **Goodsir, J.**)

1853 **Gosse, P. H.** A Naturalist's rambles on the Devonshire Coast...
By...London. xvi and 451 pp., 28 Pl.

1853 **Harting, L.** Ueber den praktischen Werth sämmtlicher bis auf die
neueste Zeit empfohlenen Verfahrungsweisen zur Erweckung der
Frühgeburt...von...In: Monatsschr. Geburtsk. i, 2, p....

* The plates have double numbers. The running numbers are marked: at 1, 1–16,
18–24.

1853 **Manès, W.** Description physique géologique et minéralogique du département de la Charente-Inférieure. Bordeaux.

1853 **Marck, W. von der.** Ueber Schwimmsteine und Feuersteine. In: Verh. naturh. Ver. Rheinland und Westphalen, x, p. 385–403.

1853 **Richter, R.** Gäa von Saalfeld. In: Einlad. Progr. Realschule Saalfeld, p. 1–32.

1853 **Roemer, F. A.** (Eine kurze Skizze seiner im letzten Sommer nach Devonshire unternommenen Reise.) In: N. Jahrb. Mineral. p. 810–818.

1853 **Schmid, E. E.** Die organischen Reste des Muschelkalkes im Saal-Thale bei Jena, von...In: N. Jahrb. Mineral. p. 9–30.

1853 **Studer, B.** Geologie der Schweiz. 2. Band...Zürich. vii and 497 pp.

1854 **Bailey, J. W.** Examination of some Deep Soundings from the Atlantic Ocean. By...In: Amer. Journ. Sc. and Arts, (2), xvii, p. 176–178.

1854 (α) **Bailey, J. W.** On some new localities of Fossil Diatomaceae in California and Oregon. In: Amer. Journ. Sc. and Arts, xvii, p. 179–180.

1854 **Carter, H. J.** Zoosperms in Spongilla. By...In: Ann. and Mag. (2), xiv, p. 334–336, Pl. xɪ*.

1854[?] **Charlesworth, E.** (Zoophytes.) In: Yorks. Ph. S. Proc. 1855, p. 73–77.

1854 **Darwin, C.** Monograph of Cirripedia, Vol. ii (see p. 302–314). London.

1854 **Edwards, H. Milne & J. Haime.** A Monograph of the British Fossil Corals. Part v....London...4°.

1854 **Ehrenberg, C. G.** Mikrogeologie...Von...Leipzig...Fol.

1854 (α) **Ehrenberg, C. G.** Über das organische Leben des Meeresgrundes in bis 10800 und 12000 Fuss Tiefe. In: Ber. Akad. Berlin, p. 54–75.

1854 (β) **Ehrenberg, C. G.** Weitere Mittheilungen über die aus grossen Meerestiefen gehobenen Grundmassen. In: Ber. Akad. Berlin, p. 191–196.

1854 (γ) **Ehrenberg, C. G.** Systematische Charakteristik der neuen mikroskopischen Organismen des tiefen atlantischen Oceans. In: Ber. Akad. Berlin, p. 236–250.

1854 (δ) **Ehrenberg, C. G.** Weitere Ermittelungen über das Leben in grossen Tiefen des Oceans. In: Ber. Akad. Berlin, p. 305–328.

1854 (ε) **Ehrenberg, C. G.** Weitere Mittheilungen über die Natur und Entstehung des Grünsandes...In: Ber. Akad. Berlin, p. 384–410.

1854 (ζ) **Ehrenberg, C. G.** Über Culturerden aus Ceylon...In: Ber. Akad. Berlin, p. 704–710.

1854 **Eichwald, K. E. von.** Die Grauwacken-Schichten von Lief- und Esth-Land. In: Bull. Soc. Natur. Moscou, xxvii, 1, p. 3–112, Pl. ɪ–ɪɪ.

1854 **Geigel, A.**

1854 **Goeppert, H. R.** Bericbt über eine...Reise...In: Verb. naturh. Ver. Rheinland und Westphalen, xi (2), i, p. 225–264, Pl. III.

1854 **Morris, J.** A Catalogue of British Fossils comprising the genera and species hitherto described; with references to their geological distribution and to the localities in which they have been found ...2nd edit., considerably enlarged. London, for and by the author....8°. viii and 372 pp.

1854 **Murchison, R. I.** Siluria...London.

1854 **Otto, E. von.** Additamente zur Flora des Quadergebirges in Sachsen. 2. Heft...Leipzig...4°. x and 53 pp.

1854 **Poey, F.** Memorias sobre la Historia natural de la isla de Cuba... Por...Toma i...Habana.

1854 **Quekett, E. J.** Lectures on Histology. Vol. ii...London.

1854 **Reuss, A. E.** Beiträge zur geognostischen Kenntniss Mährens. In: Jahrb. geol. Reichsanstalt, v, p. 659–765.

1854 **Sandberger, G. & K. L. T. Sandberger.** Systematische Beschreibung und Abbildung der Versteinerungen des Rheinischen Schichtensystems in Nassau...2. Band. Atlas. Wiesbaden...Fol.

1854 **Schafhäutl, K. F. E.** Beiträge zur näheren Kenntniss der Bayern'schen Voralpen [cont.] von...In: N. Jahrb. Mineral. p. 513–559, Pl. VII–VIII.

1854 **Sharpe, D.** On the Age of the fossiliferous sands and gravels of Farringdon and its neighbourhood. In: Quart. Journ. Geol. Soc. London, x, p. 176–198, Pl. v.

1854 **Strickland, H. E.** On the mode of growth of Halichondria Suberea. In: Rep. 23 Meet. Brit. Assoc. p. 72.

1854 **Tchihatcheff, P. von.** Dépôts paléozoiques de la Cappadou et du Bosphore par...In: Bull. Soc. Géol. France (2), xi, p. 402–416.

1855 **Ebray, T.** Note sur les Spongiaires des environs de Vierzon. In: Bull. Soc. Géol. France (2), xii, p. 1032–1037.

1855 **Ehrenberg, C. G.** Über die weitere Entwicklung der Kenntniss des Grünsandes...In: Ber. Akad. Berlin, p. 172–178.

1855 (α) **Ehrenberg, C. G.** Über neue Erkenntniss immer grösserer Organisation der Polythalamien...In: Ber. Akad. Wien, p. 272–289.

1855 (β) **Ehrenberg, C. G.** Über ein europäisches marines Polygastern-Lager...In: Ber. Akad. Berlin, p. 292–305.

1855 (γ) **Ehrenberg, C. G.** Beiträge zur Kenntniss der Flusstrübungen und der vulkanischen Auswurfsstoffe. In: Ber. Akad. Berlin, p. 561–578.

1855 **Fraas, O. F.** Beiträge zum obersten weissen Jura in Schwaben. In: Jahresh. Ver. vaterl. Naturk. Württemberg, x, p. 77–107, Pl. II.

1855 **Gosse, P. H.** A Manual of Marine Zoology for the British Isles. By... Part i. London...12°. 710 pp.

1855 **Gregory, W.** On the presence of Diatomaceae, Phytolitharia and Sponge spicules, in Soils which support Vegetation. In: Proc. Bot. Soc. Edinburgh, p. 69–72. Also in: Ann. and Mag. N. H. (2), xvi, p. 219–222.

1855 **Grube, A. E.** Beschreibung neuer oder wenig bekannter Anneliden. In: Arch. Naturgesch. xxi, 1, p. 81–136.

1855 **Kressler,** In: Polytech. Centralbl. xxi (2), ix, p. 317.

1855 **Leidy, J.** Contributions towards a Knowledge of the Marine Invertebrate Fauna of the Coasts of Rhode Island and New Jersey. In: Journ. Acad. Nat. Sc. Philadelphia (2), iii, 2, p. 135–152, Pl. x–xi.

1855 **Meyer, J. B.**

1855 **Mueller, J.** (Ueber Sphaerozoum und Thalassicolla.) In: Ber. Akad. Berlin, p. 229–253.

1855 **Peters, K. F.** Die Nerineen des oberen Jura in Oesterreich. In: Sitzungsber. Akad. Wien, xvi, p. 336–366, 4 Pl.

1855 **Roemer, C. F.** Das ältere Gebirge in der Gegend von Aachen... In: Zeitschr. d. geol. Ges. vii, p. 377–398.

1855 **Roemer, F. A.** Beiträge zur geologischen Kenntniss des nordwestlichen Harzgebirges. 3. Abtheilung. In: Palaeontogr. v, p. 1–46, Pl. i–viii.

1855 **Rose, C. B.** On the discovery of parasitic borings in fossil Fish-scales. In: Trans. Microsc. Soc. London (2), iii, p. 7–9, Pl. i.

1855 **Schaffner,...** Einige Bemerkungen über Spongilla fluviatilis, Link. In: Verh. naturh. Ver. Rheinland und Westphalen, xii (2), ii, p. 29–39.

1855 **Spada Lavini, A. & A. Orsini.** Quelques Observations géologiques sur les Apennins de l'Italie centrale, par... In: Bull. Soc. Géol. France (2), xii, p. 1202–1230, Pl. xxxii.

1855 **Terquem, O.** Paléontologie du système du lias inférieur du Grand-Duché de Luxembourg et de Hettange, du département de la Moselle. In: Mém. Soc. Géol. France (2), v, 2, p. 219–343, Pl. xii–xxvi.

1855 **Weisse, J. F.** Mikroskopische Analyse eines organischen Polirschiefers aus dem Gouvernement Simbirsk. In: Bull. Acad. St Pétersbourg, xii, coll. 273–282, Pl.... Also in: Mélanges Biol. ii, p. 237–250, Pl. i–iii.

1855 (a) **Weisse, J. F.** Resultate einer vergleichenden microskopischen Untersuchung von mehr denn dreissig verschiedenen Proben der sogenannten Schwarz-Erde (Tscherno-Sjom). In: Bull. Soc. Natur. Moscou, xxviii, 1, p. 452–459.

1856 **Achenbach, A.** Geognostische Beschreibung der Hohenzollern'schen Lande In: Zeitschr. d. geol. Ges. viii, p. 331–482, Pl. xiv.

1856 Bailey, J. W. On the Non-existence of Polarizing Silica in the Organic Kingdoms. By...In: Amer. Journ. Sc. and Arts, xxi, p. 357–358. Also in: Ann. and Mag. N. H. (2), xviii, p. 78–79; and in: Quart. Journ. Microsc. Sc. iv, p. 303–305.

1856 (a) Bailey, J. W. On the Origin of Green Sand...In: Proc. Boston Soc. v, p. 364–368. Also in: Amer. Journ. Sc. and Arts, xxii, p. 280–284.

1856 Bowerbank, J. S. On the Origin of Siliceous Deposits in the Chalk Formation. In: 26 Meeting Brit. Assoc., Sect. Geology, see 1857β.

1856 Carter, H. J. Notes on the Freshwater Infusoria of the Island of Bombay. No. 1, Organisation. By...In: Ann. and Mag. (2), xviii, pp. 115–132, 221–249, Pl. v–vii.

1856 Catullo, F. A. Dei terreni di sedimento superiore delle Venezie... Padova...4°. viii and 88 pp., 19 Pl.

1856 Dunker, W. B. R. H. Ueber mehre Pflanzenreste aus dem Quadersandsteine von Blankenburg. In: Palaeontogr. iv, p. 179–183, Pl. xxxii–xxxv.

1856 Ehrenberg, C. G. Mikrogeologie...Von...Leipzig...Fol. Fortsetz. p. 1–88.

1856 (a) Ehrenberg, C. G. Über den Grünsand und seine Erläuterung des organischen Lebens. In: Abh. Akad. Berlin, 1855, p. 85–176, Pl. i–vii.

1856 (β) Ehrenberg, C. G. Über die Meeresorganismen in 16200 Fuss Tiefe. In: Monatsber. Akad. Berlin, p. 197–201.

1856 (γ) Ehrenberg, C. G. Über das mikroskopische Leben der centralen Landflächen Mittel-Afrika's. In: Monatsber. Akad. Berlin, p. 323–338, 1 Pl.

1856 (δ) Ehrenberg, C. G. Über 2 neue südamerikanische Gebirgsmassen... In: Monatsber. Akad. Berlin, p. 425–431.

1856 Giebel, C. G. A. Das Kreidegebirge in Thüringen. In: Zeitschr. gesammt. Naturw. viii, p. 169–174.

1856 Gosse, P. H. Tenby: a Sea-side Holiday. London. xviii and 400 pp., 24 Pl.

1856 Köchlin-Schlumberger, J. Études géologiques dans le département du Haut-Rhin...In: Bull. Soc. Géol. France, xiv, p. 117–206.

1856* Leidy, J. (On Cliona.) In: Proc. Acad. Sc. Philadelphia, viii, p. 162–163. Also in: Amer. Journ. of Science and Arts (2), xxiii, p. 281–282.

1856 Lieberkühn, N. Beiträge zur Entwicklungsgeschichte der Spongillen. Von...In: Arch. Anat. and Phys. 1856, p. 1–19. Transl. in: Ann. and Mag. xvii, p. 403–413. Abstr. in: Quart. Journ. Micr. Sc. v (1857), p. 212–219.

* On the title of vol. viii is printed 1857; Leidy's article probably was published in 1856.

1856 (a) **Lieberkühn, N.** Zur Entwicklungsgeschichte der Spongillen. (Nachtrag.) Von...In: Arch. Anat. und Phys. 1856, p. 399–414, Pl. xv.

1856 (β) **Lieberkühn, N.** Zusätze zur Entwicklungsgeschichte der Spongillen. Von...(Vorgetragen in der Gesellschaft naturforschender Freunde in Berlin in der Sitzung vom 20. Mai 1856.) In: Arch. Anat. und Phys. 1856, p. 496–514, Pl. xviii, fig. 8–18.

1856 (γ) **Lieberkühn, N.** Ueber Protozoen. Aus einem Schreiben von... an C. Th. v. Siebold. In: Z. W. Z. viii, p. 307–310.

1856 **Safford, J. M.** Remarks on the genus Tetradium, with notices of the species found in Middle-Tennessee. In: Amer. Journ. Sc. and Arts (2), xxii, p. 236–238.

1856 **Sandberger, G. & K. L. T. Sandberger.** Systematische Beschreibung und Abbildung der Versteinerungen des Rheinischen Schichtensystems in Nassau...1. Band...Wiesbaden...4°. 564 pp., 1 Pl.

1856 **Schafhäutl, K. F. E.** (Ueber Versteinerungen des Kressenberges.) In: N. Jahrb. Mineral. p. 819–822.

1856 **Schlossberger, J. E.** Erster Versuch einer allgemeinen und vergleichenden Thier-Chemie. 1. Band...Leipzig und Heidelberg. 1854–1856.

1856 **Staring, W. C. H.** Natuurlijke historie van Nederland. Eerste deel... De bodem van Nederland...i. Haarlem.

1856 **Thompson, W.** The natural history of Ireland. Vol. iv...London. 546 pp.

1857 **Bailey, J. W.** Report upon the results of Microscopic Examinations of the Soundings made by Lieut. Berryman, of the U.S. Navy, on his recent voyages to and from Ireland in the "Arctic." In: Amer. Journ. Sc. and Arts (2), xxiii, p. 153–157.

1857 **Billings, E.** Notes on some of the more remarkable genera of Silurian and Devonian Fossils. In: Canadian Nat. and Geol. ii, pp. 184–198, 405–409, 17 woodcuts.

1857 (a) **Billings, E.** In: Geol. Survey Canada. Rep. 1853–1856.

1857 **Bowerbank, J. S.** On the Vital Powers of the Spongiadae. By...In: Rep. Brit. Assoc. 1856, p. 438–451. Reprinted with separate paging.

1857 (a) **Bowerbank, J. S.** On the Anatomy and Physiology of the Spongiadae. By...(abstract). In: Proc. Roy. Soc. London, viii, p. 573–577. Also in: Ann. and Mag. (2), xx, p. 298–301.

1857 (β) **Bowerbank, J. S.** On the Origin of Siliceous Deposits in the Chalk Formation. By...In: Rep. 26 Meet. Brit. Assoc., Not. & Abstr. p. 63.

1857 **Brandt, J. F. von.** Extrait d'un Mémoire ayant pour titre: "de nova Polyporum classis familia Hyalochaetidum nomine designanda." In: Bull. Acad. St Pétersbourg, xvi, coll. 65–67. Also in: Mélanges Biol. ii (1858), p. 606–608.

1857 **Byron, R.** (Ueber ein Exemplar der Posidonomya Becheri, etc.) In: N. Jahrb. Mineral. p. 57.

1857 **Carter, H. J.** On the Ultimate Structure of Spongilla, and Additional Notes on Freshwater Infusoria. By...In: Ann. and Mag. (2), xx, p. 21–41, Pl. i. Also (?) in: Journ. Bombay Branch Asiat. Soc. v, p. 574–597.

1857 **Claparède, J. L. R. A. E.** Anatomie und Entwicklungsgeschichte der Neritina fluviatilis von...In: Arch. Anat. Phys. und Wiss. Med. p. 109–248, Pl. iv–viii.

1857 **Ébray, T.** Sur l'âge du calcaire à chailles des départements du Cher, de la Nièvre et de l'Yonne. In: Bull. Soc. Géol. France (2), xiv, p. 582–585.

1857 **Edwards, H. Milne.** Leçons sur la physiologie et l'anatomie comparée de l'homme et des animaux...par...Tome ii. Paris. 8⁰.

1857 **Edwards, H. Milne & J. Haime.** Histoire naturelle des Coralliaires, ou Polypes proprement dits. Tome i...Paris. viii, xxix and 326 pp.

1857 **Egger, J. G.** Die Foraminiferen der Miocän-Schichten bei Ortenburg in Nieder-Bayern, von...In: N. Jahrb. Mineral. p. 266–311.

1857 **Ehrenberg, C. G.** Über einen vulkanischen Tuff bei Hennersdorf in Schlesien, welcher reich mit organischen Süsswasserformen gemischt ist. In: Monatsber. Akad. Berlin, p. 227–233.

1857 (a) **Ehrenberg, C. G.** Über die organischen Lebensformen in unerwartet grossen Tiefen des Mittelmeeres. In: Monatsber. Akad. Berlin, p. 538–570.

1857 **Etallon, A.** Esquisse d'une description géologique du Haut-Jura, et en particulier des environs de St Claude, par...In: Ann. Soc. d'Agricult. Lyon (3), i, p. 247–354 (with two maps). Vide pp. 268, 276, 282, 296, 302, 328, 329.

1857 **Focillon, A.** Rapport sur l'application des bateaux plongeurs de MM. Payerne et Lamiral à l'exploitation, à la culture et à l'acclimatation des animaux aquatiques, fait à la Société imperiale zoologique d'acclimatation...Par...In: Bull. Soc. Zool. d'Acclim. iv, p. 207–214.

1857 **Gray, J. E.** Synopsis of the Families and Genera of Axiferous Zoophytes or Barked Corals. By...In: Proc. Zool. Soc. London, xxv, p. 278–294, Pl. ix.

1857 **Heller, C.** Beitrag zur Fauna der Adria. In: Verh. Zool.-Bot. Ver. Wien, vi, p. 629–634, Pl. ix.

1857 **Lieberkühn, N.** Beiträge zur Anatomie der Spongien von...(Der Gesellschaft naturforschender Freunde mitgetheilt in den Sitzungen vom 6. September und 2. December 1856.)...In: Arch. Anat. und Phys. p. 376–403, Pl. xv.

1857 **Milne Edwards, H.** See Edwards.

1857 **Owen, D. D.** General Report. In: 2 Rep. Geol. Survey Kentucky, p. 7–114.

1857 **Owen, R.** Description of a new Species of Euplectella (Euplectella cucumer, O.). By...In: Trans. Linn. Soc. London, xxii, p. 117–123, Pl. xxi.

1857 **Pichler, A.** Zur Geognosie der Tyroler Alpen, von...In: N. Jahrb. Mineral. p. 689–695.

1857 **Pictet, F. J.** Traité de Paléontologie ou Histoire Naturelle des Animaux fossiles considérés dans leurs rapports zoologiques et géologiques par...Seconde édition, revue, corrigée, considérablement augmentée, accompagnée d'un Atlas de 110 planches grand in 4⁰. Tome quatrième. Paris...8⁰. 768 pp. Atlas, 4⁰, 1853–1857.

1857 **Quenstedt, F. A.** Der Jura...Tübingen, 1858 (sic). vi and 842 pp. Atlas, 4⁰*.

1857 **Renou, F.** Note sur une production d'eau douce, d'apparence spongiforme. In: Bull. Soc. Linn. Normandie, ii, p. 57–59.

1857 **Roemer, C. F.** Über Holländische Diluvial-Geschiebe, von...In: N. Jahrb. Mineral. p. 385–392.

1857 **Salter, J. W.** On the Cretaceous Fossils of Aberdeenshire; with a note on the position of the Chalk flints and Greensand, by W. Ferguson. In: Quart. Journ. Geol. Soc. London, xiii, p. 83–89, Pl. ii.

1857 **Zeuschner, L.** (Ueber den Nerineen-Kalk von Inwald bei Wadowice und den polnischen Spongiten-Kalk.) In: N. Jahrb. Mineral. p. 154–155.

1858 **Andler,...** Über die Angulaten-Schichten in der Württembergischen Jura-Formation, von...In: N. Jahrb. Mineral. p. 641–645.

1858 **Baily, W. H.** Description of fossil Invertebrata from the Crimea. In: Quart. Journ. Geol. Soc. London, xiv, p. 133–163, Pl. viii–x.

1858 **Bigsby, J. J.** On the palaeozoic basin of the state of New York. In: Quart. Journ. Geol. Soc. London, xiv, p. 335–452.

1858 **Binder, C.** Geognostisches Profil des Eisenbahn-Einschnittes von Geislingen nach Amstetten. In: Jahresh. Ver. vaterl. Naturk. Württemberg, xiv, p. 79–96, Pl. i.

1858 **Bowerbank, J. S.** Further Report on the Vitality of the Spongiadae. By...In: Rep. Brit. Assoc. p. 121–125, Pl. i. Reprint p. 3–7.

1858 (a) **Bowerbank, J. S.** On the Anatomy and Physiology of the Spongiadae. By...Part i. On the spicula. In: Phil. Trans. Roy. Soc. London, cxlviii, p. 279–332, Pl. xxiii–xxvi.

---

* On the title 1858. The book appeared in four parts: 1856, Lief. 1, p. 1–208, Pl. i–xxiv; 1857, Lief. 2, p. 209–368, Pl. xxv–xlvii; 1857, Lief. 3, p. 369–576, Pl. xlviii–lxxii; 1857, Lief. 4, pp. i–vi, 577–842, Pl. lxxiii–c.

1858 **Bronn, H. G.** Untersuchungen über die Entwickelungs-Gesetze der organischen Welt während der Bildungs-Zeit unserer Erd-Oberfläche...Stuttgart. x and 502 pp.

1858 **Dalyell, J. G.** The powers of the creator displayed in the creation; or, observations on life amidst the various forms of the humbler tribes of animated nature: with practical comments and illustrations. By...Vol. iii. pp. 87–93, 110–111, Pl. xxi–xxvi.

1858 **Desmarest, E.** Encyclopédie d'histoire naturelle...Par le Dr Chenu ...Avec la collaboration de M...Paris. 4°.

1858 **Ehrenberg, C. G.** Kurze Charakteristik der 9 neuen Genera und der 105 neuen Species des Aegaeischen Meeres und des Tiefgrundes des Mittel-Meeres. In: Monatsber. Akad. Berlin, p. 10–40.

1858 (α) **Ehrenberg, C. G.** Mittheilung über organischen Quarzsand...In: Monatsber. Akad. Berlin, p. 118–128.

1858 (β) **Ehrenberg, C. G.** Uebersicht der Bearbeitung der von den Herren Schlagintweit übergebenen Materialien zur Bestimmung des stationären mikroskopischen Lebens auf den Hochalpen des Himalaya in 18000 und 20000 Fuss Höhe. In: Monatsber. Akad. Berlin, p. 225–256.

1858 (γ) **Ehrenberg, C. G.** Über andere massenhafte mikroskopische Lebensformen der ältesten silurischen Grauwacken-Thone bei Petersburg. In: Monatsber. Akad. Berlin, p. 324–337, Pl. i.

1858 **Gervais, P. & P. J. van Beneden.** Zoologie médicale...ii...Paris. viii and 455 pp.

1858 **Gray, J. E.** On the Dysidea papillosa of Dr Johnston. By...In: Proc. Zool. Soc. London, xxvi, p. 531–532, Pl. x, fig. 3–8. Also in: Ann. and Mag. ii, p. 489.

1858 (α) **Gray, J. E.** Description of Aphroceras, a New Genus of Calcareous Spongiadae brought from Hong-kong by Dr Harland. By... In: Proc. Zool. Soc. London, xxvi, p. 113–114, Pl. x, fig. 1–2. Also in: Ann. and Mag. ii, p. 83–84.

1858 (β) **Gray, J. E.** On Aphrocallistes, a New Genus of Spongiadae from Malacca. By...In: Proc. Zool. Soc. London, xxvi, p. 114–115, Pl. xi. Also in: Ann. and Mag. N. H. ii, p. 224–225.

1858 (γ) **Gray, J. E.** Description of a New Genus of Sponge (Xenospongia) from Torres Strait. By...In: Proc. Zool. Soc. London, xxvi, p. 229–230, Pl. xii. Also in: Ann. and Mag. N. H. ii, p. 369–371.

1858 (δ) **Gray, J. E.** On Carpenteria and Dujardinia, two genera of a new form of Protozoa with attached multilocular Shells filled with Sponge, apparently intermediate between Rhizopoda and Porifera. By...In: Proc. Zool. Soc. London, xxvi, p. 266–271, 4 figs. Also in: Ann. and Mag. N. H. ii, p. 381–386.

1858 **Hall, J. & J. D. Whitney.** Report on the Geological Survey of the State of Iowa...Vol. i. Albany. 724 pp., 29 Pl. and 3 maps.

1858 **Holdsworth, E. W. H.** On Zoanthus Couchii, Johnston. By...In: Proc. Zool. Soc. London, xxvi, p. 557–560, Pl. x, fig. 3–8. Also in: Ann. and Mag. N. H. iv (1859), p. 152–156.

1858 **Horner, L.** An Account of some recent Researches near Cairo... by...Part ii. In: Phil. Trans. Roy. Soc. London, cxlviii, p. 53–92, Pl. ii–v.

1858 **Jones, T. Rymer.** The Aquarian Naturalist: a Manual for the Sea-Side. By...London. 12º. 550 pp.

1858 **Kade, G.** Ueber Petrefakten des Norddeutschen Diluviums. In: N. Jahrb. Mineral. p. 451–452.

1858 **Oppel, A.** Die Juraformation Englands, Frankreichs und des südwest-lichen Deutschlands. Schluss. In: Jahresh. Ver. väterl. Naturk. Württemberg, xiv, p. 129–291. Also published separately?.

1858 **Pacht, R.** Geognostische Untersuchungen zwischen Orel, Woronesch und Simbirsk im Jahre 1853. In: Beitr. Kenntn. russ. Reiches, xxi, p. 61–187, Pl. i–x.

1858 **Pictet, F. J. & E. Renevier.** Description des fossiles du terrain aptien de la Perte-du-Rhône. In: Matériaux paléontol. suisse...ii, p. 137–184, Pl. xix–xxii.

1858 **Raulin, F. V. & A. Leymerie.** Statistique géologique du département de l'Yonne...Auxerre.

1858 **Roemer, C. F.** Die Versteinerungen der silurischen Diluvial-Geschiebe von Gröningen in Holland von...In: N. Jahrb. Mineral. p. 257–272.

1858 **Schlegel, H.** Handleiding tot beoefening der dierkunde, door... Breda ii., xx, 628 and 2 pp., 27 Pl.

1858 **Schlossberger, J. E.** Ueber Fibroïn und die Substanz des Bade-schwamms; von...In: Ann. Chemie und Pharm. cviii [(2) xxxii], p. 62–64.

1858 **Schmidt, F.** Untersuchungen über die Silurische Formation von Ehstland, Nord-Livland und Oesel. Von...In: Archiv Naturk. Liv-, Ehst- und Kurland (1), ii, p. 1–247, 1 Pl. and 1 map.

1859 **Andrian, F. von.** Bericht über die Übersichts-Aufnahmen im Zipser und Gömörer Comitate während des Sommers 1858. In: Jahrb. geol. Reichsanst. x, p. 535–554.

1859 **Berger,...** Die Versteinerungen im Röth von Hildburghausen, gefunden von...In: N. Jahrb. Mineral. p. 168–171, Pl. iii, fig. 1–14.

1859 **Billings, E.** Fossils of the Calciferous Sandrock, including those of a deposit of White Limestone at Mingan, supposed to belong to the formation. In: Canadian Naturalist, iv, p. 345–367, 12 figs.

1859 **Binkhorst van den Binkhorst, J. F.** Esquisse géologique et palé-ontologique des couches crétacées du Limbourg...par... Bruxelles. 6, xviii and 268 pp., 1 map, 5 Pl.

1859 **Bowerbank, J. S.** On the Organization of Grantia ciliata. By...In: Trans. Microsc. Soc. London, vii, p. 79–84, Pl. v.

1859 **Brandt, J. F.** ...Symbolae ad polypos Hyalochaetides spectantes, tabulis iv illustratae. Petropoli...Fol. 24 pp.

1859 **Bronn, H. G.** Die Klassen und Ordnungen des Thierreichs, wissenschaftlich dargestellt in Wort und Bild. Von...Erster Band. Amorphozoen...Leipzig und Heidelberg. xvi and 142 pp., 12 Pl.

1859 **Busk, G.** A Monograph of the fossil Polyzoa of the Crag. London... 4º. xiii and 136 pp., 22 Pl.

1859 **Carter, H. J.** On Fecundation in the two Volvoces, and their Specific Differences; on Eudorina, Spongilla, Astasia, Euglena and Cryptoglena. By...In: Ann. Mag. N. H. (3), iii, p. 1–20, Pl. i.

1859 (α) **Carter, H. J.** On the Identity in Structure and Composition of the so-called Seed-like Body of Spongilla with the Winter-egg of the Bryozoa; and the presence of Starch-granules in each. By... In: Ann. Mag. N. H. (3), iii, p. 331–343, Pl. viii.

1859 **Coquand, H.** Synopsis des animaux et des végétaux fossiles observés dans la formation crétacée du sud-ouest de la France. In: Bull. Soc. Géol. France, (2), xvi, p. 954–1023.

1859 **Ehrenberg, C. G.** Beitrag zur Bestimmung des stationären mikroskopischen Lebens in bis 20,000 Fuss Alpenhöhe. In: Abhandl. Akad. Berlin, 1858, p. 429–456, Pl. i–iii.

1859 (α) **Ehrenberg, C. G.** Über die mit dem Proteus anguinus (Hypochthon Laurenti) zusammenlebenden mikroskopischen Thierformen in den Bassins der Magdalenengrotte in Krain. In: Monatsber. Akad. Berlin, p. 758–775, 1 Pl.

1859 (β) **Ehrenberg, C. G.** Über das mikroskopische Leben des Montblanc-Gipfels...In: Monatsber. Akad. Berlin, p. 775–783.

1859 **Eichwald, K. E. von.** Lethaea rossica ou paléontologie de la Russie, décrite et figurée. 5 et 6 Livraison. Stuttgart. i, 2, p. 269–1004, Pl. i–xxix.

1859 **Étallon, A.** Études paléontologiques sur le Haut-Jura. Rayonnés du corallien. Par...In: Mém. Soc. d'Émul. du Doubs (3), iii, p. 401–553.

1859 **Foetterle, F.** Das Gebirge des Grossherzogthums Krakau, sowie West-Galiziens. Von...In: Jahrb. geol. Reichs-Anstalt, x, p. 100–104.

1859 **Gaudry, A.** Sur les résultats de fouilles géologiques entreprises aux environs d'Amiens. In: C. R. Paris, xlix, p. 465–467.

1859 **Gray, J. E.** On the Arrangement of Zoophytes with Pinnated Tentacles. By...In: Ann. Mag. N. H. (3), iv, p. 439–444.

1859 (α) **Gray, J. E.** Description of MacAndrewia and Myliusia, two new forms of Sponges. By...In: Proc. Zool. Soc. London, xxvii, p. 437–440, Pl. xv–xvi. Also in: Ann. Mag. N. H. (3), v (1860), p. 495–498.

1859 **Greene, J. R.** A Manual of the Sub-kingdom Protozoa. With a general introduction on the principles of zoology. London. 12°. 110 pp.

1859 **Grewingk, C.** (Letter from... to Bronn, containing results of researches on silurian formations in Livland, etc.) In: N. Jahrb. Mineral. p. 62–67.

1859 **Horion, C.** Sur le terrain crétacé de la Belgique. In: Bull. Soc. Géol. France (2), xvi, p. 635–666.

1859 **Jardin, E.** Essai sur l'histoire naturelle de l'archipel de Mendana ou des Marquises. 3ᵐᵉ Partie. Zoologie. In: Mém. Soc. Cherbourg, vi, p. 161–200.

1859 **Kölliker, R. A. von.** Ueber das ausgebreitete Vorkommen von pflanzlichen Parisiten in den Hartgebilden niederer Thiere. Von... In: Z. W. Z. x, p. 215–232, Pl. xv–xvi.

1859 **Lecoq, H.** Observations sur une grande espèce de Spongille (Ephydatia lacustris?) du Lac Pavin. In: Mém. Acad. Clermont-Ferrand, i, p. 476–493. (Also in: C. R. Paris, l (1860), p. 1116–1121.)

1859 **Lieberkühn, N.** Neue Beiträge zur Anatomie der Spongien. Von... In: Arch. Anat. und Phys. pp. 353–382, 515–529, Pl. ix–xi.

1859 **Nordman, A. von & C. T. E. von Siebold.** Über einen merkwürdigen Polypenstock. In: Amtl. Ber. 34. Vers. d. Naturf. und Aerzte, p. 202–203.

1859 **Roemer, C. F.** Ueber Silurische Spongien aus dem Staate Tennessee. In: Amtl. Ber. 34. Vers. d. Naturf. und Aerzte, p. 69–70.

1859 **Salter, J. W.** Fossils from the base of the Trenton Limestone. In: Geol. Survey Canada, i, p. 1–48, Pl. i–x.

1859 **Schlossberger, J. E.** Ueber die Unterscheidung des Fibroïns von der Substanz des Badeschwamms... In: Amtl. Ber. 34. Vers. d. Naturf. und Aerzte, p. 164–165.

1859 **Städeler, G.** Untersuchungen über das Fibroïn, Spongin und Chitin, nebst Bemerkungen über den thierischen Schleim; von... In: Vierteljahrsschr. naturf. Ges. Zürich, iv, p. 117–135. Also in: Ann. Chemie und Pharm. cxi (2), xxxv, p. 12–28.

1859 **Wetherell, N. T.** On the Occurrence of Graphularia Wetherellii in Nodules from the London Clay and the Crag. By... In: Quart. Journ. Geol. Soc. London, xv, 1, p. 30–32.

1859 **Wiltshire, T.** On the Red Chalk of England. In: Geologist, ii, p. 261–278.

1859 **Wood, S. V.** On the Extraneous Fossils of the Red Crag. By... In: Quart. Journ. Geol. Soc. London, xv, 1, p. 32–45.

1860 **Bowerbank, J. S.** Note on Prof. Kölliker on Parasites in Sponges. In: Quart. Journ. Microsc. Sc. viii, p. 187–188.

1860 (α) **(Bowerbank?)** in: Catalogue of the Contents of the Museum of the Royal College of Surgeons of England. Part i. Plants and Invertebrate Animals in the dried state. London. 4°.

1860 **Capellini, G. & H. A. Pagenstecher.** Mikroskopische Untersuchungen über den innern Bau einiger fossilen Schwämme. Von...In: Z. W. Z. x, p. 364–372, Pl. xxx.

1860 **Ébray, T.** Note sur la composition géologique du sol des environs de Mâcon, et calcul des dénudations qui se sont opérées dans cette contrée; par...In: Bull. Soc. Géol. France (2), xvii, p. 507–516.

1860 **Ehrenberg, C. G.** (Beiträge zur Beurtheilung der wunderbaren japanischen Glaspflanze, der sogenannten Corallenthier-Gattung Hyalonema und der Familie der Hyalochaetiden.) In: Monatsber. Akad. Wiss. Berlin, p. 173–182.

1860 (α) **Ehrenberg, C. G.** Über zwei Staub-Meteore aus Westphalen und Syrien...In: Monatsber. Akad. Berlin, p. 137–157, Pl. i–iii.

1860 (β) **Ehrenberg, C. G.** Über den am 24–25. Januar 1859 auf das amerikanische Schiff Derby bei den Capverden gefallenen Passatstaub. In: Monatsber. Akad. Berlin, p. 203–208.

1860 (γ) **Ehrenberg, C. G.** Vorlesungen einer grösseren Zahl neuer Zeichnungen der vermeintlich aus 19,800 Fuss Meeres-tiefe gehobenen Lebensformen, so wie neue Erläuterungen dieser Grund- und Wasser-Proben. In: Monatsber. Akad. Berlin, p. 588–592.

1860 (δ) **Ehrenberg, C. G.** Über die organischen und unorganischen Mischungsverhältnisse des Meeresgrundes in 19,800 Fuss Tiefe... In: Monatsber. Akad. Berlin, p. 765–774.

1860 (ε) **Ehrenberg, C. G.** Über den Tiefgrund des Stillen Oceans zwischen Californien und den Sandwich-Inseln aus bis 15,600 Tiefe...In: Monatsber. Akad. Berlin, p. 819–833.

1860 **Étallon, A.** Rayonnés du Jura supérieur de Montbéliard. In: Soc. Émul. Montbéliard, p....

1860 (α) **Étallon, A.** Sur la Classification des Spongiaires du Haut-Jura et leur distribution dans les étages, Par...In: Actes Soc. Jur. d'Émulation pendant l'année 1858... p. 129–160, 1 Pl.

1860 (β) **Étallon, A.** Paléontostatique du Jura. Jura Graylois. Faunes du terrain jurassique moyen, Par...In: Ann. Soc. d'Agricult. Lyon, iv, p. 145–177.

1860 **Fromentel, M. E. de.** Introduction à l'étude des Éponges fossiles; par...In: Mém. Soc. Linn. Normandie (2), xi, p. 1–50, Pl. i–iv*.

1860 **Gabb, W. M.** Description of a new genus and species of Amorphozoon, from the cretaceous formation of New Jersey (Desmatocium trilobatum). In: Proc. Acad. Sc. Philadelphia, iv, p. 518.

1860 **Gaudry, A.** Sur des éponges provenant des côtes de l'Attique. In: C. R. Paris, li, p. 460.

1860 **Gorup-Besanez, E. von.** Schwämme, Schwammalgen, Badeschwämme, Spongioideae. In: Handwörterb. Chemie, vii, p. 383–386.

1860 **Gosse, P. H.** Actinologia Brittanica: a History of the British Sea Anemones and Corals...London.

* On title-page of sep. copy erroneously 1859.

1860 **Goubert, E.** Quelques mots sur l'étage éocène moyen dans le bassin de Paris; par...In: Bull. Soc. Géol. France, (2), xvii, p. 137–149, Pl. ii.

1860 **Gray, J. E.** On the genus Hyalonema. By...In: Ann. Mag. N. H. (3), v, p. 229–230.

1860 **Hibberd, S.** Science on the sea-shore. i, Flints and Sponges... ii, Physiology of Sponges...In: Recreat. Sc. i, pp. 29–31, 109–112, 9 figs.

1860 **Hogg, J.** On the distinctions of a Plant and an Animal, and on a fourth Kingdom of Nature. In: Edinburgh New Phil. Journ. (2), xii, p. 216–225, Pl.

1860 **Lecoq, H.** Observations sur les corps reproducteurs et sur l'état d'agrégation d'une grande espèce de Spongille du lac Pavin... par...In: C. R. Paris, l, p. 1165–1170.

1860 (α) **Lecoq, H.** Observations sur le degré d'animalité et sur les espèces de Spongilles, et particulièrement sur la grande espèce du lac Pavin; par...In: C. R. Paris, li, p. 5–9.

1860 **Leidy, J.** Remarks on Hyalonema mirabilis. In: Proc. Acad. Nat. Sc. Philadelphia, iv, p. 85.

1860 **Meek, F. B. & A. H. Worthen.** Descriptions of new Carboniferous Fossils from Illinois and other Western States. In: Proc. Acad. Nat. Sc. Philadelphia, iv, p. 447–472.

1860 **Meyn, L.** Siphonia praemorsa, Goldfuss, rectius Astylospongia praemorsa, Ferd. Roemer. In: Mitth. Ver. Nörd. d. Elbe Verbreit. naturw. Kenntn. Kiel, iv, p. 23–31.

1860 **Pagenstecher, H. A.** Ueber den mikroskopischen Bau einiger fossilen Schwämme. In: Verh. naturh. medic. Ver. Heidelberg, ii, p. 6–7.

1860 **Roemer, C. F.** Die silurische Fauna des westlichen Tennessee. Eine palaeontologische Monographie. Von...Breslau, 1860. 4°. viii and 100 pp., Pl. i–v.

1860 **Schlossberger, J. E.** Eine Bemerkung über das Spongien-Fibroïn. In: Zeitschr. Chem. und Pharm. iii, p. 426–427.

1860 **Schultze, M. J. S.** Sur une nouvelle espèce d'éponges (Hyalonema) prise pour un polype; par...In: C. R. Paris, l, p. 792–793.

1860 (α) **Schultze, M. J. S.** Die Hyalonemen. Ein Beitrag zur Naturgeschichte der Spongien von...Bonn, bei Adolph Marcus...4°. iii and 46 pp., 5 Pl.

1860 (β) **Schultze, M. J. S.** Über eine merkwürdige Spongie, Hyalonema. In: Verh. naturh. Ver. Rheinl. und Westphal. xvii (2), vii, p. 67–69. Also in: Froriep's Notizen, ii, p. 312–314.

1860 (γ) **Schultze, M. J. S.** Über die Japanische Glasfadenspongie. In: Verh. naturh. Ver. Rheinl. und Westphal. xvii (2), p. 85.

1860 **Smith, J. A.** On a large Cup-shaped Sponge. In: Proc. Phys. Soc Edinburgh, ii, p. 227.

1860 **Staring, W. C. H.** Natuurlijke historie van Nederland. Tweede deel
...De bodem van Nederland...Haarlem. ii, 480 pp.

1860 **Stoppani, A.** Les Pétrifications d'Esino...In: Paléontol. lombarde,
9–13 (1), 7–11, p. 81–151, Pl. xvii–xxxi. Also separately?.

1860 **Valenciennes, A.** Note sur les Spongiaires envoyés des côtes de
l'Attique par M. Albert Gaudry; par...In: C. R. Paris, li,
p. 579–581.

1860 **Woods, J. E.** On some Tertiary Rocks in the colony of South
Australia. In: Quart. Journ. Geol. Soc. London, xvi, p. 253–260.

1861 **Ancelon,** ...Mémoire sur les Spongiles de l'étang de Lindre-Basse.
In: Mém. Acad. Metz, xlii, (2) ix, p. 467–470.

1861 **Artus, W.** Ueber das Bleichen der Badeschwämme. In: Polytech.
Journ. clxii, p. 79.

1861 **Billings, E.** Palaeozoic Fossils of the Geological Survey of Canada.
Vol. i...By...Fasc. 1. Montreal. p. 1–24. Also (?) as: New
Species of Lower Silurian Fossils.

1861 **Bosquet, J.** Coup-d'œil sur la répartition géologique et géographique
des espèces d'animaux et de végétaux citées dans le tableau des
fossiles crétacés du Limbourg, inseré dans la dernière livraison
de l'ouvrage du Dr W. C. H. Staring sur le sol de la Néerlande.
Par...In: Versl. en meded. Akad. Amsterdam, xi, p. 108–120.

1861 **Bouchard-Chantereaux,** ...Observations sur les Hélices saxicaves du
Boulonnais. In: Ann. Sc. Nat. xvi, p. 197–218, Pl. iv.

1861 **Bowerbank, J. S.** (See also **MacAndrew.**) On the Anatomy and Physiology of the Spongiadae. Part ii by...(abstract). In: Proc. R.
Soc. London, xi, p. 372–375. Also in: Ann. and Mag. N. H. viii,
p. 420–422.

1861 **Brady, G. S.** Spongilla fluviatilis...In: Trans. Tyneside Naturalists'
Field Club, v, p. 151.

1861 **Bureau, E.** Observations sur le terrain dévonien de la Basse-Loire.
In: Bull. Soc. Géol. France, (2), xviii, p. 337–340.

1861 **Carpenter, W. B.** Researches on Foraminifera...By...In: Phil.
Trans. Roy. Soc. London, cl, p. 535–594, Pl. xvii–xxii.

1861 **Carter, H. J.** Notes and Corrections on the Organization of Infusoria,
etc. By...In: Ann. and Mag. (3), viii, p. 281–290.

1861 **Courtiller, A.** Description des éponges fossiles des sables du terrain
crétacé supérieur des environs de Saumur...In: Ann. Soc. Linn.
Maine et Loire, iv (= Année ix), p. 110–142, Pl. i–xl.

1861 **Ehrenberg, C. G.** Über die neueren die japanische Glaspflanze
als Spongia betreffenden Ansichten und...Erläuterungen der
Synonyme zu Hrn. Bowerbank's Spongolithen-Tafeln. In:
Monatsber. Akad. Berlin, i, p. 448–452, Pl.

1861 (a) **Ehrenberg, C. G.** Beitrag zur Übersicht der Elemente des tiefen
Meeres-grundes im Mexicanischen Golfstrome bei Florida...
Von...In: Monatsber. Akad. Berlin, i, p. 222–240.

1861 (β) **Ehrenberg, C. G.** Über die Tiefgrund-Verhältnisse des Oceans am Eingange der Davisstrasse und bei Island...In: Monatsber. Akad. Berlin, i, p. 275–315.

1861 (γ) **Ehrenberg, C. G.** Über massenhaft jetzt lebende oceanische und die fossilen ältesten Pteropoden der Urwelt...In: Monatsber. Akad. Berlin, i, p. 434–446, Pl.

1861 (δ) **Ehrenberg, C. G.** Über das mikroskopische Leben auf der Insel St Paul im Süd-Ocean. In: Monatsber. Akad. Berlin, ii, p. 1085–1102.

1861 **Fromentel, E. de.** Catalogue raisonné des Spongitaires de l'étage Neocomien. In: Bull. Soc. Yonne, xiv, p. 355–372, Pl. i–iv.

1861 **Fyfe, W. W.** Sponge-Hunting in Holy Island and Berwick Bays. In: Recreat. Sc. ii, p. 4–8, 9 figs.

1861 **Geinitz, H. B.** Dyas oder die Zechsteinformation und das Rothliegende ...Heft i...Leipzig...4°. xvi and 130 pp., 23 Pl

1861 **Giebel, C. G. A.** Tertiäre Conchylien von Latdorf im Bernburgischen. von...In: Zeitschr. gesammt. Naturwiss. xvii, p. 30–47.

1861 **Gorup-Besanez, E. von.** (Art. Spongin.) In: Handwörterb. Chem. viii, p. 153.

1861 **Gosse, P. H.** An hour among the Torbay sponges. By...In: Good Words (Macleod), 1861, p. 293–298.

1861 **Grant, R. E.** Tabular View of the primary divisions of the Animal Kingdom. London.

1861 **Gregory, F. T.** On the Geology of a part of Western Australia. In: Quart. Journ. Geol. Soc. London, xvii, p. 475–483.

1861 **Grewingk, C.** Geologie von Liv- und Kurland...In: Arch. Naturk. Liv-, Ehst- und Kurlands (1), ii, p. 479–774.

1861 **Grube, A. E.** Mittheilungen über die Aufenthaltsorte der Anneliden. In: Amtl. Ber 35. Vers. deutscher Naturf. und Aerzte, p. 78–85.

1861 (α) **Grube, A. E.** Ein Ausflug nach Triest und dem Quarnero. Beiträge zur Kenntniss der Thierwelt dieses Gebietes. Von...Berlin. viii and 175 pp., 5 Pl.

1861 **Gümbel, C. W.** Geognostische Beschreibung des Königreichs Bayern. 1. Abtheilung...Gotha. 4°. x and 950 pp.

1861 **Hall, J.** Report of the superintendent of the Geological Survey of the State of Wisconsin...Madison.

1861 **Howell, H. H. & A. Geikie.** Memoirs of the Geological Survey of Great Britain. The Geology of the neighbourhood of Edinburgh. London.

1861 **Lamiral, E.** Sur l'acclimatation des Éponges dans les eaux de la France et de l'Algérie, Par...In: Bull. Soc. d'Acclim. viii, p. 327–334.

1861 **Lecoq, H.** Observations sur une grande espèce de Spongille du Lac Pavin, Par...Clermont. 20 pp.

1861 **MacAndrew, R.** List of the British Marine Invertebrate Fauna...In: Rep. 30th Meeting Brit. Assoc. (Oxford), p. 217–236. (For various groups, the author is assisted by others: for sponges by Bowerbank. Hence quoted as: **MacAndrew-Bowerbank.**)

1861 **Martens, E. von.** Über einen neuen Polyodon aus dem Yantsekiang und über die sogenannten Glaspolypen. In: Monatsber. Akad. Berlin, i, p. 476–480.

1861 **McBain, J.** Exhibition of Sponges, with Explanatory Remarks. In: Proc. Phys. Soc. Edinburgh, ii, p. 233–240.

1861 **Pélissier, A. J. J.** Lettre adressée à M. le Président de la Société impériale d'acclimatation. In: Bull. Soc. Zool. d'Acclim. viii, p. 572.

1861 **Pengelly, W.** On the Devonian age of the world. In: Geologist, iv, p. 332–347, Pl. v.

1861 **Reuss, A. E.** Ueber die fossile Gattung Acicularia, d'Arch. In: Sitzber. Akad. Wien, xliii, 1, p. 7–10, Pl. I.

1861 **Roemer, C. F.** ...Die fossile Fauna der silurischen diluvial-Geschiebe von Sadewitz bei Oels in Nieder-Schlesien...von...Breslau... 4º. xvi and 82 pp., 6 Pl.

1861 **Seeley, H. G.** Notice of opinions on the Stratigraphical Position of the Red Limestone of Hunstanton. By...In: Ann. and Mag. N. H. vii, p. 233–244.

1861 **Soubeiran, J. L.** Rapport sur le mémoire de M. Lamiral relatif à l'acclimatation des éponges dans les eaux de la France et de l'Algérie...In: Bull. Soc. Zool. Acclim. viii, p. 433–437.

1861 **Trautschold, H.** Recherches géologiques aux environs de Moscou. Couche jurassique de Mniovniki. In: Bull. Soc. Nat. Moscou, xxxiv, 1, p. 64–94, Pl. IV–VIII.

1862 **Balsamo Crivelli.** See Crivelli.

1862 **Bowerbank, J. S.** On the Anatomy and Physiology of the Spongiadae ...By...In: Phil. Trans. Roy. Soc. London, clii, pp. 747–829, 1087–1135, Pls. XXVII–XXXV, LXXII–LXXIV.

1862 (a) **Bowerbank, J. S.** Supplement to Part i. "On the Anatomy and Physiology of the Spongiadae" by...In: Phil. Trans. Roy. Soc. London, clii, p. 830–836, Pl. XXXVI.

1862 **Carpenter, W. B.** Introduction to the Study of the Foraminifera. By...London.

1862 **Coquand, H.** Description physique, géologique, paléontologique et minéralogique du département de la Charente. Besançon et Marseille, 1858–1862. 2 vols.

1862 **Crivelli, G. Balsamo.** Descrizione delle spugne esistenti nel Gabinetto di Storia Naturale del R. Universita di Pavia. In: Atti Instit. Lomb. iii, p. 337. Prelim. account of 1863, Crivelli.

1862 **Dawson, J. W.** On the Flora of the Devonian Period in North-Eastern America. In: Quart. Journ. Geol. Soc. London, xviii, p. 296–328, Pl. xii–xvii, also Silliman, xxxv, 1863, p. 311–319.

1862 **Ehrenberg, C. G.** Erläuterung eines neuen wirklichen Passatstaubes aus dem atlantischen Dunkelmeere...In: Monatsber. Akad. Berlin, p. 202–222, Pl.

1862 (a) **Ehrenberg, C. G.** Über die rothen Meteorstaubfälle im Anfang des Jahres 1862 in den Gasteiner und Rauriser Alpen und bei Lyon. In: Monatsber. Akad. Berlin, p. 511–535.

1862 (β) **Ehrenberg, C. G.** Zweite Mittheilung über die mikroskopischen Lebensformen als Nahrung des Höhlen-Salamanders. In: Monatsber. Akad. Berlin, p. 579–598.

1862 **Étallon, A.** Études paléontologiques sur le Haut-Jura. Par...Monographie du corallien. In: Mém. Soc. d'Émul. Doubs (3), vi, p. 53–244.

1862 (a) **Étallon, A.** Paléontostatique du Jura...Par...In: Mém. Soc. d'Émul. Doubs (3), vi, p. 245–260.

1862 **Geinitz, H. B.** Dyas...Heft 2...Leipzig. 4°. viii pp. and p. 131–342, Pl. xxiv–xlii.

1862 **Gümbel, C. W.** Die Streitberger Schwammlager und ihre Foraminiferen-Einschlüsse. In: Jahresh. Ver. vaterl. Naturk. Württemberg, xviii, p. 192–238, Pl. iii–iv.

1862 **Hall, J.** Notice of some new species of fossils from a locality of the Niagara Group in Indiana...In: Trans. Albany Instit. iv, p. 195–228.

1862 **Hall, J. & J. Whitney.** Report of the Geological Survey of the State of Wisconsin; by...Vol. i. Albany. 453 pp.

1862 **Hauer, F. von.** Über das Vorkommen der Trias-Kalksteine im Vertesgebirge und im Bakonyer Walde. In: Jahrb. Geol. Reichs-Anstalt, xii, p. 164–166.

1862 **Lamiral, E.** Rapport sur un essai d'acclimatation des éponges de Syrie dans les eaux françaises de la Méditerranée...In: Bull. Soc. d'Acclim. ix, p. 641–653.

1862 (a) **Lamiral, E.** Lettre adressée par...à M. le Président de la société impériale d'acclimatation, au sujet des Éponges dans la Méditerranée. In: Bull. Soc. d'Acclim. ix, p. 817.

1862 **MacCoy, F.** A Synopsis of the Silurian fossils of Ireland...by... London. 4°. 72 pp. and 4 Pl. Ed. ii of MacCoy, 1846.

1862 **Mackie, S. J.** What are the Ventriculites? In: Geologist, v, p. 161–167.

1862 **Roemer, C. F.** Ueber die Diluvial-Geschiebe von nordischen Sedimentär-Gesteinen...In: Zeitschr. d. geol. Gesellsch. xiv, p. 575–637.

1862 (a) **Roemer, C. F.** Notiz über die Auffindung einer senonen Kreide-
Bildung bei Bladen...In: Zeitschr. d. geol. Gesellsch. xiv,
p. 765–769.

1862 **Schmidt, (E.) O.** Die Spongien des adriatischen Meeres. Von...
Leipzig. 4°. viii and 88 pp., 7 Pl.

1862 (a) **Schmidt, E. O.** Die Einführung der künstlichen Schwammzucht
in Dalmatien. Von...In: Triester Zeitung, xii, no. 60.

1862 **Schultze, M. J. S.** Ueber Polytrema miniaceum, Blainv. In: Verh.
naturh. Ver. Rheinland und Westphalens, xix, (2) ix, p. 13–14.

1862 **Senoner, A.** Sulle spugne del Mare Adriatico...In: Bull. Naut. e
Geograf. ii, p. 36–38.

1862 **Terquem, O. & E. Piette.** Le Lias inférieur de la Meurthe...par...
In: Bull. Soc. Géol. France (2), xix, p. 322–394, Pl. viii bis.

1862 **Wallich, G. C.** The North-Atlantic Sea-bed;...Part i. London. 4°.

1862 **White, M. C.** Discovery of Microscopic Organisms in the Siliceous
Nodules of the Palaeozoic Rocks of New York. In: Amer. Journ.
Sc. and Arts (2), xxxiii, p. 385–386.

1863 **Balsamo Crivelli.** See Crivelli.

1863 **Beck, E. G.**
In: Monatsber. Geburtsk. xxii, p. 386–...

1863 **Billings, E.** Geological Survey of Canada. Report of progress from
its commencement to 1863. Montreal. 983 pp., 498 figs.

1863 **Bowerbank, J. S.** A Monograph of the Spongillidae. By...In: Proc.
Zool. Soc. London, p. 440–472, Pl. xxxviii.

1863 **Crivelli, G. Balsamo.** Di alcuni Spongiari del Golfo di Napoli.
Memoria del Professore...In: Atti Soc. Ital. v, p. 284–302, Pl.
iv–vi.

1863 (a) **Crivelli, G. Balsamo.** Descrizione delle Spugne esistenti nel
Gabinetto di Storia naturale della regia Università di Pavia con
premessi alcuni cenni sulle raccolte zoologiche dello stesso.
Memoria di...In: Mem. Instit. Lomb. ix, p. 325–341, Pl. x–xi.
Sep. copy, p. 1–19, Pl. i–ii.

1863 **Dana, J. D.** On two Oceanic Species of Protozoans related to the
Sponges. By...In: Amer. Journ. Sc. and Arts, (2), xxxv, p. 386–
388. Also in: Ann. and Mag. N. H. (3), xii, p. 54–55, fig. 1–2.

1863 **Drescher, R.** Ueber die Kreide-Bildungen der Gegend von Löwen-
berg. In: Zeitschr. d. geol. Ges. xv, p. 291–366, Pl. viii–ix.

1863 **Ehrenberg, C. G.** Über das unsichtbar wirkende Leben im Mittel-
meere und in den sich an dasselbe ostwärts nach Central-Asien
hin anschliessenden Meeren und Seeen. In: Monatsber. Akad.
Berlin, p. 291–295.

1863 (α) **Ehrenberg, C. G.** Über die nicht natürliche Gestaltung der Glaspflanze Hyalonema Sieboldi Gray. In: Monatsber. Akad. Berlin, p. 300–305.

1863 (β) **Ehrenberg, C. G.** Beitrag zur Kenntniss der unterseeischen Agulhas-Bank an der Südspitze Afrikas als eines sich kundgebenden grünsandigen Polythalamien-Kalkfelsens. In: Monatsber. Akad. Berlin, pp. 379–394, Map.

1863 (γ) **Ehrenberg, C. G.** Über Hyalonema Sieboldii. In: Sitzungsber. Ges. naturf. Freunde Berlin, p. 13.

1863 **Geinitz, H. B.** Über zwei neue dyadische Pflanzen, von...In: N. Jahrb. Mineral. p. 525–530.

1863 **Hall, J.** Note on the geological range of the genus Receptaculites in American Paleozoic strata. In: 16 Annual Rep. State Cabinet New York, p. 67–69.

1863 (α) **Hall, J.** Note on the occurrence of Astylospongia in the Lower Heidelberg Rocks. In: 16 Annual Rep. State Cabinet New York, p. 69–70.

1863 (β) **Hall, J.** Observations upon some spiral growing Fucoidal remains of the Palaeozoic Rocks of New York. In 16 Annual Rep. State Cabinet New York, p. 76–83.

1863 (γ) **Hall, J.** Observations upon the genera Uphantaenia and Dictyophyton; with notices of some species from the Chemung Group of New York and the Waverley Sandstone of Ohio. In: 16 Annual Rep. State Cabinet New York, p. 84–91.

1863 **Harting, P.** Bijdrage tot de kennis der mikroskopische fauna en flora van de Banda-zee, naar aanleiding van een onderzoek van eenige door diepzeeloodingen van 900 tot 4000 vademen uit die zee opgebragte gronden. Door...In: Verhand. K. Akad. Wetensch. Amsterdam, x, p. 1–34, Pl. 1–3.

1863 **Heller, C.** Die Crustaceen des südlichen Europa. Crustacea podophtalmia...Wien. xi and 356 pp., 10 Pl.

1863 **Hoernes, M.** (Über Coeloptychien (Spongien) aus der Oberen Kreide von Vordorf unfern Braunschweig.) In: Jahrb. geol. Reichs–Anstalt, xiii, p. 40–41.

1863 **Johnson, J. Y.** Description of a New Siliceous Sponge from the Coast of Madeira. By...In: Proc. Zool. Soc. London, p. 257–259. Also in: Ann. and Mag. N. H. (3), xiii, p. 257–258.

1863 **Kölliker, R. A. von.** Ueber den Bau der Spongien. In: Würzburger naturw. Zeitschr. iv, p. xiv.

1863 **Lamiral, E.** Second rapport sur un essai d'acclimatation des éponges de Syrie dans les eaux françaises de la Méditerranée, Par... In: Bull. Soc. d'Acclim. x, p. 8–15.

1863 **Lieberkühn, N.** Ueber Bewegungserscheinungen bei den Schwämmen Von...In: Arch. Anat. und Phys. p. 717–730, Taf. xix.

1863 Lorenz, J. R. Physicalische Verhältnisse und Vertheilung der Organismen im Quarnerischen Golfe. Wien. xii and 382 pp., 5 Pl.

1863 Loriol, P. de. Description des animaux invertebrés fossiles contenus dans l'étage néocomien moyen du mont Salève. Par...Bâle-Genève. 4°. 214 pp., 22 Pl.

1863 Meyer, C. J. A. Three Days at Farringdon. Position of Sponge-Gravel. In: Geologist, vii, 1864, p. 5–11.

1863 Moesch, C. (Vorläufiger Bericht über die Ergebnisse der im Sommer 1862 ausgeführten Untersuchung im weissen Jura der Kantone Solothurn und Bern.) In: Verh. Schweiz. naturf. Ges. p. 156–168.

1863 Noguès, A. F. Note sur une Grauwacke dévonienne fossilifère des Pyrénées, par...In: C. R. Acad. Paris, lvi, p. 1122–1123.

1863 Oppel, A. Ueber jurassische Cephalopoden (Fortsetzung); von...In: Palaeontol. Mittheil. Mus. bayer. Staates, i, 2, p. 163–266.

1863 Roemer, C. F. Über Vorarbeiten zur Herstellung einer Geologischen Karte von Ober-Schlesien; von...In: N. Jahrb. Mineral. p. 334–342.

1863 (α) Roemer, C. F. Geognostische Bemerkungen auf einer Reise nach Constantinopel, von...In: N. Jahrb. Mineral. p. 513–524.

1863 Rummel, F. Beiträge zur Kenntniss der Trias Unterfrankens; von... In: N. Jahrb. Mineral. p. 786–800.

1863 Schafhäutl, K. F. E. Süd-Bayerns Lethaea geognostica...Leipzig. Fol. xvii and 487 pp., 98 Pl., 2 maps.

1863 Schmidt, E. O.
In: Grazer Zeitung, no. 161.

1863 Schultze, M. J. S. Ueber Polytrema miniaceum, ein Polythalamie. Von...In: Arch. Naturgesch. xxix, 1, p. 81–102, Pl. viii.

1863 Stimpson, W. Notes on the "Glass Coral" of Japan (Hyalonema mirabilis). In: Amer. Journ. Sc. and Arts (2), xxxv; p. 458–459.

1863 Thurmann, J. Lethaea Bruntrutana...Œuvre posthume de...terminée et publiée par A. Étallon: 3 Partie. In: N. Denkschr. schweiz. Ges. xx [(2) x], p. 357–500, Pl. L–LXII.

1864 Alberti, F. A. von. Ueberblick über die Trias, mit Berücksichtigung ihres Vorkommens in den Alpen...Stuttgart. 355 pp.

1864 Beltremieux, E. Faune du département de la Charente-Inférieure. La Rochelle. 1°.

1864 Bocage, J. V. Barboza du. Note sur la Découverte d'un Zoophyte de la famille Hyalochaetides sur la côte du Portugal. Par...In: Proc. Zool. Soc. London, p. 265–269, Pl. xxii.

1864 Boettger, R. Ueber das verschiedene Verhalten rother Pflanzenpigmente zur Schwammsubstanz und ein darauf gegründetes Verfahren echten Rothwein von künstlich gefärbten zu unterscheiden. Von... In: Journ. prakt. Chemie, xci, p. 246–247.

1864 **Bowerbank, J. S.** A Monograph of the British Spongiadae. By...
Vol. i. London: Published for the Ray Society...8°. xx and
290 pp., 37 Pl. Abstr. and Rev. in: Athenaeum (1865), p. 186–
187.

1864 (a) **Bowerbank, J. S.** Description of two American Sponges. By...
In: Canadian Naturalist, New Ser. vol. i, p. 304–307.

1864 **Burgersdijk, L. A. J.** Een blik op het maaksel en de levensverschijn-
selen der Sponsen...In: Jaarb. Zoöl. Gen. Amsterdam, p. 141–
161.

1864 **Carpenter, W. B.** ("On the Structure and Affinities of Eozoon Canadense.")...By...In: Proc. R. Soc. London, xiii, p. 545–549.
Also in: Ann. and Mag. N. H. (3), xv, p. 325–329.

1864 **Deslongchamps, E. E.** Études sur les étages jurassiques inférieurs de
la Normandie. In: Mém. Soc. Linn. Normandie, xiv, 1, p...

1864 **Duchassaing de Fonbressin, P. & G. Michelotti.** Spongiaires de la
Mer caraïbe, par...Mémoire publié par la Société hollandaise des
sciences à Harlem...In: Natuurk. Verh. Mij. Haarlem, xxi,
p. 1–124, Pl. i–xxv.

1864 **Ehrenberg, C. G.** (Über eine neue Erweiterung der Kenntniss des das
caspische Meer characterisirenden reichen Lebens.) In: Monatsber. Akad. Berlin, p. 182–183.

1864 **Étallon, A.** Études paléontologiques sur le Jura graylois, par...In:
Mém. Soc. d'Émul. Doubs (3), viii, p. 221–506.

1864 **Fromentel, M. E. de.** Polypiers coralliens des environs de Gray...
In: Mém. Soc. Linn. Normandie, xiv, 4.

1864 **Geinitz, H. B.** Zwei Arten von Spongillopsis: S. dyadica, S. carbonica.
In: N. Jahrb. Mineral. p. 517–519.

1864 **Göppert, H. R.** Die fossile Flora der Permischen Formation. In:
Palaeontogr. xii, 1, p. 1–56, Pl. i–xxxi.

1864 **Gray, J. E.** Notes on Pustularia rosea, Gray, and Hyalonema. By...
In: Ann. and Mag. N. H. (3), xiii, p. 111.

1864 **Gressly, A.** (Spongille trouvée dans le lac de Neuchâtel.) In: Bull.
Soc. Sc. Nat. Neuchâtel, vi, pp. 286, 304.

1864 **Grube, A. E.** Die Insel Lussin und ihre Meeresfauna...Breslau.
v and 116 pp.

1864 **Guillaume.** ...Notice sur les éponges du lac de Neuchâtel...Par...
In: Bull. Soc. Sc. Nat. Neuchâtel, vi, p. 405–411, 1 Pl.

1864 **Heller, C.** Horae Dalmatinae...Von...In: Verh. zool. botan. Ges.
Wien, xiv, p. 17–64.

1864 **Huxley, T. H.** Lectures on the Elements of Comparative Anatomy
in the Classification of Animals and in the Vertebrate Skull.
London. 290 pp.

1864 **Jacquet, E.** Description géologique des falaises de Biarritz, Bidart, Guétary et Saint-Jean-de-Luz. In: Actes Soc. Linn. Bordeaux, xxv (3), v, p. 1–58, Map.

1864 **Kölliker, A.** Icones histiologicae oder Atlas der vergleichenden Gewebelehre. Herausgegeben von...Erste Abtheilung. Der feinere Bau der Protozoen. Mit ix Tafeln und 15 Holzschnitten. Leipzig, Engelmann. 1864. 4º. Vide: pp. iv, 5, 6, 46–74, 82–84, fig. 4–15, Pl. vii, fig. 6–13, Pl. viii–ix.

1864 **Merian, P.** Über die Stellung des Terrain à Chailles in der Schichtenfolge der Juraformation von...In: N. Jahrb. Mineral. p. 520–529.

1864 **Meyer, C. J. A.** Notes on the Branchiopoda from the Pebble-bed of the Lower Greensand of Surrey...By...In: Geol. Mag. i, p. 249–257, Pl. xi–xii.

1864 **Niles, W. H.** (On Pasceolus Halli, Billings, and its relation to the other Cystideans.) In: Proc. Boston Soc. x, p. 19–20.

1864 **Roemer, F. A.** Die Spongitarien des norddeutschen Kreidegebirges ...von...In: Palaeontogr. xiii, pp. i–iv, 1–64, Pl. i–xix.

1864 **Salter, J. W.** On some new fossils from the Lingula-flags of Wales. In: Quart. Journ. Geol. Soc. London, xx, p. 233–241, Pl. xiii.

1864 **Sandberger, K. L. F.** Beobachtungen in der Würzburger Trias...In: Würzburger naturw. Zeitschr. v, p. 201–231.

1864 **Sars, M.** Bemaerkninger over det dyriske Livs Udbredning i Havets Dybder. In: Forhandl. Vidensk. Selsk. Christiania, p. 53–68.

1864 **Schmidt, (E.) O.** Supplement der Spongien des adriatischen Meeres. Enthaltend die Histologie und systematische Ergänzungen... Leipzig. 4º. vi and 48 pp., 4 Pl. Abstr. and rev. in: Gött. Gel. Anz. p...; Rec. Zool. Literat. ii (1866), pp. 788, 794–795.

1864 **Seeley, H. G.** On the Fossils of the Hunstanton Red Rock. By...In: Ann. and Mag. N. H. (3), xiv, p. 276–280.

1864 **Slegt, H. D.** Mededeeling namens...door G. F. de Bruijn Kops omtrent sponsbanken nabij het eiland Kompol, op den Z. W. hoek van Borneo. In: Natuurk. Tijdschr. Nederl. Indië, xxvii (6), ii, p. 438–439.

1864 **Tate, R.** On the Correlations of the cretaceous formation of the North-east of Ireland. In: Abstr. Proc. Geol. Soc. London, p.... Also in: Phil. Mag. and Journ. Sc. xxviii, p. 562.

1864 **Ulex, G. L.** Ueber die Verbreitung des Kupfers im Thierreich. In: Pharm. Zeitschr. Russland, iv, p. 322–328. Also in: Journ. prakt. Chemie, xcv (1865), p. 367–374.

1864 **Verneuil, P. E. P. de.** Note sur les fossiles recueillis en 1863 par M. de Tchihatchef aux environs de Constantinople. In: Bull. Soc. Géol. France (2), xxi, p. 147–156.

1864 **Waagen, W.** Der Jura in Franken, Schwaben und der Schweiz...
In: Jahresber. Ver. vaterl. Naturk. Württemberg, xix, 1863,
p. 117–350.

1864 **Wallich, G. C.** On the Process of Mineral Deposit in the Rhizopods
and Sponges...By...In: Ann. and Mag. N. H. (3), xiii, p. 72–82,
fig. 1–5.

1865 **Baily, W. H.** The Cambrian rocks of the British Islands, with especial
reference to the occurrence of this Formation and its Fossils in
Ireland. In: Geol. Mag. ii, p. 385–400.

1865 **Baird, W.** On new Tubicolous Annelids, in the Collection of the
British Museum. Part 2. By...In: Journ. Linn. Soc. London,
viii, p. 157–160, Pl. v.

1865 **Bocage, J. V. Barboza du.** Sur l'Habitat du Hyalonema lusitanicum.
Par...In: Proc. Zool. Soc. London, p. 662–663.

1865 **Benecke, E. W.** Über Trias und Jura in den Südalpen von...In:
Geogn. paläontol. Beitr. i, 1, pp. iii–iv, 1–204, Pl. i–xi and 6 figs.

1865 **Billings, E.** Geological Survey of Canada. Palaeozoic fossils. Vol. i.
Containing descriptions and figures of new or little known species
of organic remains from the Silurian Rocks. Montreal, 1865
(read: 1861–1865), 426 pp.*

1865 **Carpenter, W. B.** Additional Note on the Structure and Affinities of
Eozoön Canadense. By...In: Quart. Journ. Geol. Soc. London,
xxi, p. 59–66, Pl. viii–ix and 1 woodcut. Also in: Canadian
Naturalist (2), ii, p. 111–119.

1865 **Dawson, J. W.** On the Structure of certain Organic Remains in the
Laurentian Limestones of Canada. By...In: Quart. Journ.
Geol. Soc. London, xxi, p. 51–59, Pl. vi–vii. Also in: Canadian
Naturalist (2), ii, p. 99–111.

1865 **Duncan, P. M.** A Description of the Echinodermata from the Strata
on the South-eastern Coast of Arabia, and the Bagh on the
Nerbudda...By...In: Quart. Journ. Geol. Soc. London, xxi,
p. 349–363, fig. 1–4.

1865 **Eck, H.** Über die Formationen des bunten Sandsteins und des
Muschelkalks in Oberschlesien und ihre Versteinerungen. Berlin.
viii and 150 pp., 1 table and 2 Pl.

1865 **Ferry, H. de.** Notes sur les crustacés et les spongitaires de la base de
l'étage Bathonien des environs de Mâcon. In: Bull. Soc. Linn.
Normandie, ix, p. 365–375, 2 Pl.

1865 **Fritzsche, C. J.** Bemerkungen zu der Abhandlung des Hrn. Goebel:
"Untersuchung des Carnallits von Maman..." Von...In: Bull.
Acad. St Pétersbourg, ix, coll. 56–64. Also in: Journ. f. prakt.
Chemie, xcvii (1866), p. 30–37.

---

* Fasc. 1 (p. 1–24) appeared in 1861 (cf. 1861); fasc. 2–3 (p. 25–168) in 1862;
fasc. 4–5 (p. 169–420) in 1865.

1865 **Goebel, A.** Untersuchung des Carnallits von Maman in Persien und über die wahre Ursache der rothen Färbung mancher natürlichen Salze, von...In: Bull. Acad. St Pétersbourg, ix, coll. 1–26. Also in: Journ. f. prakt. Chemie, xcvii (1866), p. 6–29.

1865 **Gosse, P. H.** A year at the shore by...with thirty-six illustrations by the author, printed in colours by Leighton brothers...London.

1865 **Gümbel, C. W.** Die Nummuliten-führenden Schichten des Kressenbergs...von...In: N. Jahrb. Mineral. p. 129–170.

1865 **Hyatt, A.** Remarks on the Beatriceae, a new division of Mollusca. In: Amer. Journ. Sc. and Arts (2), xxxix, p. 261–266.

1865 **Julien, A. A.** On two varieties of sponge-spicules. In: Amer. Journ. Sc. and Arts (2), xl, p. 379–383, 4 figs.

1865 **Kestevan, W. E.** (On the spongeous origin of Flint fossils: a correction.) In: Geol. Mag. ii, p. 329–330.

1865 **Laube, G. C.** Die Fauna der Schichten von St. Cassian...1. Abtheilung. Spongitarien...In: Denkschr. Akad. Wien, xxiv, p. 223–296, Pl. I–X.

1865 **Lieberkühn, N.** Beiträge zur Anatomie der Kalkspongien. Von... In: Arch. Anat. und Phys. p. 732–748, Pl. XIX.

1865 **Martin, J.** Zone à Avicula contorta ou étage rhaetien. Par...In: Mém. Acad. Dijon, xii, p. 1–270, Pl. I- III.

1865 **Müller, F.** Ueber Darwinella aurea, einen Schwamm mit sternförmige Hornnadeln. Von...In: Arch. Mikrosk. Anat. i, p. 344–353, Pl. XXI.

1865 **Oppel, A.** Geognostische Studien in dem Ardèche Département von ...In: Palaeont. Mitth. Mus. bayer. Staates, i, 3, p. 305–322.

1865 **Salter, J. W. & W. F. Blanford.** Palaeontology of Niti in the Northern Himalaya...

1865 **Schafhäutl, K. F. E.** Die Nummuliten-führenden Schichten des Kressenberges...In: N. Jahrb. Mineral. p. 769–788, Pl. V.

1865 **Schultze, M. J. S.** (Ueber ein Exemplar von Hyalonema Sieboldi aus Japan und einen Schwamm mit Nadeln aus Hornsubstanz.) In: Verhandl. naturh. Ver. preuss. Rheinl. und Westphalens, xxii [(3), ii], p. 6–7.

1865 **Spratt, T. A. B.** Travels and Researches in Crete. By...Vol. i. London.

1865 **Tate, R.** On the Correlation of the Cretaceous Formations of the North-east of Ireland. By...In: Quart. Journ. Geol. Soc. London, xxi, p. 15–44, Pl. III–v.

1865 **Terquem, O. & E. Piette.** Le Lias inférieur de l'est de la France... par...In: Mém. Soc. Géol. France (2), viii, mém. No. i, p. 1–175, Pl. I–XVIII.

1865 **Winchell, A.** Descriptions of new species of Fossils, from the Marshall Group of Michigan...By...In: Proc. Acad. Nat. Sc. Philadelphia (2), ix, p. 109–133.

81

1865 **Winchell, A. & O. Marcy.** Enumeration of Fossils collected in the Niagara Limestone at Chicago...In: Mem. Boston Soc. i, 1, p. 81–113, Pl. II-III.

1866 **Archiac, E. J. A. d', P. Fischer & P. E. P. de Verneuil.** Asie Mineure, description physique de cette contrée par M. Pierre de Tchihatcheff...Paléontologie de l'Asie Mineure. Paris. 424 pp. Atlas with 20 Pl.

1866 **Balsamo-Crivelli, G.** See Crivelli.

1866 **Beltrémieux, E.** Faune fossile du département de la Charente Inférieure. In: Ann. Acad. La Rochelle, xii, p....

1866 **Beneden, P. J. van.** (Coelenteraten-Natur Spongien.) In: Amtl. Ber. 40. Vers. d. Naturf. und Aerzte, p. 263.

1866 (a) **Beneden, P. J. van.** Recherches sur la faune littorale de Belgique. (Polypes), par...In: Mém. Acad. Belgique, xxxvi, p. 1-207, Pl. I-XIX.

1866 **Billings, E.** Geological Survey of Canada. Catalogue of the Silurian fossils of the Island of Anticosti...Montreal. 93 pp., 28 figs.

1866 **Bowerbank, J. S.** A Monograph of the British Spongiadae. By... vol. ii. London: published for the Ray Society by Robert Hardwicke...8⁰. 388 pp.

1866 (a) **Bowerbank, J. S.** On Hyalonema mirabilis, in reply to Dr Gray. By...In: Ann. and Mag. N. H. (3), xviii, p. 397-401.

1866 **Clark, H. J.** (On the Nature of Sponges.) In: Proc. Boston Soc. xi (1866-1868), p. 16-17.

1866 (a) **Clark, H. J.** Conclusive proofs of the animality of the ciliate Sponges and of their affinities with the Infusoria Flagellata; by... In: Amer. Journ. Sc. and Arts (2), xlii, p. 320-324. Also in: Ann. and Mag. (3), xix (1867), p. 13-18. Abstr. and rev. in: Rec. Zool. Literat. iii (1867), p. 642-644.

1866 **Claus, C. F. W.** (Ueber die Coelenteraten-Natur der Spongien.) In: Amtl. Ber. 40. Vers. d. Naturf. und Aerzte, p. 263.

1866 **Cohn, F. J.** (Ueber organische Einschlüsse im Carnallit von Stassfurt.) In: 43. Jahresber. schles. Ges. p. 54-56.

1866 **Crivelli, G. Balsamo.** Di alcuni organi speciali osservati in una spugna...In: Rendiconti R. Istit. Lomb. iii, p. 333-336, 1 Pl.

1866 **Dittmar, A. von.** Zur Fauna der Hallstädter Kalke...Von...In: Geognost. paläont. Beitr. i, 2, p. 319-397, Pl. XII-XX.

1866 **Ehrenberg, C. G.** Über Hyalonema lusitanicum (und über die thierische oder Pflanzen-Natur der Schwämme.) In: Monatsber. Akad. Berlin, p. 823-837. Transl. into English by Dallas, in Ann. and Mag. N. H. (3), xix (1867), p. 419-427.

1866 (a) **Ehrenberg, C. G.** (Vorläufige Mittheilung) Über einen Phytolitharien-Tuff als Gebirgsart im Toluca-Thale von Mexiko. In: Monatsber. Akad. Berlin, p. 158-168, 1 Pl.

B. S.   6

1866 **Eichwald, K. E. von.** Lethaea rossica ou Paléontologie de la Russe...
par...Stuttgart...Vol. ii. (1), (Livraison ix) 224 pp., 15 Pl. in
Atlas.

1866 (a) **Eichwald, K. E. von.** Ueber die Neocomschichten Russlands.
Von...In: Zeitschr. d. geol. Ges. xviii, p. 245–280, Pl. ii.

1866 **Grave.** ...Sur le tissu sarcodique de l'Éponge. Note de...In: C. R.
Paris, lxiii, p. 54–55.

1866 **Gray, J. E.** Note on the "Glass-Rope" Hyalonema. By...In: Ann.
and Mag. N. H. (3), xviii, p. 287–296.

1866 (a) **Gray, J. E.** Notes on Dr Bowerbank's Paper on Hyalonema.
By...In: Ann. and Mag. N. H. (3), xviii, p. 485–486.

1866 (β) **Gray, J. E.** Venus's Flower-basket (Euplectella speciosa). By...
In: Ann. and Mag. N. H. (3), xviii, p. 487–490.

1866 **Grotrian.** ...Ueber Coeloptychien der Oberen Kreide von Vordorf.
In: Amtl. Ber. 40. Vers. d. Naturf. und Aerzte, p. 148–149.

1866 **Guppy, R. J. L.** On the Relations of the Tertiary Formations of the
West Indies. By...In: Quart. Journ. Geol. Soc. London, xxii,
1, p. 570–593, Pl. xxvi. [Sponge, p. 584 and fig. 19.]

1866 **Haeckel, E. H. P. A.** Generelle Morphologie der Organismen...
Berlin. i. xxxii and 574 pp.; ii. clx and 462 pp.

1866 **Hesse, E.** Observations sur des Crustacés rares ou nouveaux des
côtes de France...In: Ann. Sc. Nat. (5), vi, p. 51–87, Pl. iv.

1866 **Hughes, W. R.** Notes on the Development of a Deep-Sea Sponge.
In: Rep. 35th Meeting Brit. Assoc. p. 86–87.

1866 **Lieberkühn, N.** Über den Furchungsprocess der Spongillen-Eier. In:
Sitz. Bes. Ges. naturf. Freunde Berlin, p. 20.

1866 **Loriol, P. de.** Description des fossiles de l'Oolithe corallienne...
Genève. 4º. 100 pp., 6 Pl.

1866 **Ludwig, R. A. B. S.** Corallen aus paläolothischen Formationen.
Von...In: Palaeontographica, xiv, p. 133–244, Pl. xxxi–lxxii*.

1866 **Mackie, S. J.** An Illustrated Catalogue of British Fossil Sponges...
i. London.

1866 **Meek, F. B. & A. H. Worthen.** Descriptions of Invertebrates from
the Carboniferous System. In: Geol. Survey of Illinois, ii, p. 145–
411, Pl. xiv–xx, xxiii–xxxii.

1866 (a) **Meek, F. B. & A. H. Worthen.** Descriptions of Palaeozoic fossils
from the Silurian, Devonian, and Carboniferous Rocks of Illinois
and other Western States. In: Proc. Chicago Acad. i, p. 11–23.

1866 **Müller, A.** Ueber Süsswasserschwämme oder Spongillen. In:
Schriften phys. ökon. Ges. Königsberg, vii. Sitzungsber. p. 13–14.

1866 **Nadler, G.** Ueber den angeblichen Jodgehalt der Luft und verschiedener Nahrungsmittel. Von...In: Journ. prakt. Chemie,
xcix, p. 183–206.

---

* The first part (in which no sponges are mentioned) appeared 1865.

1866 **Oppel, A.** Über die Zone des Ammonites transversarius von...
Beendet und herausgegeben von Dr. W. Waagen. In: Geognost
paläontol. Beitr. i, 2, p. 205–318.

1866 **Plant, J.** Notes relating to the Discovery of Primordial Fossils in
the Lingula-Flags in the Neighbourhood of Tyddyngwladis
Silverlead Mine. By...(abstract). In: Quart. Journ. Geol. Soc.
London, xxii, p. 505–506.

1866 **Quenstedt, F. A.** Handbuch der Petrefaktenkunde. 2. Auflage...
Tübingen, 1867*. 982 pp., 86 Pl. and 185 woodcuts.

1866 **Reuss, A. E.** Die Bryozoen, Anthozoen und Spongiarien des braunen
Jura von Balin bei Krakau. Von...In: Sitzungsber. Akad.
Wien, liii, (i), p. 229–231.

1866 **Roemer, C. F.** Geognostische Beobachtungen im Polnischen Mittel-
gebirge. Von...In: Zeitschr. d. geol. Ges. xviii, p. 667–690, Pl. xiii.

1866 **Roemer, F. A.** Beiträge zur geologischen Kenntniss des nordwestlichen Harzgebirges. Von...5. Abth. In: Palaeontogr. xiii,
p. 201–235, Pl. xxxiii–xxxv.

1866 **Schmidt, E. O.** Zweites Supplement der Spongien des adriatischen
Meeres. Enthaltend die Vergleichung der adriatischen und
britischen Spongiengattungen...Von...Leipzig. 4º. iv and
24 pp., 1 Pl.

1866 (a) **Schmidt, E. O.** Vorläufiger Bericht über die Untersuchung der
Bowerbank'schen Spongien. In: Sitzungsber. Akad. Wien, liii,
p. 147–151.

1866 **Seebach, K. von.** Die Zoantharia perforata der paläozoischen Periode.
Von...In: Nachrichten Ges. und Univ. Göttingen, p. 235–243.

1866 **Seeley, H. G.** Notice of Torynocrinus and other new and little-known
Fossils from the Upper Greensand of Hunstanton...By...In:
Ann. and Mag. N. H. (3), xvii, p. 173–183.

1866 **Suess, E.** On the Existence of Hyalonema in a Fossil State. By...
In: Ann. and Mag. N. H. (3), xviii, p. 404.

1866 **Waagen, W.** Nachträge zu seiner Abhandlung über die Classification
der Schichten des oberen Jura. In: N. Jahrb. Mineral. p. 570–574.

1866 **Whitaker, W.** On the "Lower London Tertiaries" of Kent. By...
In: Quart. Journ. Geol. Soc. London, xxii, 1, p. 404–435 Pl. xxii.

1866 **Würtenberger, F. J. & L. Würtenberger.** Der weisse Jura im Klettgau
und angrenzenden Randengebirg. In: Verh. naturw. Ver. Carls-
ruhe, 2, p. 11–68.

1866 **Zeuschner, L.** Über die verschiedenen Formationen, auf denen sich
der polnische Jura abgesetzt hat, von...In: N. Jahrb. Mineral.
p. 788–800.

1867 **Bigsby, J. J.** A brief Account of the "Thesaurus Siluricus," with a
few facts and inferences. By...In: Proc. R. Soc. London, xv,
p. 372–385. Also in: Ann. and Mag. N. H. (3), xix, p. 289–302.

* Read 1865–1866; 1st part (p. 1–320, 24 Pl.) appeared 1865, the rest 1866.

1867 **Bocage, J. V. Barboza du.** On Hyalonema lusitanicum. By...In: Ann. and Mag. N. H. (3), xx, p. 123–127.

1867 **Bowerbank, J. S.** On Hyalonema mirabile. By...In: Proc. Zool. Soc. London, p. 18–34, Pl. ɪv–ᴠ.

1867 (a) **Bowerbank, J. S.** Additional observations on Hyalonema mirabile. By...In: Proc. Zool. Soc. London, p. 350–351.

1867 (β) **Bowerbank, J. S.** On Alcyoncellum speciosum. By...In: Proc. Zool. Soc. London, p. 351–359.

1867 (γ) **Bowerbank, J. S.** On Hyalonema lusitanicum. By...In: Proc. Zool. Soc. London, p. 901–907.

1867 **Clark, H. J.** On the Spongiae Ciliatae as Infusoria Flagellata; or, Observations on the Structure, Animality, and Relationship of Leucosolenia botryoides, Bowerbank. By...In: Mem. Boston Soc. i (1866–1869)*, p. 305–340, Pl. 9–10. Also in: Ann. and Mag. (4), i (1868), pp. 133–142, 188–215, 250–264, Pl. ᴠ–ᴠɪɪ.

1867 **Cohn, F. J.** Beiträge zur Physiologie der Phycochromaceen und Florideen. Von...In: Arch. Mikr. Anat. iii, p. 1–60, Pl. ɪ–ɪɪ.

1867 **Dumortier, E.** Études paléontologiques sur les dépôts jurassiques du bassin du Rhône. Deuxième partie: Lias inférieur...Paris. iv and 256 pp., 50 Pl.

1867 **Ehrenberg, C. G.** Weitere Entwicklung des Hyalonema lusitanicum und der Spongiaceen. In: Monatsber. Akad. Berlin, p. 843–857.

1867 (a) **Ehrenberg, C. G.** Nachtrag zur Kenntniss der organischen kieselerdigen Gebilde. In: Monatsber. Akad. Berlin, p. 298–318.

1867 (β) **Ehrenberg, C. G.** Über die organischen Kieselgebilde und die den Baströhren der Pflanzen vergleichbaren Spongolithen mit besonderer Rücksicht auf die Hyalonema-Gebilde. In: Sitz. Ber. Ges. naturf. Freunde Berlin, p. 9–10.

1867 **Etheridge, R.** On the Physical Structure of West Somerset and North Devon...By...In: Quart. Journ. Geol. Soc. London, xxiii, p. 568–698, 8 figs.

1867 **Favre, A.** Recherches géologiques dans les parties de la Savoie, du Piémont et de la Suisse voisines du Mont-Blanc...Paris. 3 vols. 150 pp. and Atlas, 32 Pl.

1867 **Fischer, P.** Note sur quelques Spongiaires fossiles de la Craie, appartenant au groupe des Géodies. In: Actes Soc. Linn. Bordeaux, xxvi (= 3, vi), p. 233–237, Pl. ɪɪɪ.

1867 **Geinitz, H. B.** Beiträge zur älteren Flora und Fauna, von...In: N. Jahrb. Mineral. p. 273–290.

1867 **Gray, J. E.** Notes on the Arrangement of Sponges, with the Description of some New Genera. By...In: Proc. Zool. Soc. London, p. 492–558, Pl. xxvɪɪ, xxvɪɪɪ; Rev. Record of Zool. Lit. iv. (1868), p. 669–674.

---

* The number containing Clark's paper appeared Sept. 1867.

1867 (α) **Gray, J. E.** Notes on Zoanthinae, with the Descriptions of some New Genera. By...In: Proc. Zool. Soc. London, p. 233–240.

1867 (β) **Gray, J. E.** On Placospongia, a New Generic Form of Spongiadae in the British Museum. By...In: Proc. Zool. Soc. p. 127–129. Woodcut.

1867 (γ) **Gray, J. E.** Notes on Hyalonema lusitanicum, and on the Genus in general. By...In: Proc. Zool. Soc. London, p. 117–125.

1867 (δ) **Gray, J. E.** Observations on Dr Bowerbank's Paper on Hyalonema lusitanicum. By...In: Proc. Zool. Soc. London, p. 1001–1003.

1867 (ε) **Gray, J. E.** Additional Notes on Euplectella speciosa. By...In: Ann. and Mag. N. H. (3), xix, p. 138–139.

1867 **Gümbel, C. W.** Skizze der Gliederung der oberen Schichten der Kreideformation (Pläner) in Böhmen, von...In: N. Jahrb. Mineral. p. 795–809.

1867 **Hall, F. M.** On the Relative Distribution of Fossils throughout the North Devon Series. By...In: Quart. Journ. Geol. Soc. London, xxiii, p. 371–381.

1867 **Hancock, A.** Note on the Excavating Sponges; with Descriptions of four new Species. By...(Pl. vii and viii.) In: Ann. and Mag. (3), xix, p. 229–242. Also in: Nat. Hist. Trans. Northumberland and Durham, i (1867), p. 337–353, Pl. xvi–xvii.

1867 **Hébert, E.** Le terrain crétacé des Pyrénées; par...In: Bull. Soc. Géol. France (2), xxiv, p. 323–380, Pl. v.

1867 **Hogg, J.** The Name Porifera. In: Athenaeum, p. 160.

1867 **Judd, J. W.** On the strata which form the Base of Lincolnshire Wolds. By...In: Quart. Journ. Geol. Soc. London, xxiii, p. 227–251.

1867 **Lieberkühn, N.** Ueber das contractile Gewebe der Spongien. Von... (Vorgetragen in der Gesellschaft naturforschender Freunde zu Berlin am 17. Juli 1866.) In: Arch. Anat. und Phys. p. 74–86, Taf. iii–iv.

1867 **Marcusen, J.** Zur Fauna des Schwarzen Meeres. Vorläufige Mittheilung von... In: Arch. f. Naturgesch. xxxiii, p. 357–363.

1867 **Marsh, O. C.** Notice of a New Genus of Fossil Sponges from the Lower Silurian: Brachiospongia. In: Amer. Journ. Sc. and Arts (2), xliv, p. 88.

1867 **Meek, F. B.** Note on the genus Palaeacis, Haime, 1860 (= Sphenopoterium M. and W. 1866). In: Amer. Journ. Sc. and Arts (2), xliv, p. 419–420.

1867 **Moore, C.** On Abnormal Conditions of Secondary Deposits...By...In: Quart. Journ. Geol. Soc. London, xxiii, p. 449–568, Pl. xiv–xvii.

1867 **Müller, F.** Ueber Balanus armatus und einen Bastard dieser Art und des Balanus improvisus var. assimilis Darw. Von...In: Arch. f. Naturgesch. xxxiii, p. 329–356, Taf. vii–ix. Transl. in: Ann. and Mag. (4), i (1868), p. 393–412, Pl. xx. Abstr. and rev. in: Rec. Zoll. Liter. iv (1868), pp. 621, 622, 675.

1867 **Norman, A. M.** Report of the Committee appointed for the purpose of Exploring the Coasts of the Hebrides by means of the Dredge. Part ii. In: Rep. 36th Meeting Brit. Assoc. p. 193–206.

(1867? **Pomel, A.** Paléontologie...Cf. 1872.)

1867 **Quenstedt, F. A.** Begleitworte zur geognostischen Specialkarte von Württemberg.

1867 **Renevier, E.** Notices géologiques et paléontologiques sur les Alpes Vaudoises...v. Complément de la Faune de Cheville. Par... In: Bull. Soc. Vaud. ix, p. 389–482.

1867 **Reuss, A. E.** Die Bryozoen, Anthozoen und Spongiarien des Braunen Jura von Balin bei Krakau. Von...In: Denkschr. Akad. Wien, xxvii, p. 1–26, Pl. i–iv. Also in: Q. J. Geol. Soc. xxiii, p. 23–25.

1867 **Rosen, F.** Ueber die Natur der Stromatoporen und über die Erhaltung der Hornfaser der Spongien im fossilen Zustande. Inaug. Diss. Dorpat. 98 pp.

1867 **Sandberger, K. L. F.** Die Gliederung der Würzburger Trias...In: Würzb. naturw. Zeitschr. vi, p. 131–155, Pl. viii–ix.

1867 **Schafhäutl, K. F. E.** Weitere Beiträge zur näheren Kenntniss der bayerischen Alpen von...In: N. Jahrb. Mineral. p. 257–272, Pl. i–ii.

1867 **Schmarda, L. K.** Die maritime Production der österreichischen Küstenländer...In: Oesterr. Revue, v, p. 45-89. Also separately?.

1867 **Schmidt, (E.) O.** Spongologische Mittheilungen. Von...In: Arch. Mikr. Anat. iii, p. 390–392.

1867 (a) **Schmidt, E. O.** Die Künstliche Zucht des Badeschwammes. In: Agronom. Zeitung, p 759–...

1867 **Schultze, M. J. S.** On Hyalonema. By...In: Ann. Mag. N. H. (3), xix, p. 153–160.

1867 (a) **Schultze, M. J. S.** Ueber Hyalonema. Von...In: Arch. Mikr. Anat. iii, p. 206–214. German transl. from Schultze, 1867.

1867 **Selenka, E.** Ueber einige neue Schwämme aus der Südsee. Von... In: Z. W. Z. xvii, p. 565–571, Pl. xxxv.

1867 **Semper, C.** Einige Worte über Euplectella aspergillum Owen und seine Bewohner. Von...In: Arch. Naturgesch. xxxiii, p. 84–89. Transl. in: Ann. and Mag. N. H. (4), ii (1868), p. 26–30.

1867 **Slack, H. J.** Venus' flower-basket: Euplectella speciosa. In: Intellect. Observer, xi, p. 161–166, 1 Pl.

1867 **Thomson, C. W.** On the "Glass-rope" Hyalonema. By...In: Intellect. Observer, xi, p. 81–94, 1 coloured Pl.

1867 **Würtenberger, L.** Die Schichtenfolge des Schwarzen und Braunen Jura im Klettgau von...In: N. Jahrb. Mineral. p. 39–59.

1867 **Zittel, K. A. & W. Vogelgesang.** Beiträge zur Statistik der inneren Verwaltung des Grossherzogthums Baden...4°. 62 pp., 2 Pl.

87

1868 **Andrews, W.** (On rare Irish Sponges.) In: Proc. Dublin N. H. Soc. v,
p. . Also in: Ann. and Mag. N. H. (4), i, p. 307–308.

1868 **Bate, C. Spence.** Report of the Committee appointed to explore the
Marine Fauna and Flora of the South Coast of Devon and
Cornwall. No. 2...Reporter...In: Rep. 37th Meet. Brit. Assoc.
p. 275–287, Pl. i–iii. (Sponges by **Bowerbank.**)

1868 **Baudelot, E.** Observation sur les œufs de l'éponge d'eau douce. In:
Bull. Soc. Sc. Nat. Strasbourg, i, p. 127–128.

1868 **Benecke, E. W.** Ueber einige Muschelkalk-Ablagerungen der Alpen.
Von...In: Geognost. paläont. Beitr. ii, p. 3–67, Pl. i–iv.

1868 **Bigsby, J. J.** Thesaurus Siluricus, the Flora and Fauna of the Silurian
Period. London. 4º. 214 pp.

1868 **Bocage, J. V. Barboza du.** On Hyalonema boreale. By...In: Ann.
and Mag. N. H. (4), ii, p. 36–38.

1868 **Bowerbank, J. S.** Observations on Dr Gray's "Notes on the Arrangement of Sponges, with the Description of some New Genera."
By...In: Proc. Zool. Soc. London, p. 118–137.

1868 (a) **Bowerbank, J. S.** (Sponges.) In: Rep. 37th Meet. Brit. Assoc.
p. 285–286. See 1868 **Bate,** suprà.

1868 **Carter, H. J.** On a Variety of Spongilla Meyeni from the River Exe,
Devonshire. By...In: Ann. and Mag. (4), i, p. 247–250.

1868 **Claus, C. F. W.** Ueber Euplectella aspergillum (R. Owen). Ein
Beitrag zur Naturgeschichte der Kieselschwämme. Von...
Marburg. N. G. Elwert'sche Universitäts-Buchhandlung. 4º.
28 pp. and 4 Pl.

1868 **Czernay, A.** Soobschtschenie o besposwonocz nich schiwotnich
Charkowskoj gubernii. In: Trudi 1 sjesda russk. estestv. p. 183–
184.

1868 **Czerniawsky, V.** Predvaritelnoe soobschtschenie po faunje Czernago
moria. In: Trudi 1 sjesda russk. estestv. p. 9–10.

1868 (a) **Czerniawsky, V.** Materialia ad zoographiam ponticam compara-
tam. In: Trudi 1 sjesda russk. estestv. p. 19–136, 8 Pl.

1868 **Dames, W.** Ueber die in der Umgebung Freiburgs in Nieder-Schlesien
auftretenden devonischen Ablagerungen. Von...In: Zeitschr.
d. geol. Ges. xx, p. 469–508, Pl. x–xi.

1868 **Dewalque, G.** Prodrome d'une description géologique de la Belgique,
par...Bruxelles-Liège. 442 pp.

1868 **Edgell, H. A. W.** Fish remains in the Lower Devonian of South Devon
and Cornwall. In: Geol. Mag. v, p. 247–248.

1868 **Fischer, P.** Recherches sur les éponges perforantes fossiles...In:
Nouv. Arch. Mus. tom. iv, p. 117–172, Pl. 24–25.

1868 **Funck, N.** Notice sur le Euplectella aspergillum, R. Owen. In: Ann.
Soc. Malacol. Belgique, iii, p. xii–xiv. Also in: Soc. Sc. Nat.
Luxembourg, x (1869), p. 266–267.

1868 **Goodsir, J.** The Anatomical Memoirs of...Edinburgh, i, p. 405–419 (Tethea).

1868 **Gray, J. E.** Observations on Sponges and on their Arrangement and Nomenclature. By...In: Ann. and Mag. N. H. (4), i, p. 161–173.

1868 (α) **Gray, J. E.** Note on Theonella, a New Genus of Coralloid Sponges from Formosa. By...In: Proc. Zool. Soc. London, p. 565–566, fig. 1–3.

1868 (β) **Gray, J. E.** Notes on the Ceratelladae, a family of Keratose Sponges. By...In: Proc. Zool. Soc. London, p. 575–579, fig. 1–2.

1868 (γ) **Gray, J. E.** Note on Xylospongia cookii, a New Genus of Palmated Sponges in the collection of the British Museum. By...In: Proc. Zool. Soc. London, p. 637.

1868 (δ) **Gray, J. E.** On the peculiar Structure and Function of the spicules of Hyalonema. By...In: Ann. and Mag. N. H. (4), i, p. 292–295.

1868 (ε) **Gray, J. E.** Note on Hyalonema boreale, Lovén. By...In: Ann. and Mag. N. H. (4), i, p. 484.

1868 (ζ) **Gray, J. E.** Note on a new Japanese Coral (Isis Gregorii), and on Hyalonema. By...In: Ann. and Mag. N. H. (4), ii, p. 263–264.

1868 (η) **Gray, J. E.** On a new Free Form of Hyalonema Sieboldii, and its manner of growth. By...Ibid. p. 264–276.

1868 (θ) **Gray, J. E.** On Tetilla euplocamos and Hyalonema boreale. By...In: Ann. and Mag. N. H. (4), p. 319–320.

1868 (ι) **Gray, J. E.** Note on Hyalonema Schultzei, Semper. By...In: Ann. and Mag. N. H. (4), ii, p. 373–377.

1868 (κ) **Gray, J. E.** Euplectella. In: Ann. and Mag. N. H. (4), ii, p. 388.

1868 (λ) **Gray, J. E.** On the Name Alcyoncellum. By...Ibid. p. 390–391.

1868 (μ) **Gray, J. E.** (Letter from..., containing a reply to Dr Bowerbank's remarks on his "Notes on Sponges.") In: Proc. Zool. Soc. London, p. 532–533.

1868 **Gümbel, C. W.** Geognostische Beschreibung des Königreichs Bayern. 2. Abth...Gotha. 4⁰. viii and 968 pp.

1868 **Herklots, J. A. & W. Marshall.** Notice sur deux espèces nouvelles d'éponges de la Famille des Lophospongiae. Par...In: Arch. Néerland. iii, p. 435–438.

1868 **Hogg, J.** Is the Freshwater Sponge (Spongilla) an animal? In: Popular Science Rev. vii, p. 134–141.

1868 **Holl, H. B.** On the Older Rocks of South Devon and East Cornwall. By...In: Quart. Journ. Geol. Soc. London, xxiv, p. 400–454, fig. 1–5, Pl. xviii.

1868 **Hyatt, A.** Sponges. By...In: Amer. Naturalist, ii, p. 303–306.

1868 **Jeffries, B. J.** On Euplectella speciosa. In: Proc. Boston Soc. xii, p. 76.

89

1868 **Jenzsch, G.** Über eine mikroskopische Flora und Fauna krystallinischer Massengesteine (Eruptivgesteine). Von...Leipzig. 31 pp.

1868 **Judd, J. W.** On the Speeton Clay. By...In: Quart. Journ. Geol. Soc. London, xxiv, p. 218–250, fig. 1–4.

1868 **Kostytschef, P. & O. Marggraf.** Ueber die chemische Zusammensetzung der in dem Apatitsandstein der russischen Kreideformation vorkommenden versteinerten Schwämme. In: Bull. Acad. St Pétersbourg, xiii, coll. 19–20. Also in: Journ. prakt. Chemie, cv, p. 63–64.

1868 **Lankester, E. R.** Preliminary Notice of some Observations with the Spectroscope on Animal Substances. In: Journ. Anat. and Phys. (2), ii, p. 114–116.

1868 (a) **Lankester, E. R.** On the Discovery of the Remains of Cephalaspidian Fishes in Devonshire and Cornwall; and on the Identity of Steganodictyum, M'Coy, with genera of those Fishes. By... In: Quart. Journ. Geol. Soc. London, xxiv, p. 546–547.

1868 **Leidy, J.** Description of a new Sponge: Pheronema Annae. In: Proc. Acad. Nat. Sc. Philadelphia, p. 9–11.

1868 **Lieberkühn, N.** Über die contractilen Zellen der Spongillen. In: Sitzungsber. Ges. Beförd. ges. Naturw. Marburg, p. 42–43.

1868 **Loriol, P. de.** Monographie des couches de l'étage Valangien des Carrieres d'Arzier...In: Matériaux paléontol. Suisse (4), x-xi, p. 1–110, Pl. I-IX.

1868 **Loven, S.** Om en märklig i Nordsjön lefvande art af Spongia. Af... In: Öfvers. Kon. Vetensk. Akad. Förhandl. xxv,p.105–121, Pl.2. English transl. in: Ann. and Mag. N. H. (4), ii, p. 81–91, Pl. VI.

1868 **Martens, E. von.** Ueber einige ostasiatische Süsswasserthiere. Von ...In: Arch. Naturg. xxxiv, p. 1–67, Pl. I.

1868 (a) **Martens, E. von.** Der Glasschwamm. In: Naturforscher, i, p. 178–179.

1868 **Meek, F. B.** Preliminary notice of a remarkable new genus of Corals, probably typical of a new Family. By...In: Amer. Journ. Sc. and Arts (2), xlv, p. 62–64. Also in: Ann. and Mag. N. H. (4), p. 225–227.

1868 (a) **Meek, F. B.** Note on Ethmophyllum and Archaeocyathus. In: Amer. Journ. Sc. and Arts (2), xlvi, p. 144.

1868 **Meek, F. B. & A. H. Worthen.** Paleontology of Illinois. In: Geol. Survey Illinois, iii, p. 291–565, Pl. I-XX.

1868 **Miklucho-Maclay, N.** Beiträge zur Kenntniss der Spongien. i....von ...In: Jen. Zeitschr. iv, p. 221–240, Pl. IV-V.

1868 **Norman, A. M.** On the British Species of Alpheus, Typton, and Axius, and on Alpheus Edwardsii of Audouin. By...In: Ann. and Mag. N. H. (4), ii, p. 173–178.

1868 **Parfitt, E.** On the Marine and Freshwater Sponges of Devonshire. In: Trans. Devonshire Assoc. p. 443–462.

1868 **Peach, C. W.** The Fossil Fishes of Cornwall. In: Geol. Mag. v, p. 568–569. Also in: Rep. 38th Meet. Brit. Assoc. (1869), p. 76.

1868 **Pourtales, L. F. de.** Contributions to the Fauna of the Gulf Stream at great depths. By...In: Bull. Mus. Compar. Zool. Harvard Coll. i, p. 103–141.

1868 **Rennes, A. J. M.** Des Éponges. Classification, preparation, Commerce. Paris...24 pp.

1868 **Sars, M.** Fortsatte Bemaerkninger over det dyriske Livs Udbredning i Havets Dybder. In: Forhandl. Vidensk. Selskab. Christiania, xi, 1869, p. 246–275. Transl. in: Ann. and Mag. N. H. (4), iii (1869), p. 423–441.

1868 **Schloenbach, C. U.** Ueber die norddeutschen Galeriten-Schichten und ihre Brachiopoden-Fauna. In: Sitzungsber. Akad. Wien, lviii, p. 181–224, Pl. i–iii.

1868 **Schmidt, (E.) O.** Die Spongien der Küste von Algier. Mit Nachträgen zu den Spongien des adriatischen Meeres. (Drittes Supplement.) ...von...Mit fünf Kupfertafeln. Leipzig...Engelmann, 1868. 4⁰. iv and 44 pp. and 5 Pl. Rev. in: C. R. lxvii, p. 141–142; Rec. Zool. Literat. v (1869), pp. 575, 576, 586–588.

1868 **Semper, C.** (Einige neue Kieselschwämme der Philippinen.) In: Verh. phys. med. Ges. Würzburg, i, p. xxix–xxx. Transl. in: Ann. and Mag. N. H. (4), ii, p. 372–373.

1868 **S(e)n(one)r, A.** Die Schwammzucht und -fischerei im adriatischen Meere. In: Zool. Garten, ix, p. 101–103.

1868 **Thomson, C. W.** On the "Vitreous" Sponges. By...In: Ann. and Mag. N. H. (4), i, p. 114–132, Pl. iv.

1868 **Williamson, W. C.** The Common Fresh-water Sponge (Spongilla fluviatilis). In: Popular Sc. Review, vii, p. 1-9, Pl. i.

1868 **Wright, E. P.** Notes on Irish Sponges, Part i. A List of the Species. By...In: Proc. Irish Acad. x, p. 221–228. N.B.—On the title-page of vol. x is printed 1870, but Wright's article appeared in 1868. [Separate copy paged 49–56.]

1868 **(a) Wright, E. P.** Notes on Deep-sea Dredging. By...In: Ann. and Mag. N. H. (4), ii, p. 423–427.

1868 **Wright, E. P. & C. W. Thomson.** On Hyalonema, Gray. In: Ann. and Mag. N. H. (4), ii, p. 320.

1868 **Würtenberger, L.** Einige Beobachtungen im Weissen Jura des oberen Donauthales von...In: N. Jahrb. Mineral. p. 540–547.

1868 **Zaddach, E. G.** Das Tertiär-Gebirge Sammlands. In: Schriften physik. ökon. Ges. Königsberg, viii, p. 85–197, Pl. vi–xvii.

1869 **Agassiz, L. J. R.** Report upon Deep-Sea Dredgings in the Gulf Stream, during the Third Cruise of the U.S. Steamer Bibb... by...In: Bull. Mus. Comp. Zool. Harvard Coll. i, p. 363–386.

1869 **Bocage, J. V. Barboza du.** Éponges siliceuses nouvelles de Portugal et de l'île Saint-Iago (archipel de Cap-vert), par...In: Jorn. de Scienzas Acad. Lisboa, ii, p. 159–161, Pl. x–xi.

1869 **Bowerbank, J. S.** On Dr Gray's Genus Theonella. By...In: Proc. Zool. Soc. London, p. 389–390.

1869 (a) **Bowerbank, J. S.** On the Generic Name Alcyoncellum, and in reply to Dr Gray's "Observations on Sponges and on their Arrangement and Nomenclature,"...By...In: Ann. and Mag. N. H. (4), iii, p. 84–87.

1869 (β) **Bowerbank, J. S.** A Monograph of the Siliceo-fibrous Sponges. By...Parts i, ii. In: Proc. Zool. Soc. London, pp. 66–100, 323–351, Pl. iii–vi, xxi–xxv.

1869 **Brady, G. S. & D. Robertson.** Notes on a week's Dredging in the West of Ireland. By...In: Ann. and Mag. N. H. (4), iii, p. 353–374, Pl. xviii–xxii.

1869 **Carpenter, W. B.** On the Temperature and animal Life of the Deep-Sea. In: Notices Proc. R. Inst. v, p. 503–510.

1869 (a) **Carpenter, W. B.** Preliminary Report by...of Dredging Operations on the Seas to the North of the British Islands, carried on in Her Majesty's Steam-vessel "Lightning." In: Proc. R. Soc. London, xvii, p. 168–200.

1869 **Carpenter, W. B. & H. B. Brady.** [Description of Parkeria and Loftusia, two gigantic types of Arenaceous Foraminifera. (Abstract).] In: Proc. R. Soc. London, xvii, p. 400–404.

1869 (a) **Carpenter, W. B. & H. B. Brady.** Description of Parkeria and Loftusia, two gigantic types of Arenaceous Foraminifera. By... In: Phil. Trans. Roy. Soc. London, clix, p.721–745, Pl.lxxii–lxxx.

1869 **Carter, H. J.** A Descriptive Account of four Subspherous Sponges, Arabian and British, with General Observations. By...In: Ann. and Mag. (4), iv, p. 1–28, Pl. i–ii.

1869 (a) **Carter, H. J.** Description of a Siliceous Sand-Sponge found on the South-east Coast of Arabia. By...In: Ann. and Mag. N. H. (4), iii, p. 15–17. Woodcut 1–4.

1869 (β) **Carter, H. J.** Calcareous Sponges. By...In: Ann. and Mag. N. H. (4), iii, p. 466.

1869 (γ) **Carter, H. J.** On Grayella cyathophora, a new Genus and Species of Sponges. By...In: Ann. and Mag. N. H. (4), iv, p. 189–197, Pl. vii.

1869 **Czerniawsky, V.** Ze predvaritelnoe soobschtschenie po faunje Czernago moria. In: Trudi 2 sjesda russk. estestv. p. 1–3.

1869 **Ehrenberg, C. G.** Über mächtige Gebirgsschichten vorherrschend aus mikroskopischen Bacillarien unter und bei der Stadt Mexiko. In: Abhandl. Akad. Berlin, ii, p. 1–66, Pl. i–iii.

1869 (a) **Ehrenberg, C. G.** Über den am 24. März dieses Jahres mit Nord-Ost-Sturm gefallenen rothen Passatstaub in den Dardanellen... In: Monatsber. Akad. Berlin, p. 308–320, Pl.

1869 **Eisig, H.** Nereis hircinicola (Nova Species). Von...In: Z. W. Z. xx, p. 103–105, Pl. xi, fig. 3–4.

1869 **Gray, J. E.** Note on Ianthella, a new Genus of Keratose Sponges. By...In: Proc. Zool. Soc. London, p. 49–51.

1869 (a) **Gray, J. E.** On the Manner of Growth of Hyalonema. By...In: Ann. and Mag. N. H. (4), iii, p. 192–196.

1869 **Gregory,**... (On the Habits of Hyalonema.) In: Ann. and Mag. N. H (4), iii, p. 172.

1869 **Haeckel, E. (H. P. A.).** Ueber den Organismus der Schwämme und ihre Verwandtschaft mit den Corallen. Von...In: Jen. Zeitschr. v*, p. 207–235. Transl. in: Ann. and Mag. (4), v, (1870), pp. 1–13, 107–120.

1869 (a) **Haeckel, E. (H. P. A.).** Prodromus eines Systems der Kalkschwämme. Von...In: Jen. Zeitschr. v†, p. 236–254. Transl. in: Ann. and Mag. (4), v, p. 176–191.

1869 **Hancock, A. & T. Atthey.** On some curious Fossil Fungi from the Black Shale of the Northumberland Coalfield. By...In: Ann. and Mag. N. H. (4), iv, p. 221–228, Pl. ix–x.

1869 **Hull, E.** On the Evidences of a Ridge of Lower Carboniferous Rocks crossing the Plain of Cheshire beneath the Trias...In: Quart. Journ. Geol. Soc. London, xxv, p. 171–184.

1869 **Huxley, T. H.** An Introduction to the Classification of Animals. By...London.

1869 **Jaccard, A.** Description géologique du Jura vaudois et neuchâtelois ...par...In: Matér. Carte Géol. Suisse (Beiträge Geol. Karte Schweiz), Livr. vi, p. 1–342, Pl. i–viii.

1869 **Kunth, A.** Beiträge zur Kenntniss fossiler Korallen. Von... i. Korallen des schlesischen Kohlenkalkes. In: Zeitschr. d. geol. Ges. xxi, p. 183–220, Pl. ii–iii.

1869 **Loriol, P. de & V. Gilliéron.** Monographie paléontologique et stratigraphique de l'étage Urgonien inférieur du Landeron...In: N. Denkschr. Schweiz. Ges. (Nouv. Mém. Helvét.), xxiii [(3), iii], mém. 5, Pl. i–viii.

1869 **Ludwig, R. A. B. S.** Ueber die Gliederung der devonischen Formation im Dillenburgischen und Biedenkopfischen Theile des Westerwalds, von...In: N. Jahrb. Mineral. p. 658–685.

1869 **Malmgren, A. J.** Om förekomsten af djurlif på stora hafsdjup. In: Öfvers. finska vetensk. soc. förhandl. xii, p. 40–49.

* On the title-page of this volume is printed 1870. As, however, the article appeared translated in the January number of the "Annals" (1870) authors are right to put Haeckel's article in 1869. Moreover, Haeckel states himself (1872, i, p. 33) that the article appeared in 1869.

† Appeared, according to Haeckel (1872, i, p. 33), in Sept. 1869.

1869 **Martens, E. von.** Ueber Thiernamen. Von... In: Zool. Garten, x, p. 50–56.

1869 **Möller, V. von.** Carte géologique du versant occidental de l'Oural. St Pétersbourg. Fol.

1869 **Moore, T. J.** On the Habitat of the Regadera (Watering-pot) or Venus's Flower-basket (Euplectella aspergillum, Owen). By... In: Ann. and Mag. N. H. (4), iii, p. 196–199.

1869 **Norman, A. M.** Notes on a few Hebridean Sponges, and on a new Desmacidon from Jersey. By... In: Ann. and Mag. (4), iii, p. 296–299. Abstr. and rev. in: Rec. Zool. Lit. vi (1870), pp. 673, 680.

1869 (a) **Norman, A. M.** Shetland Final Dredging Report. Part ii. On the Crustacea, Tunicata, Polyzoa, Echinodermata, Actinozoa, Hydrozoa, and Porifera. By... In: Rep. 38th Meeting Brit. Assoc. p. 247–336.

1869 **Ooster, W. A.** Pétrifications remarquables des Alpes suisses. Le Corallien de Wimmis... Genève et Bâle. 4°. ix and 52 pp.

1869 **Rath, G. vom.** Aus Norwegen von... In: N. Jahrb. Mineral. p. 385–444.

1869 **Reichert, K. B.** Über Hyalonemen, welche Dr. Reger aus Jeddo mitgebracht hat. In: Sitzungsber. Ges. naturf. Freunde Berlin, p. 25.

1869 **Roy de Méricourt,... le.** Considérations sur l'hygiène des pêcheurs d'éponges, Par... In: Ann. d'Hygiène (2), xxxi, p. 274–286.

1869 **Safford, J. M.** Geology of Tennessee. Nashville. 550 pp., 9 Pl.

1869 **Schimper, W. P.** Traité de paléontologie végétale... Tome premier. Paris-Londres-Madrid. vi, iv and 738 pp.

1869 **Schlüter, C. A.** (Ueber die jüngsten Schichten der unteren Senon-Bildungen und deren Verbreitung.) In: Verh. naturh. Ver. pr. Rheinl. und Westphalens, xxv (3), v, p. 92–93.

1869 **Schmidt, (E.) O.** Vorläufige Mittheilungen über die Spongien der grönländischen Küste. Von... In: Mitth. naturw. Ver. Steiermark, ii, p. 89-97. Ed. 2, 1870. Abstr. and rev. in: Rec. Zool. Lit. vi (1870), p. 673–675; Zeitschr. ges. Naturw. xxxv (1870), p. 165.

1869 (a) **Schmidt, (E.) O.** Das natürliche System der Spongien. Von... In: Mitth. naturw. Ver. Steiermark, ii, p. 261–269. Ed. 2, 1870. Abstr. and rev. in: Zeitschr. ges. Naturw. xxxvii (1871), p. 155–157.

1869 (β) **Schmidt, E. O.** Wirbellose Thiere in Brehm's Thierleben. Bd. vi.

1869 **Thomson, C. Wyville.** On Holtenia, a Genus of Vitreous Sponges. By... In: Phil. Trans. Roy. Soc. London, clix, p. 701–720, Pl. LXVII–LXXI.

1869 (a) **Thomson, C. W.** On the Depths of the Sea. By... In: Ann. and Mag. N. H. (4), iv, p. 112–124, fig. 1–2. Also in: Journ. Dubl. Soc. v, 1870, p. 316–326.

1869 (β) **Thomson, C. W.** On Holtenia, a Genus of Vitreous Sponges. By ...(abstract). In: Proc. R. Soc. London, xviii, p. 32–35, fig. Also in: Ann. and Mag. (4), iv, p. 284–287.

1869 **Vaillant, L.** Note sur la vitalité d'une éponge de la famille des Corticatae, la Tethya lyncurium, Lamarck. Par...In: C. R. Acad. Paris, lxviii, p. 86–88. (Repr. p. 1–2.)

1869 **Verrill, A. E.** On the parasitic habits of Crustacea. By...In: Amer. Naturalist, iii, p. 239–250, fig. 41–42.

1869 **Weiss, C. E.** Über die Gliederung der Trias in der Umgegend von Saarbrücken. In: N. Jahrb. Mineral. p. 215–219.

1869 **Wiltshire, T.** On the Red Chalk of Hunstanton. By...In: Quart. Journ. Geol. Soc. London, xxv, i, p. 185–191.

1869 **Wright, B. M.** Sponges. By...In: Amer. Naturalist, iii, p. 449–455. With 1 woodcut.

1869 **Wright, E. P.** (Note on Placospongia melobesioides, Gray.) In: Quart. Journ. Microsc. Sc. ix, p. 322.

1870 **Allman, G. J.** Note on Polytrema miniaceum. By...In: Ann. and Mag. N. H. (4), v, p. 327–373.

1870 **Beneden, P. J. van.** Sur le commensalisme dans le règne animal. Par...In: Bull. Acad. Belgique, xxix, p. 179–182.

1870 **Bocage, J. V. Barboza du.** Sur l'existence de la "Holtenia Carpenteri" Wyville Thomson dans les côtes du Portugal. In: Jorn. Sc. Acad. Lisboa, iii, p. 69–70.

1870 (α) **B(ocage, J. V.) B(arboza du).** A vida animal nas grandes profundidades do oceano. In: Jorn. Sc. Acad. Lisboa, iii, p. 71–72.

1870 **Carpenter, W. B.** On the Temperature and Animal Life of the Deep Sea. In: Notices Proc. R. Instit. vi, p. 63–82. Also in: Nature, i, pp. 488–490, 540–542, 563–566.

1870 **Carpenter, W. B. & J. Gwyn Jeffreys.** Report on deep sea researches carried on during the months of July, August and September 1870, in H.M. Surveying ship "Porcupine." By,..In: Proc. R. Soc. London, xix, p. 146–221.

1870 **Carter, H. J.** Note on the Sponges Grayella, Osculina, and Cliona. By...In: Ann. and Mag. (4), v, p. 73–83. Review: Zool. Rec. vii (1871), p. 503.

1870 (α) **Carter, H. J.** On the Ultimate Structure of Marine Sponges. By ...In: Ann. and Mag. (4), vi, p. 329–341. Abstr. in: Zool. Rec. vii (1871), p. 503.

1870 (β) **Carter, H. J.** On two New Species of Subspherous Sponges, with Observations. By...In: Ann. and Mag. N. H. (4), vi, p. 176–182, Pl. xiii.

1870 (γ) **Carter, H. J.** On Haliphysema ramulosa (Bowerbank) and the Sponge-spicules of Polytrema. By...In: Ann. and Mag. N. H. (4), v, p. 389–392.

1870 **Clark, H. J.** Polarity and Polycephalism: an essay on individuality. In: Amer. Journ. Sc. and Arts (2), xlix, p. 69–75.

1870 **Curioni, G.** Osservazioni geologiche sulla Val-Trompia. In: Mem. Istit. Lombardo, xii (3), iii, p. 41–100.

1870 **Czernay, A.** Ob iglach gubok, wstrjeczajuschtschichsia we ilje oser Limana i Czajki Zmiewskago sjesda Charkowskoj gubernii. In: Trudi Charkowsk. Obschtsch. i, p. 1–6, Pl. ix.

1870 **Duchassaing de Fonbressin, P.** Revue des Zoophytes et des Spongiaires des Antilles. Paris. 82 pp.

1870 **Ehlers, E.** Die Esper'schen Spongien in der zoologischen Sammlung der K. Universität Erlangen...Erlangen, E. Th. Jacob. 4°. 36 pp. Abstr. and rev. in: Zool. Rec. ix (1874), pp. 473–474, 477–478.

1870 **Ehrenberg, C. G.** Über die wachsende Kenntniss des unsichtbaren Lebens als felsbildende Bacillarien in Califorien. In: Abhandl. Akad. Berlin, p. 1–74, Pl. i–iii.

1870 **Fuchs, C. W. C.** Die alten Sediment-Formationen und ihre Metamorphose in den französischen Pyrenäen von...In: N. Jahrb. Mineral. p. 719–752, Pl. vii.

1870 **Gegenbaur, C.** Grundzüge der vergleichenden Anatomie. Von... Leipzig. xii and 892 pp. Ed. ii.

1870 **Geinitz, H. B.** Ueber Versteinerungen aus einer sandigen Ablagerung der Kreideformation von Château de Meauene bei Lude...In: Sitzungsber. Isis, Dresden, p. 149–150.

1870 **Gray, J. E.** Note on a new Genus of Sponge from West Australia. By... In: Ann. and Mag. N. H. (4), vi, p. 272.

1870 (α) **Gray, J. E.** Notes on Anchoring Sponges...By...In: Ann. and Mag. N. H. (4), vi, p. 309–312.

1870 (β) **Gray, J. E.** Axos Cliftoni. By...In: Ann. and Mag. N. H. (4), vi, p. 346.

1870 **Greppin, J. B.** Description géologique du Jura bernois...In: Beiträge geol. Karte Schweiz. xx and 357 pp.

1870 **Grube, A. E.** Mittheilungen über St. Malo und Roscoff, und die dortige Meeres-besonders die Annelidenfauna. Von...In: Abh. Schles. Gesellsch. (1869–1872), p. 75–146, Pl. i–ii.

1870 (α) **Grube, A. E.** [Ueber Exemplare von der sogenannten Glaspflanze (Hyalonema Sieboldii) und von der Regadera (Euplectella aspergillum).] In: 47. Jahresber. schles. Ges. p. 45–48.

1870 **Gümbel, C. W.** Vorläufige Mittheilungen über Tiefseeschlamm von... In: N. Jahrb. Mineral. p. 753–767.

1870 **Haeckel, E. (H. P. A.).** Beiträge zur Plastidentheorie. Von...In: Jen. Zeitschr. v (1870), p. 492–550, Taf. xvii–xviii.

1870 **Harting, P.** Mémoire sur le genre Potérion. Par...In: Natuurk Verh. Utr. Gen. ii, 40 pp. and 4 Pl.

1870 **Heymann, H.** (Ueber mitteldevonische Petrefacten aus den Phosphoritlagerstätten von Nassau.) In: Verhandl. naturh. Vereins pr. Rheinl. und Westphal. xxvi (3), vi, p. 222–224.

1870 **Jeffreys, J. G.** Food of Oceanic Animals. In: Nature, i, p. 315.

1870 **Judd, J. W.** Additional Observations on the Neocomian Strata of Yorkshire and Lincolnshire...By...In: Quart. Journ. Geol. Soc. London, xxvi, i, p. 326–347, Pl. xxiii.

1870 **Kent, W. S.** On two new Siliceous Sponges taken in the late Dredging-Expedition of the Yacht "Norna" off the Coasts of Spain and Portugal. By...In: Ann. and Mag. (4), vi, p. 217–224, Pl. xv. Abstr. and rev. in: Zool. Rec. vii (1871), p. 505.

1870 (α) **Kent, W. S.** On a New Anchoring Sponge, Dorvillia agariciformis. By...In: Monthly Microsc. Journ. iv, p. 293–295, Pl. lxvi, fig. 1–19.

1870 (β) **Kent, W. S.** Häckel on the Relationship of the Sponges to the Corals. By...In: Ann. and Mag. N. H. (4), v, p. 204–218, 4 figs.

1870 (γ) **Kent, W. S.** On the Hexactinellidae, or Hexradiate Spiculed Siliceous Sponges taken in the "Norna" Expedition off the coast of Spain and Portugal. With Description of New Species, and Revision of the Order. By...In: Monthly Microsc. Journ. iv, p. 241–252, Pl. lxiii–lxv.

1870 (δ) **Kent, W. S.** Notice of a new Vitreous Sponge, Pheronema (Holtenia) Grayi. By...In: Ann. and Mag. N. H. (4), vi, p. 182–186.

1870 (ε) **Kent, W. S.** Professor Häckel and Mr E. Ray Lankester on the Affinities of the Sponges. By...In: Ann. and Mag. N. H. (4), vi, p. 250–255.

1870 **Kriesch, J.** Egy ritka Kovaszivacs és a szivacsok rövid természetrajza. In: Termesz. Közl. ii, p. 432–435.

1870 **Lankester, E. Ray.** Professor Häckel and Mr Kent on the Zoological Affinities of the Sponges. By...In: Ann. and Mag. N. H. (4), vi, p. 86–92.

1870 (α) **Lankester, E. Ray.** On Comparative Longevity in Man and the Lower Animals. By...London.

1870 **Leidy, J.** Remarks on some curious Sponges. By...In: Amer. Naturalist, iv, p. 17–22, fig. 11–12.

1870 **Lieberkühn, N.** Ueber Bewegungserscheinungen der Zellen. Von... In: Schriften Gesellsch. Beförd. Naturw. Marburg, ix, p. 335–385, Pl. i–v. Abstr. in: Ann. and Mag. (4), vi, p. 497–498; Arch. Sc. Phys. and Nat. xxxix, p. 158–160; Zool. Rec. vii (1871), p. 503; Zeitschr. ges. Naturw. xl (1872), p. 420–426.

1870 **Lindström, G.** A Description of the Anthozoa Perforata of Gotland. By...In: Kon. Svenska vetensk. akad. Handl. ix, 6, p. 1–12, 1 Pl.

1870 **Ljungman, A. V.** Några geologiska jakttagelser gjorda under en resa i mellersta Bohuslän sommaren 1870. Upsala. 22 pp.

1870 **Meek, F. B. & A. H. Worthen.** Descriptions of new species and genera of Fossils from the palaeozoic rocks of the Western States. By... In: Proc. Acad. Nat. Sc. Philadelphia (2), xiv, p. 22–56.

1870 **Metzger, A.** Über die wirbellosen Meeresthiere der ostfriesischen Küste, ein Beitrag zur Fauna der deutschen Nordsee. In: 20. Jahresber. naturh. Ges. Hannover, p. 22–36.

1870 **Miklucho-Maclay, N.** Ueber einige Schwämme des nördlichen Stillen Oceans und des Eismeeres, welche im Zoologischen Museum der Kaiserlichen Akademie der Wissenschaften in St. Petersburg aufgestellt sind. Ein Beitrag zur Morphologie und Verbreitung der Spongien. Von... In: Mém. Acad. Imp. St Pétersbourg (7), xv, No. 3 (24 pp. and 2 Pl.).

1870 (α) **Miklucho-Maclay, N.** Bemerkungen zur Schwammfauna des Weissen Meeres und des Arktischen Oceans. Von N. M. v. Maclay. In: Bull. Acad. St Pétersbourg, xv, coll. 203–205.

1870 (β) **Miklucho-Maclay, N.** Eine zoologische Exkursion an das Rothe Meer. 1869. Von... In: Petermann's Mittheil. xvi, p. 124–126.

1870 (γ) **Miklucho-Maclay, N.** O gubkach sjewernoi i sjewerowoototschnoi Assii. In: Trudi St Petersb. obschtschv. i, 1, p. 67.

1870 **Möbius, K.** Ueber Austern- und Miesmuschelzucht und die Hebung derselben an den norddeutschen Küsten... Berlin. 67 pp., 19 figs. and Map.

1870 **Moore, C.** Australian Mesozoic Geology and Palaeontology. By... In: Quart. Journ. Geol. Soc. London, xxvi, 1, p. 226–261, Pl. x–xviii.

1870 **Nathorst, A. G.** Om lagerföljden inom Cambriska formationen vid Andrarum i Skåne. In: Öfvers. vetensk. akad. förhandl. xxvi, 1869, p. 61–65.

1870 **Parfitt, E.** Fossil sponge spicules in the Greensand of Haldon and Blackdown. By... In: Rep. and Trans. Devonshire Assoc. iv, p. 138–144.

1870 **Perty, J. A. M.** Ueber die Klasse der Schwämme oder Spongien. Von... In: Mittheil. naturf. Ges. Bern, p. xix–xxv.

1870 **Roemer, (C.) F.** Geologie von Oberschlesien... Breslau. 4°. xiv and 610 pp., Atlas, 50 Pl.

1870 **Sandberger, K. L. F.** Neue Petrefacten in der fränkischen Trias und dem mittleren Oolithe Oberbadens. Von... In: N. Jahrb. Mineral. p. 604–606.

1870 **Schmidt, (E.) O.** Grundzüge einer Spongien-Fauna des atlantischen Gebietes von... Leipzig. Fol. iv and 88 pp., 6 Pl. Abstr. and rev. in: Zool. Rec. vii (1871), pp. 501–504, 506–508; N. Jahrb. Mineral. (1871), p. 216–218.

1870 (a) **Schmidt, E. O.** Beiträge zur Descendenztheorie und zur Systematik der Spongien. Von...In: Ausland, xliii, pp. 30–35, 179–181, 246–247.

1870 **Senoner, A.** (Schwammzuchtin Dalmatien.) In: Zool. Gart. xi, p. 33–34.

1870 **Sharp, S.** The Oolites of Northamptonshire. By...In: Quart. Journ. Geol. Soc. London, xxvi, 1, p. 354–391, fig. 1–5.

1870 **Sicard, A.** Études pratiques sur la vitalité des jeunes éponges et leur croissance. Par...In: Bull. Soc. Acclim. (2), vii, p. 424–431.

1870 **Stewart, C.** On a New Sponge, Tethyopsis columnifer. By...In: Quart. Journ. Micr. Sc. x, p. 281–282.

1870 **Thomson, C. W.** On Holtenia, a Genus of Vitreous Sponges. By... In: Phil. Trans. clix, p. 701–720, Pl. LXVII–LXXXI.

1870 (a) **Thomson, C. W.** Food of Oceanic Animals. In: Nature, i, p. 315–316

1870 **Tietze, E.** Mittheilungen über den Niederschlesischen Culm und Kohlenkalk. In: Verhandl. geol. Reichsanst. iv, p. 118–123.

1870 **Trautschold, H.** Palaeontologischer Nachtrag...über die geognostischen Verhaeltnisse des Kreises Meschtschowsk im Gouvernem. Kaluga, von...In: Bull.Soc.Nat.Moscou, xlii, 2, p. 230–233, Pl. IV.

1870 **Vaillant, L.** Note sur la disposition des pores ou orifices afférents dans la Cliona celata, Grant. Note de...In: Comptes Rendus, lxx, p. 41–43. Transl. in: Ann. and Mag. (4), v, p. 146–148.

1870 **Wright, E. P.** Notes on Sponges. 1, On Hyalonema mirabilis, Gray. 2, On Aphrocallistes Bocagei, sp. nov. 3, On a new Genus and Species of Deep Sea Sponge. By...In: Q. J. M. S. x, p. 1–9, Pl. I–III.

1870 (a) **Wright, E. P.** On the minute structure of Dehitella atrorubens, Gray. By...In: Quart. Journ. Microsc. Sc. x, p. 90–92.

1871 [**B(ennett), A. W.**] Affinities of Coccoliths and of Sponges. In: Amer. Nat. v, p. 726–727. (Review of a paper by Carter.)

1871 **Binder, C.** Sind die festen Kalkbänke mit Spongiten und mit Terebr. lacunosa bei Geislingen Weisser Jura β oder γ? In: Jahrb. Ver. vaterl. Naturk. Württemberg, xxvii, p. 293–300.

1871 **Carpenter, W. B.** On the latest Scientific Researches in the Mediterranean. In: Notices Proc. R. Inst. vi, p. 236–259.

1871 **Carter, H. J.** On two undescribed Sponges and two Esperiadae from the West Indies; also on the Nomenclature of the Calcisponge Clathrina, Gray. By...In: Ann. and Mag. (4), vii, p. 268–283, Pl. XVII. Review: Zool. Rec. viii (1873), pp. 482, 485.

1871 (a) **Carter, H. J.** A Description of two new Calcispongiae, to which is added Confirmation of Prof. James Clark's Discovery of the True Form of the Sponge-cell (Animal), and an Account of the Polype-like Pore-area of Cliona corallinoides contrasted with Prof. E. Haeckel's view on the Relationship of the Sponges to the Corals. By...In: Ann. and Mag. (4), viii, p. 1–27, Pl. I–II.

1871 (β) **Carter, H. J.** Discovery of the Animal of the Spongiadae confirmed. By...In: Ann. and Mag. (4), vii, p. 445. Also in: Amer. Nat. v, p. 441-442.

1871 (γ) **Carter, H. J.** A Descriptive Account of three Pachytragous Sponges growing on the Rocks of the South Coast of Devon. By...In: Ann. and Mag. N. H. (4), vii, p. 1-15, Pl. iv.

1871 (δ) **Carter, H. J.** On Fossil Sponge-spicules of the Greensand compared with those of existing Species. By...In: Ann. and Mag. N. H. (4), vii, p. 112-141, Pl. vii-x.

1871 (ε) **Carter, H. J.** Description and Illustrations of a new Species of Tethya, with Observations on the Nomenclature of the Tethyadae. By...In: Ann. and Mag. N. H. (4), viii, p. 99-105, Pl. iv.

1871 (ζ) **Carter, H. J.** Parasites of the Sponges. By...In: Ann. and Mag. N. H. (4), viii, p. 330-332.

1871 **Chimmo, W.** Natural History of Deep-sea Soundings between Galle and Java. [Not published; title in Proc. Linn. Soc. London, p. xxx.]

1871 **Clark, H. J.** Note on the Infusoria flagellata and the Spongiae ciliatae. By...In: Amer. Journ. Sc. and Arts (3), i, p. 113-114. Also in: Ann. and Mag. N. H. (4), vii, p. 247-248.

1871 (α) **Clark, H. J.** The American Spongilla a Craspedote Flagellate Infusorian. By...In: Amer. Journ. Sc. and Arts (3), ii, p. 426-436, Pl. ii. Also in: Ann. and Mag. N. H. (4), ix (1872), p. 71-81, Pl. xi, and in Monthly Microsc. Journ. vii (1872), p. 104-114, Pl. xi.

1871 **Cohn, F. (J.).** [Ueber das Vorkommen von Kieselschwamm-nadeln in einem dichten grauen Kalkstein.] In: 48 Jahresber. schles. Ges. p. 63-64.

1871 **Cubitt, C. E.** Linear Projection considered in its Application to the Delineation of Objects under microscopic Observation. By...In: Monthly Microsc. Journ. v, p. 205-212, Pl. lxxxii-lxxxiv.

1871 **Cunningham, R. O.** Notes on the Natural History of the Strait of Magellan and West Coast of Patagonia...Edinburgh.

1871 **Dawson, J. W.** Hydrous Silicates injecting the pores of Fossils. By...In: Nature, iv, p. 162-163.

1871 **Dumortier, E.** Sur quelques gisements de l'oxfordien inférieur de l'Ardèche. Par...Paris-Lyon. iv and 85 pp.

1871 (**Ehlers, E.** Aulorhipis elegans, n. g. et sp.) In: Sitz. Ber. Phys. Med. Soc. Erlangen, iii, p. 61-62. English transl. in: Ann. Mag. N. H. (4), vii, p. 302-303.

1871 (α) **Ehlers, E.** Aulorhipis elegans, eine neue Spongienform; nebst Bemerkungen über einzelne Punkte aus der Organisation der Spongien. Von...In: Z. W. Z. xxi, p. 540-567, Pl. xlii.

1871 **Ehrenberg, C. G.** Uebersicht der seit 1847 fortgesetzten Untersuchungen über das von der Atmosphäre unsichtbar getragene reiche organische Leben. Von...In: Abhandl. Akad. Berlin, p. 1–150, Pl. i–ii. Abstr. in Z. Ges. Nat. xxxviii (N.S. iv), p. 206–207.

1871 **Flemming, W.** Bemerkungen über die neue Gray'sche Hornschwammgattung Janthella. Von...In: Verhandl. phys. med. Ges. Würzburg, ii, p. 1–7, Pl. i.

1871 **Geinitz, H. B.** Das Elbthalgebirge in Sachsen. Von...Erster Theil. Der untere Quader. i. Die Seeschwämme des unteren Quaders... In: Palaeontogr. xx, p. 1–42, Pl. i–x.

1871 (α) **Geinitz, H. B.** Ueber fossile Seeschwämme. In: Sitzungsber. Isis, Dresden, p. 88–90.

1871 **Giebel, C. G. A.** (Ueber Janthella verrucosa.) In: Zeitschr. gesammte Naturw. iv (2), xxxviii, p. 380.

1871 **Gray, J. E.** Note on Spongia linteiformis and S. lycopodium, Esper. By...In: Ann. and Mag. N. H. (4), viii, p. 142.

1871 (α) **Gray, J. E.** Note on Spongia linteiformis, Esper. By...In: Ann. and Mag. (4), viii, p. 370.

1871 **Haeckel, E. H. P. A.** Ueber die sexuelle Fortpflanzung und das natürliche System der Schwämme. Von...Jen. Zeitschr. vi, p. 641–651.

1871 **Harkness, R. & H. Hicks.** On the Ancient Rocks of the St David's Promontory, South Wales, and their Fossil Contents. By...In: Quart. Journ. Geol. Soc. London, xxvii, 1, p. 384–402, Pl. xv–xvi.

1871 **Harting, P.** Leerboek van de grondbeginselen der dierkunde...door ...Derde deel...Eerste afdeeling...Tiel, 1870*.

1871 **Johnson, M. H.** On flint. By...In: Proc. Geologists' Assoc. (2), ii, p. 251–269.

1871 **Kayser, E.** Studien aus dem Gebiete des Rheinischen Devon. 2. Die Devonischen Bildungen der Eifel. Von...In: Zeitschr. d. geol. Ges. xxiii, p. 289–376, Pl. vi.

1871 **Kent, W. S.** Note on Dorvillia agariciformis. By...In: Ann. and Mag. N. H. (4), vii, p. 37–38.

1871 (α) **Kent, W. S.** On a new Genus of Sponges from North Australia. By...In: Proc. Zool. Soc. London, p. 615–616, Pl. xlviii.

1871 (β) **Kent, W. S.** Notes on Prof. James Clark's Flagellate Infusoria, with description of new species. In: Monthly Microsc. Journ. vi, p. 261–265, Pl. cv.

1871 (γ) **Kent, W. S.** Affinities of the Sponges. By...In: Nature, iv, p. 184.

1871 (δ) **Kent, W. S.** Zoological results of the 1870 dredging expedition of the yacht "Norna" off the coast of Spain and Portugal. In: Nature, iv, p. 456–458. Also in: Amer. Journ. Sc. and Arts, cii (3), 2, p. 385–387.

* Read 1867–1871.

1871 **Kölliker, R. A. von.** Beiträge zur Kenntniss der Polypen. Von...
In: Verhandl. phys. med. Ges. Würzburg, ii, p. 11–30, Pl. III–IV.

1871 **Linnarsson, J. G. O.** Geognostika och palaeontologiska iakttagelser
öfver Eophytonsandstenen i Vestergötland. In: Kongl. Svenska
vetensk. akad. handl. ix. No. 7, p. 1–9, Pl. I–V.

1871 (α) **Linnarsson, J. G. O.** Om några försteningar från Sveriges och
Norges "Primordialzon." In: Öfvers. vetensk. akad. forhandl.
xxviii, p. 789–796, Pl. XVI.

1871 **Lütken, C. F.** De saakalde Glas-koraller eller Glas-svampe. In:
Tidsskr. popul. Fremstillngr. iii, p. 1–39.

1871 **Marchi, P.** Scienza del popolo. 2 Ser. Vol. ii. Spugne e coralli...
Milano. 32mo. 40 pp.

1871 **Meyer, C. J. A.** On Lower Tertiary Deposits recently exposed at Ports-
mouth. By...In: Q. J. G. S. London, xxvii, 1, p. 74–89, fig. 1–4.

1871 **Parfitt, E.** On a Species of Arenaceous Foraminifer (?) from the
Carboniferous Limestone of Devonshire. By...In: Ann. and
Mag. N. H. (4), vii, p. 158–161, Pl. XI.

1871 (α) **Parfitt, E.** On the Boring of Molluscs, Annelids and Sponges into
Rocks, Wood and Shells. In: Rep. and Trans. Devonsh. Assoc.
iv, p. 456-466.

1871 (β) **Parfitt, E.** Affinities of the Sponges. By...In: Nature, iv, p. 201–202.

1871 **Pengelly, W.** Notes on the existence of precretaceous Sponges. In:
Rep. and Trans. Devonshire Assoc. iv, p. 536–539.

1871 **Prestwich, J.** On the Structure of the Cragbeds of Suffolk and
Norfolk, with some Observations on their Organic Remains.
By...In: Quart. J. Geol. Soc. London, xxvii, 1, p. 115–146, Pl. VI.

1871 **Quenstedt, F. A.** Über den Unteren Weissen Jura. Von...In:
N. Jahrb. Mineral. p. 859–869.

1871 **Roper, F. C. S.** On Dysidea fragilis, a species of Sponge. By...In:
3rd Annual Rep. and Proc. Eastbourne Soc.

1871 **Schlüter, C. A.** Die Spongitarien-Bänke der unteren Mukronaten- und
oberen Quadraten-Schichten, und über Lepidospongia rugosa
insbesondere. In: Verhandl. nat. Ver. preuss. Rheinlande und
Westphalens*, xxvii; (3) vii; Sitzgsb. p. 139–141.

1871 **Sicard, A.** Études pratiques sur la revivification des madrépores et
des éponges...In: Rép. Travaux Soc. Statist. Marseille (7), xxx.

1871 **Thomson, C. W.** The Continuity of the Chalk. By...In: Nature, iii,
p. 225–227.

1871 **Verrill, A. E.** On the Distribution of Marine Animals on the Southern
Coast of New England. By...In: Amer. J. Sc. (3), ii, p. 357–362.

1871 **W.** Affinities of the Sponges. By...In: Nature, iv, p. 224.

* Roy. Soc. Cat. of Sc. Papers gives ref.: "Bonn, Sitz. Ber. Niederrhein. Gesell.
1870, pp. 139–141." But the only title-page is as above, and "Verhandl." and
"Sitzgsb." are alike included in the subject-index on the following page.—G. P. B.

1871 **Waller, J. G.** On the so-called boring or burrowing sponge: Cliona celata, Grant. By...In: Journ. Quekett Microsc. Club, ii, p. 269–277, 1 Pl.

1871 **Whiteaves, J. F.** Deep-sea dredging in the gulf of St Lawrence. In: Nature, v, p. 8–9.

1871 **Willemoes-Suhm, R. von.** Biologische Beobachtungen über niedere Meeresthiere. Von...2. Ueber einen jungen Kalkschwamm... In: Zeitschr. wiss. Zool. xxi, pp. 382, 396, Taf. xxxi, fig. 4.

1871 **Young, J. & J. Armstrong.** On the carboniferous Fossils of the West of Scotland...by. In: Trans. Geol. Soc. Glasgow, iii, Suppl. p...

1871 **Zaddach, E. G.** (Über verschiedene Präparate aus den Sammlungen des Kgl. zoologischen Museums.) In: Schriften physik. ökon. Ges. Königsberg, xi, p. 18–20.

1872 **Binder, C.** Erwiderung auf das Schreiben des Hrn. Prof. Dr. v Quenstedt in Tübingen. Von...In: N. Jahrb. Mineral. p. 411–414.

1872 **Bowerbank, J. S.** Contributions to a General History of the Spongiadae. By...Part i–iii. In: Proc. Zool. Soc. London, 1872, pp. 115–129, 196–202, 626–635, Pls. v, vi, x, xi, xlvi–xlix.

1872 (α) **Bowerbank, J. S.** Observations on Mr Carter's paper "On two new Sponges from the Antarctic Sea, and on a new Species of Tethya from Shetland; together with Observations on the Reproduction of Sponges commencing from Zygosis of the Sponge-animal." By...In: Ann. and Mag. N. H. (4), x, p. 58–61.

1872 **Buxton, A. F.** On Sponges. By...In: 2nd Rep. Cambridge Nat. Sc. Club.

1872 **Carter, H. J.** On two new Sponges from the Antarctic Sea, and on a new Species of Tethya from Shetland; together with Observations on the Reproduction of Sponges commencing from Zygosis of the Sponge-animal. By...In: Ann. and Mag. N. H. (4), ix, p. 409–435, Pl. xx–xxii.

1872 (α) **Carter, H. J.** Proposed Name for the Sponge-animal, viz. "Spongozoon"; also on the Origin of Thread-cells in the Spongiadae. By...In: Ann. and Mag. N. H. (4), x, p. 45–51.

1872 (β) **Carter, H. J.** Additional Information on the Structure of Tethya dactyloidea, Cart. By...In: Ann. and Mag. N. H. (4), ix, p. 82–84, Pl. x.

1872 (γ) **Carter, H. J.** Description, with Illustrations, of a new Species of Aplysina from the N.W. Coast of Spain. By...In: Ann. and Mag. N. H. (4), x, p. 101–110, Pl. vii.

1872 (δ) **Carter, H. J.** Descriptions of two new Sponges from the Philippine Islands. By...In: Ann. and Mag. N. H. (4), x, p. 110–113.

1872 (ε) **Carter, H. J.** Answer to Dr Bowerbank's "Observations on Mr Carter's paper, etc." in the last Number of the Annals. By...In: Ann. and Mag. N. H. (4), x, p. 141.

1872 **Dewalque, G.** Un spongiaire nouveau, du système eifelien. Par...
In: Bull. Acad. R. Belgique (2), xxxiv, p. 23–26, 1 Pl. Also in:
Journ. Zool. ii (1873), p. 292–294.

1872 **Ehrenberg, C. G.** (Mikrogeologische Studien als Zusammenfassung
seiner Beobachtungen des kleinsten Lebens der Meeres-Tiefgründe aller Zonen und dessen geologischen Einfluss.) In:
Monatsber. Akad. Berlin, p. 265–322.

1872 **Eimer, Th.** Nesselzellen und Samen bei Seeschwämmen. Von...In:
Arch. f. mikrosk. Anat. viii, p. 281–294. With 2 woodcuts.

1872 (α) **Eimer, T.** (Über die Verwandschaft der Schwämme mit den
Korallen.) In: Verhandl. phys. med. Ges. Würzburg, iii, p. ii.

1872 (β) **Eimer, T.** (Über Untersuchungen an Seeschwämmen.) In:
Verhandl. phys. med. Ges. Würzburg, iii, p. xiii–xvi. Also with
discussion in: Z. Ges. Nat. xl (N.S. vi), p. 202–205.

1872 **Fischer, P. & L. de Folin.** Note sur les draguages exécutés dans la
fosse du Cap. Breton durant l'année 1871...In: C. R. Acad.
Paris, lxxiv, p. 750–753.

1872 **Geinitz, H. B.** Das Elbthalgebirge in Sachsen. Von...Zweiter Theil.
Der mittlere und obere Quader. i. Seeschwämme, Korallen,
Seeigel, Seesterne und Haarsterne. In: Palaeontogr. xx, 2, pp. i–
viii, 1–19, Pl. i–vi.

1872 **Gibson, J.** The natural history and commerce of Sponges. In:
Pharm. Journ. ii, pp. 865–867, 904–906.

1872 **Godet, P.** (Notice sur Euplectella aspergillum.) In: Bull. Soc. Sc.
Nat. Neuchatel, ix, p. 178–182.

1872 **Gray, J. E.** Notes on the Classification of the Sponges. By...In:
Ann. and Mag. (4), ix, p. 442–461. Rev. in: Zool. Rec. ix, p. 473.

1872 (α) **Gray, J. E.** On the name Tethya and its Varieties of Spelling.
By...In: Ann. and Mag. N. H. (4), x, p. 150–151.

1872 (β) **Gray, J. E.** On the Animal of the Glass-rope. By...In: Ann.
and Mag. N. H. (4), ix, p. 324–325.

1872 (γ) **Gray, J. E.** Notice of a new Netted Sponge (Meyerella) from the
Philippines. By...In: Ann. and Mag. N. H. (4), x, p. 76.

1872 (δ) **Gray, J. E.** On a new Genus of Hexradiate and other Sponges
discovered in the Philippine Islands by Dr A. B. Meyer. By...
In: Ann. and Mag. N. H. (4), x, p. 134–139.

1872 (ε) **Gray, J. E.** Marine Sponges in the British Museum. By...In:
Ann. and Mag. N. H. (4), x, p. 152.

1872 **Greeff, E.** Madeira und die canarischen Inseln in naturwissen-
schaftlicher besonders zoologischer Beziehung. Marburg. 4°.

1872 **Grimm, O.** Ueber Synura urella, Ehrb., und Uroglena volvox, Ehrb.,
und den wahrscheinlichen genetischen Zusammenhang der Catal-
lacten mit den Schwämmen von...In: Nachr. K. Ges. Wiss.
G. A. Univ. Göttingen, p. 539–540.

1872 **Gulliver, G.** (Note on Spongiadae.) In: Quart. Journ. Microsc. Sc. xii, p. 106–108.

1872 **Haeckel, E. (H. P. A.).** Die Kalkschwämme. Eine Monographie in zwei Bänden Text und einem Atlas mit 60 Tafeln Abbildungen von... Berlin, 1872. (Part translated A. & M. N. H. (4) xi, pp. 241–262, 421–430.)

1872 **Hicks, H.** On some Undescribed Fossils from the Menevian Group. By...In: Quart. Journ. Geol. Soc. London, xxviii, 1, p. 173–185, Pl. v–vii.

1872 **Holl, H. B.** Notes on fossil Sponges. In: Geol. Mag. ix, pp. 309–315, 343–352. Also in: Cincinnati Quart. Journ. Sc. i (1874), p. 65–80.

1872 (α) **Holl, H. B.** The Epitheca in Fossil Sponges. In: Monthly Microsc. Journ. viii, p. 141–142.

1872 **Kent, W. S.** Note on Tethya muricata, Bowerbank, and Dorvillia agariciformis, Kent. By...In: Ann. and Mag. N. H. (4), x, p. 209–212.

1872 (α) **Kent, W. S.** The zoological results of the Dredging Expedition of the Yacht "Norna" off the Coast of Spain and Portugal in 1870. By...In: Rep. 41st Meet. Brit. Assoc. p. 132–133.

1872 **Kitton, F.** The Glass-Rope Sponge, Hyalonema mirabilis (Gray). In: Hardwicke's Science Gossip, p. 35–38, fig. 19–31.

1872 (α) **Kitton, F.** Remarks on Palythoa investing the Glass-rope Sponge. In: Hardwicke's Science Gossip, p. 56.

1872 (β) **Kitton, F.** The Glass-rope Sponge. In: Hardwicke's Science Gossip, p. 106–107.

1872 (γ) **Kitton, F.** On the Spongeous Origin of Flints. In: Trans. Norfolk and Norwich Nat. Soc. p....

1872 **Koninck, L. G. de.** Nouvelles recherches sur les animaux fossiles du terrain carbonifère de la Belgique. Première partie. In: Mém. Acad. R. Belgique, xxxix, p. 1–178, Pl. i–xv.

1872 **Lacaze-Duthiers, (F. J.) H. de.** Sur la nature des éponges. In: Archives Zool. Expérim. i, p. lxv–lxvii.

1872 **Moulins, Ch. des.** Fragments zoologiques. i. Questions obscures relatives à l'Hydractinia echinata, Flem., et à l'Alcyonium domuncula, Lamk. tous deux logeurs de Pagures...Par...In: Actes Soc. Linn. Bordeaux, xxviii [(3), viii*], p. 325–356.

1872 **(Pagenstecher, H. A.)** Votrag des Herrn Prof...."Zur Kenntniss der Schwamme" von 15. Dezember 1871...1. Geschichtliche Einleitung. In: Verh. naturh. med. Ver. Heidelberg, vi, p. 1–66.

1872 **Parfitt, E.** Mode in which Cliona burrows. In: Monthly Microsc. Journ. vii, p. 123–125.

1872 (α) **Parfitt, E.** Does Cliona burrow? In: Monthly Microsc. Journ. vii, p. 186–188.

* On the title page is printed 1871; the article of des Moulins is, however, signed "23. Janvier 1872."

1872* **Pomel, A.** Paléontologie ou Description des animaux fossiles de la province d'Oran. Par...Zoophytes. 5ᵉ fasc. Spongiaires. Oran. 4°. v and 256 pp., 36 Pl.

1872 **Quenstedt, F. A.** Über die geognostische Specialkarte von Württemberg und den Weissen Jura Beta. In: N. Jahrb. Min. p. 198–202.

1872 **Roemer, C. F.** (Ueber die Auffindung einer Jurassischen Diluvial-Geschiebes bei Strehlen südlich von Breslau.) In: 49. Jahresber. schles. Ges. p. 41–43.

1872 (*a*) **Roemer, C. F.** (Ueber seine Reise nach England.) In: N. Jahrb. Mineral. p. 67–73.

1872 **Sars, G. O.** On Some Remarkable Forms of Animal Life from the great deeps off the Norwegian Coast. i. Partly from posthumous manuscripts of the late professor Dr Michael Sars by... Christiania. 4°. 82 pp. and 6 Pl.

1872 **Schlüter, C. A.** Ueber die Spongitarien-Bänke der oberen Quadraten-und unteren Mukronaten-Schichten des Münsterlandes. Bonn. 38 pp., 1 Pl.

1872 **Schmidt, E. O.** Ueber die Entwicklung der Kieselkörper der Spongien. In: Tagebl. 45. Versamml. d. Naturf. und Aerzte, p. 139. Abstr. in Z. Ges. Nat. xl (N.S. vi), p. 314.

1872 **Schultze, M. J. S.** (Ueber Poterion Neptuni.) In: Verhandl. naturh. Ver. preuss. Rheinl. und Westph. xxviii; (3) viii; Sitzgsb. p. 15. [See footnote, p. 101.]

1872 **Senoner, A.** (Thiergärten in Italien und künstliche Schwammzucht auf Lesina.) In: Zool. Garten, xiii, p. 283–284.

1872 **Sinzow, J.** Ueber Jura- und Kreideversteinerungen im Gouvernement Saratow. In: Materialy dla Geologii Rossii, iv, p. ...

1872 **Slack, H. J.** The nature of Sponges. By...In: Popular Sc. Rev. xi, p. 167–176, Pl. LXXXIV.

1872 **Smith, G.** The Euplectella. In: Hardwicke's Science Gossip, p. 180–181.

1872 **Sollas, W. J.** Some Observations on the Upper Greensand Formation of Cambridge. By...In: Quart. Journ. Geol. Soc. London, xxviii, 1, p. 397–400 (abridged).

1872 (*a*) **Sollas, W. J.** On the Ventriculitae of the Cambridge Upper Greensand. By...In: Quart. Journ. Geol. Soc. London, xxix, p. 63–70.

1872 **Uljanin...** Catalogus animalium ponticorum, p. 103. (Quotation from Czerniawsky.)

1872 (*a*) **Uljanin, W.** Materialy dlia fauny Czernago moria. (Russian.) Abstr. in: Zool. Rec. ix (1874) p. 472.

1872 **Waller, J. G.** Observations on the Fresh-Water Sponges. In: Journ. Quekett Microsc. Club, iii, p. 42–48.

* According to Rauff (1893, p. 61) the complete paper appeared in 1872, although the preface is dated 1866. "Vermuthlich wurde es stückweise gedruckt..."

1872 **Whiteaves, J. F.** Notes on a Deep-sea Dredging Expedition round the Island of Anticosti, in the Gulf of St Lawrence. By...In: Ann. and Mag. N. H. (4), x, p. 341-354. Also in: Canadian Naturalist, vi, p. 351-354, and vii (1873), p. 86-100.

1873 **Ankum, H. J. van.** (Sponges from Helgoland.) In: Tijdschr. Ned. Dierk. Ver. i, p. 29- .

1873 (α) **Ankum, H. J. van.** (Sponges from Helgoland.) In: Tijdschr. Ned. Dierk. Ver. i, p. 45-46.

1873 **Barrande, J.** Systême silurien du centre de la Bohême. i. partie... Prague et Paris. 4°. xxx and 647 pp., 35 Pl.

1873 **Bowerbank, J. S.** Contributions to a General History of the Spongiadae. By...Parts iv and v. In: Proc. Zool. Soc. London, pp. 3-25, 319-333, Pls. i-iv, xxviii-xxxi.

1873 (α) **Bowerbank, J. S.** Report on a Collection of Sponges found at Ceylon by E. W. H. Holdsworth, Esq. By...In: Proc. Zool. Soc. London, p. 25-32, Pl. v-vii.

1873 (β) **Bowerbank, J. S.** Reply to Dr J. E. Gray's Observations on certain Species of Sponges described in the Proceedings of the Zoological Society for 1873. By...In: Ann. and Mag. N. H. (4), xii, p. 488-491.

1873 **Carter, H. J.** Transformation of an entire Shell into Chitinous Structure by the Polype Hydractinia, with short Descriptions of the Polypidoms of five other Species. By...In: Ann. and Mag. (4), xi, p. 1-15, Pl. i.

1873 (α) **Carter, H. J.** On two new Species of Gummineae, with Special and General Observations. By...In: Ann. and Mag. (4), xii, p. 17-30, Pl. i.

1873 (β) **Carter, H. J.** On the Hexactinellidae and Lithistidae generally, and particularly on the Aphrocallistidae, Aulodictyon, and Farreae, together with Facts elicited from their Deciduous Structures, and Descriptions respectively of Three New Species. By...In: Ann. and Mag. N. H. (4), xii, pp. 349-373, 437-472, Pl. xiii-xvii.

1873 (γ) **Carter, H. J.** Description of Labaria hemisphaerica, Gray, a new Species of Hexactinellid Sponge, with Observations on it and the Sarcohexactinellid Sponges generally. By...In: Ann. and Mag. N. H. (4), xi, p. 275-286.

1873 (δ) **Carter, H. J.** Points of Distinction between the Spongiadae and the Foraminifera. By...In: Ann. and Mag. (4), xi, p. 351-356.

1873 **Chapman, H. C.** Criticism on an Observation of Professor Thomson on Certain Sponges, etc. In: Amer. Naturalist, vii, p. 485-486.

1873 **Dawson, J. W.** Annual Address of the President (Nat. Hist. Soc. Montreal), 1872. In: Canadian Naturalist and Q. Journ. Sc. vii, 1875 (see p.   ).

1873 **Eckhel, G. von.** Der Badeschwamm in Rücksicht auf die Art seiner Gewinnung, die Geographische Verbreitung und locale Variation. Von...Triest. 42 pp., 2 Pl., 1 map. Transl. Italian, ibid.

1873 **Ehlers, E.** Zur Kenntniss der Fauna von Nowaja-Semlja. In: Sitz. Ber. phys. med. Soc. Erlangen, v, p. 7–10.

1873 **Ehrenberg, C. G.** Mikrogeologische Studien über das kleinste Leben der Meeres-Tiefgründe aller Zonen und dessen geologischen Einfluss. Von...In: Abhandl. Akad. Berlin, 1872, Phys. Klasse, p. 131–397, Pl. I–XII.

1873 **Etheridge, R.** ...Memoirs of the geological Survey of Scotland...

1873 **Ewald, J. von.** Ueber ein neues Coeloptychium aus der oberer Senoner Kreide von Haldem in Westphalen. In: Sitzungsber. Ges. naturf. Freunde, Berlin, p. 38–40.

1873 **Fisher, O.** On the Phosphatic Nodules of the Cretaceous Rock of Cambridgeshire. By...In: Quart. Journ. Geol. Soc. London, xxix, 1, p. 52–63, Pl. VI.

1873 **Ford, S. W.** On some new species of Fossils from the Primordial or Potsdam Group of Rensselaer County, N. Y. By...In: Amer. Journ. Sc. and Arts, cv (3), v, p. 211–215.

1873 **Giard, A.** Contributions a l'histoire naturelle des Synascidies par... In: Arch. Zool. Expér. ii, p. 481–514, Pl. XIX.

1873 **Giebel, C. G. A.** (Bemerkungen über Thalamopora cribrosa.) In: Zeitschr. ges. Naturw. xli (2), vii, p. 361–362.

1873 **Gray, J. E.** Notes on the Siliceous Spicules of Sponges, and on their Division into Types. By...In: Ann. and Mag. N. H. (4), xii, p. 203–217.

1873 (α) **Gray, J. E.** Natal Sponges. By...In: Ann. and Mag. N. H. (4), xii, p. 264.

1873 (β) **Gray, J. E.** A Sponge on Hyalonema. By...In: Ann. and Mag. N. H. (4), xii, p. 76.

1873 (γ) **Gray, J. E.** On the Genus Oceanapia, Norman (Rhizochalina of Oscar Schmidt). By...In: Ann. and Mag. N. H. (4), xii, p. 266.

1873 (δ) **Gray, J. E.** Sponges from Ceylon. By...In: Ann. and Mag. N. H. (4), xii, p. 266–267.

1873 (ε) **Gray, J. E.** Manufactured Glassrope. By...In: Ann. and Mag. N. H. (4), xii, p. 347.

1873 (ζ) **Gray, J. E.** On two new Free Sponges from Singapore. By...In: Ann. and Mag. N. H. (4), xi, p. 234–235.

1873 **Grube, A. E.** Ueber einige bisher noch unbekannte Bewohner des Baikalsee's. In: 50. Jahresber. schles. Ges. p. 66–68.

1873 **Haeckel, E. (H. P. A.)** (Calcaria in Möbius, 1873, pp. 99, 149.)

1873 **Hall, J. & R. P. Whitfield.** ...Descriptions of new species of fossils from the Devonian rocks of Iowa. By...In: 23rd Annual Rep. Condition State Cabin. New York, p. 223–243, Pl. IX–XIII.

1873 **Hicks, H.** On the Tremadoc Rocks in the Neighbourhood of St David's, South Wales, and their Fossil Contents. By...In: Quart. Journ. Geol. Soc. London, xxix, 1, p. 39–52, Pl. iii–v.

1873 **Holdsworth, E. W. H.** Note on the Occurrence of Xenospongia patelliformis, Gray, on the Coast of Ceylon. By...In: Proc. Zool. Soc. London, p. 32–33.

1873 **Jentink, F. A.** (Occurrence of Spongilla in town canals.) In: Tijdschr. Ned. Dierk. Ver. i, p. 44.

1873 **Judd, J. W.** The Secondary Rocks of Scotland. By...First paper... In: Quart. Journ. Geol. Soc. London, xxix, 1, p. 97–197, Pl. vii–viii.

1873 **Kritschagin, N.** Otczet o faunisticzeskich issljedovaniach proisvedennich ljeton 1872 g. na wostocznom beregu Czernago moria. In: Sapiski kiewsk. obschtschv. iii, p....

1873 **L., S.** How to clean the Euplectella. By...In: Amer. Naturalist, vii, p. 496.

1873 (a) **L., S.** Who first determined the true position of Hyalonema? In: Amer. Naturalist, vii, p. 565.

1873 **Lee, H.** (On Sponges in the Brighton Aquarium.) In: Land and Water for Sept. 1873, p. 445. Abstr. 1874 Bowerbank, J. S., vol. iii, p. 338–9.

1873 **Lindström, G.** Några anteckninger om Anthozoa tabulata. In: Öfers. vetensk. Akad. Förhandl. xxx, 4, p. 3–20. English transl. in: Ann. and Mag. N. H. (4), xviii (1876), p. 1–17.

1873 **Loretz, H.** Geognostische Beobachtungen in der Alpiner Trias der Gegend von Niederdorf, Sexten und Cortina in Süd-Tirol. Von ...In: Neues Jahrb. Mineral. p. 271–291.

1873 (a) **Loretz, H.** Zur Geognosie der Gegend von Niederdorf, Sexten und Cortina in Südtirol. In: N. Jahrb. Mineral. p. 612–626.

1873 **Marck, W. von der.** Neue Beiträge zur Kenntniss der fossilen Fische und anderer Thierreste aus der jüngsten Kreide Westfalens, sowie Aufzählung sämmtlicher seither in der Westfälischen Kreide aufgefunden Fischreste. Von...In: Palaeontogr. xxii (2), ii, p. 55–74, Pl. i–ii.

1873 **Marschall, A. F. von.** Nomenclator zoologicus...Vindobonae... 4 and 482 pp.

1873 **Möbius, K. A. (O. Schmidt & E. Haeckel).** Die wirbellosen Thiere der Ostsee. In: Jahresber. Comm. Untersuch. deutschen Meere. Kiel, i, p. 97–154.

1873 **Nicholson, H. A.** On some new Species of Stromatopora. By...In: Ann. and Mag. N. H. (4), xii, p. 89–95, Pl. iv.

1873 **Pfeiffer, L.** Nomenclator botanicus...Vol. i, Pars prior. Casselis... 4°. x and 808 pp.

1873 **Salter, J. W.** A catalogue of the collection of Cambrian and Silurian Fossils contained in the geological museum of the University of Cambridge, by....Cambridge...4º. xlviii and 204 pp.

1873 **Schmidt, (E.) O.** (Silicispongiae in Möbius, 1873, pp. 99, 147, 148.)

1873 **Senoner, A.** (Ueber die künstliche Schwammzucht in Lesina.) In: Zool. Garten, xiv, p. 272.

1873 **Sharp, S.** The Oolites of Northamptonshire. By...In: Quart. Journ. Geol. Soc. London, xxix, 1, p. 225–300, Pl. IX–X.

1873 **Smith, L. Toulmin.** Wyville Thomson and the Ventriculidae. In: Nature, viii, p. 484–485.

1873 **Sollas, W. J.** On the Ventriculitae of the Cambridge Upper Greensand. By...In: Quart. Journ.Geol. Soc. xxix, 1, p. 63–70, 6 figs.

1873 (α) **Sollas, W. J.** On the Coprolites of the Upper Greensand Formation, and on Flints. By...In Quart. J. Geol. Soc. xxix, p. 76–81.

1873 (β) **Sollas, W. J.** On the Foraminifera and Sponges of the upper Greensand of Cambridge. In: Geol. Mag. x, p. 268–274. Abstr. in: Proc. Camb. Phil. Soc. ii, 1876, p. 299–300.

1873 **Stoliezka, F.** The Corals or Anthozoa, with notes on the Sponges, Foraminifera, Arthrozoa and Spondylozoa, by...In: Mem. Geol. Survey India. Palaeontol. indica. (8), iv, 4, p. 1–70, Pl. I–XII.

1873 **Thomson, C. Wyville.** The Depths of the Sea...By...London. xx and 527 pp., 84 woodcuts, 7 Pl. In the same year there seems to have appeared a second edition. French transl. in 1875.

1873 (α) **Thomson, C. W.** The Challenger Expedition. In: Nature, vii, p. 385–388. Also in: Amer. Journ. Sc. and Arts (3), v, p. 401–405.

1873 (β) **Thomson, C. W.** Notes from the "Challenger." In: Nature, viii, pp. 28–30, 246–249, 5 figs.

1873 **Tribolet, M. de.** Recherches géologiques et paléontologiques dans le Jura Neuchatelois. Première partie...In: Mém. Soc. Neuchâtel, v, p. 37–141, Pl. I.

1873 **Verrill, A. E.** Report upon the Invertebrate Animals of Vineyard Sound and the adjacent waters, with an account of the physical characters of the region. By...In: U.S. Commission of Fish and Fisheries. Part i, Rep. Condition Sea Fisheries S. Coast New England...p. 295–778, figs. 1–4, Pl. I–XXXVIII.

1873 (α) **Verrill, A. E.** Brief Contributions to Zoölogy, from the museum of Yale College. No. xxv, Results of recent Dredging Expeditions on the Coast of New England. No. 3, by...In: Amer. Journ. Sc. (3), vi, p. 435–441.

1873 (β) **Verrill, A. E.** Review of Sars' book "On some remarkable Forms," etc...In: Amer. Journ. Sc. vi, p. 470–471.

1873 **Wetherell, J. W.** On some Fossils from the Margate Chalk. By... In: Proc. Geol. Assoc. (2), III, 1874, p. 192–197.

1873 **Zwiedinek von Südenhorst, J.** Syrien und seine Bedeutung für den Welthandel...Wien...144 pp.

1874 **Agassiz, A.** Embryology of the Ctenophora. By...In: Mem. Amer. Acad. Cambridge (2), x, p. 355–398, Pl. i–v.

1874 **Allman, G. J.** A new order of Hydrozoa. In: Nature, x, p. 251.

1874 **Ankum, H. J. van.** Kalklichaampjes bij Echinometra lucunter Ag. door...In: Tijdschr. Nederl. Dierk. Ver. i, p. 188–196, Pl. ix–x.

1874 **Barrois, C.** Note sur la craie de l'île de Wight, par...In: Bull. Soc. Géol. France (3), ii, p. 428–433.

1874 **Bayan, F.** Sur la succession des assises et des faunes dans les terrains jurassiques supérieurs, par...In: Bull. Soc. Géol. France (3), ii, p. 316–343, Pl. x–xi.

1874 **Billings, E.** On some new or little known fossils from the Silurian and Devonian Rocks of Ontario. In: Canadian Naturalist (2), vii, p. 230–240, 2 figs.

1874 **Bleicher,...** Matériaux pour servir à l'histoire du terrain cretacé inférieur de l'Hérault, par...In: Bull. Soc. Géol. France (3), ii, p. 21–27.

1874 **Bowerbank, J. S.** A Monograph of the British Spongiadae. By... Vol. iii. London: printed for the Ray Society. 8º.

1874 (α) **Bowerbank, J. S.** Contributions to a General History of the Spongiadae. By...Part vi...In: Proc. Zool. Soc. London, p. 298–305, Pl. xlvi–xlvii.

1874 **Carter, H. J.** Descriptions and Figures of Deep-sea Sponges and their Spicules from the Atlantic Ocean, dredged up on board H.M.S. "Porcupine," chiefly in 1869; with Figures and Descriptions of some remarkable Spicules from the Agulhas Shoal and Colon, Panama. By...In: Ann. and Mag. (4), xiv, pp. 207–221, 245–257, Pl. xiii–xv.

1874 (α) **Carter, H. J.** On the Nature of the Seed-like Body̅ of Spongilla; on the Origin or Mother Cell of the Spicule; and on the Presence of Spermatozoa in the Spongida. By...In: Ann. and Mag. (4), xiv, p. 97–111, Pl. x.

1874 (β) **Carter, H. J.** Development of the Marine Sponges from the earliest Recognizable Appearance of the Ovum to the Perfected Individual. By...In: Ann. and Mag. (4), xiv, pp. 321–337, 389–406, Pl. xx xxii.

1874 (γ) **Carter, H. J.** Further Instances of the Sponge-Spicule in its Mother Cell. By...In: Ann. and Mag. (4), xiv, p. 456–458, Pl. xxi, fig. 26–27.

1874 (δ) **Carter, H. J.** On Halisarca lobularis, Schmidt, off the South Coast of Devon, with Observations on the Relationship of the Sponges to the Ascidians, and Hints for Microscopy. By...In: Ann. and Mag. N. H. (4), xiii, p. 433–440.

1874 (ε) **Carter, H. J.** On the Spongozoa of Halisarca Dujardinii. By...
In: Ann. and Mag. N. H. (4), xiii, p. 315–316.

1874 **Courtiller, A.** Éponges fossiles des sables du terrain crétacé supérieur
des environs de Saumur,...Paris. 54 pp., 109 Pl. Ed. ii from
Courtiller, 1861.

1874 **Dames, W.** (Ueber Astylospongia und Aulocopium von Gotland.)
In: Zeitschr. d. geol. Ges. xxvi, p. 613–614.

1874 **Darbishire, R. D.** Note on a Deposit of Middle-Pleistocene Gravel in
the Worden-Hall Pits, near Leyland, Lancashire...In: Quart.
Journ. Geol. Soc. London, xxx, 1, p. 38–40.

1874 **Davey, E. C.** The "Sponge-gravel" beds at Coxwell, near Faringdon
...In: Trans. Newbury Field Club, ii, p...

1874 **Dumortier, E.** Études paléontologiques sur les dépots jurassiques du
Bassin du Rhone. Quatrième partie. Lias Supérieur...Paris.
iv and 339 pp., 62 Pl.

1874 **Dybowsky, B. N.** Beiträge zur näheren Kenntniss der in dem Baikal-
See vorkommenden niederen Krebse aus der Gruppe der Gam-
mariden...(Beiheft zum x Bande der Horae societatis entomo-
logicae Rossicae.) St. Petersburg. 4º. 190 pp., 14 Pl.

1874 **Ehlers, E.** Beiträge zur Kenntniss der Vertikalverbreitung der
Borstenwürmer im Meere. Von...In: Z. W. Z. xxv, p. 1–102,
Pl. i–iv.

1874 **Ehrenberg, C. G.** Das unsichtbar wirkende Leben der Nordpolarzone
am Lande und in den Meerestiefgründen bei 300-mal verstärkter
Seh Kraft...In: Zweite d. Nordpolarfahrt, ii, p. 437–467, Pl. i–iv.

1874 **Eisen, G.** Om Aulorrhipis och dess förmodade slägtskap med spongi-
orna. Af...In: Bihang Svenska Vet. Akad. Handl. ii, p. (1–16),
Pl. i–ii.

1874 **Geinitz, H. B.** Das Elbthalgebirge in Sachsen von...Zweiter Theil...
In: Palaeontogr. xx, 2, pp. i–viii, 1–245, Pl. i–xlvi.

1874 **Gray, J. E.** On the Arrangement of Sponges. By...In: Ann. and
Mag. N. H. (4), xiii, p. 284–290.

1874 (α) **Gray, J. E.** The Habitat of Labaria hemisphaerica. By...In:
Ann. and Mag. N. H. (4), xiii, p. 188.

1874 **Gümbel, C. W.** Ueber Conodictyum bursiforme Étallon, einer Fora-
minifere aus der Groupe der Dactyloporideen. Von...In:
Sitzungsber. Akad. München, iii, p. 282–294, Pl. i.

1874 **Hadlow, H.** The Hyalonema mirabilis. In: Trans. Asiat. Soc. Japan,
i, p. 10–19.

1874 **Haeckel, E. (H. P. A.).** Die Gastraea-Theorie, die phylogenetische
Classification des Thierreichs und die Homologie der Keimblätter. Von...In: Jen. Zeitschr. viii, p. 1–55, Taf. i. Transl.
in: Quart. Journ. Microsc. Sc. xiv, pp. 142–165, 233–247, Pl. vii.

1874 (α) **Haeckel, E. (H. P. A.).** Kalk- und Gallertspongien. Bearbeitet
von...In: Zweite deutsche Nordpolfahrt, ii, p. 434–436.

1874 **Hébert, E.** Comparaison de la Craie des côtes d'Angleterre avec celle de France, par...In: Bull. Soc. Géol. France (3), ii, p. 416–428.

1874 **Helmhacker, R.** Geognostische Beschreibung eines Theiles der Gegend zwischen Benesov und der Sázava. In: Arch. naturwiss. Landesdurchforsch. Böhmen, ii, 2, 1, p. 409–446, 2 Pl.

1874 **Higgin, T.** On the Structure of the Skeleton of Euplectella aspergillum. By...In: Ann. and Mag. N. H. (4), xiii, p. 44–48, Pl. III.

1874 (α) **Higgin, T.** (On Euplectella aspergillum.) In: Proc. Lit. and Phil. Soc. Liverpool, xxviii, p. xlvi–xlviii. [Also **Moore, J. J.** in: ibid. p. xliv.]

1874 **Higgins, H. H.** (Synopsis of Invertebrates.) In: ibid. p. xlv and Appendix.

1874 **Hyatt, A.** (Remarks on Hollow-fibred Horny Sponges.) In: Proc. Boston Soc. xvii, p. 204–215.

1874 **Lankester, E. Ray.** The Mode of occurrence of chlorophyll in Spongilla. In: Quart. Journ. Microsc. Sc. (2), xiv, p. 400–401.

1874 **Leidy, J.** (Remarks on Sponges.) In: Proc. Acad. Nat. Sc. Philadelphia, p. 144–145.

1874 **M'Intosh, W. C.** On the Invertebrate Marine Fauna and Fishes of St Andrews. By...In: Ann. and Mag. N. H. (4), xiii, p. 140–145.

1874 **Metschnikoff, E.** Zur Entwickelungsgeschichte der Kalkschwämme. Von...In: Z. W. Z. xxiv, p. 1–14, Taf. I.

1874 **Meyer, A. B.** Note on the Habitat of Psetalia globulosa and Labaria hemisphaerica, Gray. By...In: Ann. and Mag. N. H. (4), xiii, p. 66–67.

1874 **Meyer, C. J. A.** On the Cretaceous Rocks of Beer Head and the adjacent Cliff-sections...In: Quart. Journ. Geol. Soc. London, xxx, p. 369–393, fig. 1.

1874 **Meyn, L.** Silurische Schwämme und deren eigenthümliche Verbreitung, ein Beitrag zur Kunde der Geschiebe. In: Zeitschr. d. geol. Ges. xxvi, p. 41–58.

1874 **Miller, S. A.** Genus Pasceolus (Billings). In: Cincinnati Quart. Journ. Sc. i, p. 4–5.

1874 (α) **Miller, S. A.** Pasceolus Darwini (S. A. Miller). In: Cincinnati Quart. Journ. Sc. i, p. 5–6, fig. 1–2.

1874 (β) **Miller, S. A.** Pasceolus Claudei (S. A. Miller). In: ibid. p. 6–7, fig. 3.

1874 **Moesch, C.** Der südliche Aargauer Jura und seine Umgebungen... In: Beitr. geol. Karte Schweiz, 10, xi, 128, 53 und xl pp., 2 Pl.

1874 **Motte, F. la.** Cenni sulla pesca e sull' allevamento della spugna da bagno e del corallo rosso nel golfo adriatico. Di...Translation from the German (F. Breisach). In: Boll. agrar. Dalmazia...p. 1–20.

1874 **Nicholson, H. Alleyne.** On the Affinities of the Genus Stromatopora, with Descriptions of two new species. By...In: Ann. and Mag. N. H. (4), xiii, p. 4–14, fig. 1–3.

1874 **Price, F. G. H.** On the Gault of Folkestone. By...In: Quart. Journ. Geol. Soc. London, xxx, 3, p. 342–366, Pl. xxv.

1874 **Rutot, A.** Note sur la découverte de deux Spongiaires ayant provoqué la formation des grès fistuleux et des tubulations sableuses de l'étage Bruxellien des environs de Bruxelles. In: Ann. Soc. Malacol. Belgique, ix, p. 55–68, Pl. iii.

1874 **Schmidt, (E.) O.** Kieselspongien. Bearbeitet von...In: Zweite d. Nordpolarfahrt, ii, p. 429–433, Pl. i.

1874 **Schultze, M. J. S.** (Ueber einige ausgezeichnete Exemplare von Schwämmen: Hyalonema und Euplectella.) In: Verhandl. naturh. Ver. preuss. Rheinl. und Westphalens, xxx (3), x, p. 65–67. [See footnote p. 101.]

1874 **Shone, W.** Discovery of Foraminifera, etc., in the Boulder-clays of Cheshire. By...In: Quart. Journ. Geol. Soc. London, xxx, p. 181–184.

1874 **Thomson, C. W.** On Dredgings and Deep-sea Soundings in the South Atlantic...By...In: Proc. R. Soc. London, xxii, p. 423–428. Also in: Ann. and Mag. N. H. (4), xiv, p. 231–237.

1874 (α) **Thomson, C. W.** Preliminary Notes on the Nature of the Sea-bottom procured by the Soundings of H.M.S. "Challenger"... By...In: Proc. R. Soc. London, xxiii, p. 32–49. Also in: Nature, xi, pp. 95–97, 116–119.

1874 **Toucas, A.** Note sur la géologie des Environs de Toulon, par...In: Bull. Soc. Géol. France (3), ii, p. 457–463, fig. 20.

1874 **Verrill, A. E.** Brief Contributions to Zoölogy from the Museum of Yale College...Results of recent Dredging Expeditions on the Coast of New England...by...In: Amer. Journ. Sc. (3), vii, pp. 38–46, 131–138, 405–414, 498–505, Pl. iv–viii.

1874 (α) **Verrill, A. E.** (Explorations of Casco Bay.) In: Proc. Amer. Assoc. 22nd Meet. 2, p. 340–395, Pl. i–vi.

1874 **Whiteaves, J. F.** On recent Deep-Sea Dredging operations in the Gulf of Lawrence. By...In: Amer. Journ. Sc. (3), vii, p. 210–219. Also in: Canadian Naturalist, vii (1875), p. 257–267.

1874 **Wright, E. P.** Report on the Structure and Mode of Life of Hyalonema lusitanica, Bocage. By...In: Proc. Irish Acad. i, p. 549–552.

1874 **Wright, J.** On the Discovery of Microzoa in the Chalk Flints of the North of Ireland....In: 44th Meet. Brit. Assoc. Belfast, p. 95–96.

1875 **Allman, G. J.** On the Structure and Systematic Position of Stephanoscyphus mirabilis, the type of a new Order of Hydrozoa. In: Trans. Linn. Soc. London, i, p. 61–66, Pl. xiv.

1875 **Barrois, C.** Description géologique de la craie de l'île de Wight. In: Ann. Sc. Géol. vi, p. 1–30.

1875 (α) **Barrois, C.** La Zone à Belemnites plenus...In: Ann. Soc. Géol. du Nord, ii, p. 146–193.

1875 **Bennett, G.** On the Euplectella aspergillum, Owen, or "Venus's Flower Basket," a species of Sponge belonging to the Alcyonoid family; and a notice of the Hyalonema or "Glass Rope" Sponge. In: Papers, Proc. and Rep. R. Soc. Tasmania, p. 59–65.

1875 **Bonney, T. G.** Cambridgeshire Geology...Cambridge...pp. 82+vii.

1875 **Bowerbank, J. S.** Contributions to a General History of the Spongiadae. By...Part vii. In: Proc. Zool. Soc. London, p. 281–296

1875 (α) **Bowerbank, J. S.** Further Observations on Alcyoncellum speciosum, Quoy et Gaimard, and Hyalonema mirabile, Gray. By... In: Proc. Zool. Soc. London, p. 607–610.

1875 (β) **Bowerbank, J. S.** A Monograph of the Siliceo-fibrous Sponges. By...Parts iii, iv, v. In: Proc. Zool. Soc. London, pp. 272–281, 503–509, 558–565, Pls. xxxix–xl, lvi–lvii, lxi–lxii.

1875 **Carter, H. J.** Notes Introductory to the Study and Classification of the Spongida. By...In: Ann. and Mag. N. H. (4), xvi, pp. 1–40, 126–145, 177–200, Pl. iii and woodcuts.

1875 (α) **Carter, H. J.** On the Genus Rossella (a Hexactinellid Sponge), with the Descriptions of three Species. By...In: Ann. and Mag. N. H. (4), xv, p. 113–122, Pl. x. [See also (β) Higgin.]

1875 **Cornet, F. L.** Compte-rendu de l'excursion du 31. août aux environs de Ciply, par...In: Bull. Soc. Géol. France (3), ii, p. 567–577, fig. 3–5.

1875 **Dawson, J. W.** The Dawn of Life...London. 239 pp., 8 Pl.

1875 **Dohrn, A.** Mittheilungen aus und über die zoologische Station von Neapel...von...In: Z. W. Z. xxv, p. 457–480.

1875 **Eckhel, G. von.** Nuove communicazioni sopra le spugne. In: Boll. soc. adriat. i, p. 100–104.

1875 **Ehrenberg, C. G.** Fortsetzung der mikrogeologischen Studien als Gesammt-Uebersicht der mikroskopischen Palaeontologie gleichartig analysirter Gebirgsarten der Erde,...In: Abhandl. Akad. Berlin, p. 1–225, Pl. i–xxx.

1875 **Fischer, P.** Sur la présence dans les mers actuelles d'un type de Sarcodaires des terrains secondaires. In: C. R. Acad. Paris, lxxxi, p. 1131–1133. Also in: Journ. Zool. iv, p. 530–533. English transl. in: Ann. and Mag. N. H. (4), xvii (1876), p. 103–104.

1875 **Geinitz, H. B.** Das Elbthalgebirge in Sachsen, von...Zweiter Theil... In: Palaeontogr. xx, 2, p. 190–245, Pl. xxxvii–xlvi.

1875 **Gosselet, J.** Le terrain dévonien des environs de Stolberg. In: Ann. Soc. Géol. du Nord, iii, p. 8–16.

1875 **Grimm, O. von.** Briefliche Mittheilungen an C. Th. v. Siebold ueber eine zoologische Untersuchungs-Expedition nach dem Kaspischen Meere. Von...In: Z. W. Z. xxv, p. 323–326. French transl. in: Arch. Sc. Phys. et Nat. liv, p. 427–432; English transl. in: Ann. and Mag. N. H. (4), xvii (1876), p. 176–178.

1875 **Gümbel, C. W.** Beiträge zur Kenntniss der Organisation und systematischen Stellung von Receptaculites. Von...In: Abh. Bayer. Akad. xii (Denkschr. xliv), I, p. 167–215, Pl. A.

1875 **Haeckel, E. (H. P. A.).** Die Gastrula und die Eifurchung der Thiere. Von...(Fortsetzung der "Gastraea-Theorie" u.s.w. Bd. viii dieser Zeitschr. S. 1–57*.) In: Jen. Zeitschr. ix, p. 402–508, Taf. XIX–XXV.

1875 **Hall, J.** The Niagara and Lower Helderberg Group...In: 27th Annual Rep. Regents, Univ. New York, p. 117–131, Pl...

1875 **Hall, J. & R. P. Whitfield.** Descriptions of Invertebrate Fossils, mainly from the Silurian system. In: Rep. Geol. Survey Ohio, ii, p. 65–161, Pl. I–IX.

1875 **Hicks, H.** On the Succession of the Ancient Rocks in the Vicinity of St David's, Pembrokeshire...By...In: Quart. Journ. Geol. Soc. London, xxxi, p. 167–193, Pl. VIII–XI.

1875 **Higgin, T.** On two Hexactinellid Sponges from the Philippine Islands in the Liverpool Free Museum. By...In: Ann. and Mag. N. H. (4), xv, p. 377–390, Pl. XXI–XXII.

1875 (α) **Higgin, T.** Note by...In: Ann. and Mag. N. H. (4), xvi, p. 77.

1875 (β) **Higgin, T.** On a new Sponge of the Genus Luffaria, from Yucatan, in the Liverpool Free Museum. By...With remarks by H. J. Carter. In: Ann. and Mag. N. H. (4), xvi, p. 223–227, Pl. VI.

1875 (γ) **Higgin, T.** On some new Sponges from the Philippine Islands. In: Proc. Lit. and Phil. Soc. Liverpool, xxix, p. li–lii.

1875 (δ) **Higgin, T.** Sponges, their Anatomy, Physiology and Classification. By...In: Proc. Lit. and Phil. Soc. Liverpool, xxix, p. 193–216, Pl. I–III.

1875 **Huguenin,** ... Note sur la zone à Ammonites tenuilobatus de Crussol (Ardèche), par...In: Bull. Soc. Géol. France (3), ii, p. 519–527.

1875 **Hyatt, A.** Revision of the North American Poriferae; with Remarks upon Foreign Species. Part i. By...In: Mem. Boston Soc. N. H. ii, p. 399–408, Pl. XIII.

1875 **Jackson, W. H.** On a new Peritrichous Infusorian (Cyclochaeta spongillae). In: Quart. Journ. Microsc. Sc. (2), xv, p. 243–249, Pl. XII.

1875 **Jeffreys, J. G. & A. M. Norman.** ...Submarine-cable Fauna. By... In: Ann. and Mag. N. H. (4), xv, p. 169–176, Pl. XII.

1875 **Jones, T. Rupert.** On Quartz, Chalcedony, Agate, Flint, Chert, Jasper, and other forms of silica geologically considered, by...In: Proc. Geol. Assoc. iv, p. 439–459.

1875 **Jukes-Browne, A. J.** On the Relations of the Cambridge Gault and Greensand. By...In: Quart. Journ. Geol Soc. London, xxxi, p. 256–314, Pl. XIV–XV.

* Read 55.

116

1875 **Kayser, E.** Ueber die Billings'sche Gattung Pasceolus und ihre Verbreitung in paläozoischen Ablagerungen. Von...In: Zeitschr. d. geol. Ges. xxvii, p. 776–783, Pl. xx.

1875 **Küstermann, H.** Hyalonema Sieboldi, Gray. Von...In: Arch. mikrosk. Anat. xi, p. 286–291, Pl. xvi.

1875 **Loretz, H.** Einige Petrefacten der Alpinen Trias aus den Südalpen. Von...In: Zeitschr. d. geol. Ges. xxvii, p. 784–841, Pl. xxi–xxiii.

1875 **Lortet,**... Quelques points de l'organisation des éponges fibreuses de Syrie. In: 4. Session Assoc. Franc. Nantes.

1875 **Lütken, C. F.** A revised Catalogue of the Spongozoa of Greenland. By...In: Manual Nat. Hist. Greenland.

1875 **McIntosh, W. C.** The marine Invertebrates and Fishes of St Andrews. By...Edinburgh-London. 4º. 86 pp., 9 Pl. Reprint, as regards sponges, of 1874, McIntosh.

1875 **Marshall, W.** Untersuchungen über Hexactinelliden. Von...In: Z. W. Z. xxv, Suppl. p. 142–243, Pl. xi–xvii.

1875 **Maurer, F.** Paläontologische Studien im Gebiet des rheinischen Devon ...Von...In: N. Jahrb. Mineral. p. 596–618.

1875 **Meek, F. B. & A. H. Worthen.** Palaeontology of Illinois...In: Geol. Survey, Illinois, vi, p. 491–532, Pl. xxiii–xxxii.

1875 **Mellin, J. C.** St Helena...By...London...65 Pl.

1875 **Meyer, A. B.** On Hyalonema cebuense. By...In: Ann. and Mag. N. H. (4), xvi, p. 76–77.

1875 **Nicholson, H. A.** Report upon the Palaeontology of the Province of Ontario...Toronto. 96 pp., 45 woodcuts, 4 Pl.

1875 (α) **Nicholson, H. A.** Descriptions of the Amorphozoa from the Silurian and Devonian Formations. In: Rep. Geol. Survey, Ohio, ii, p. 243–255.

1875 **Packard, A. S.** Life Histories of the Protozoa and Sponges. By... In: Amer. Naturalist, ix, p. 87–108, fig. 29–50.

1875 **Phillips, J.** Illustrations of the Geology of Yorkshire...London. 4º. Ed. iii.

1875 **Pillet, L. & E. de Fromentel.** Description géologique et paléontologique de la colline de Lemens, Chambéry. Chambéry....pp.

1875 **Schlüter. C. A.** Der Emscher Mergel. In: Verhandl. naturh. Ver. preuss. Rheinlande und Westphalens, xxxi (4), i, p. 89–98.

1875 **Schmidt, (E.) O.** Spongien bearbeitet von...In: Jahresber. Comm. Unters. d. Meere, ii, iii, p. 115–120, Pl. i.

1875 (α) **Schmidt, (E.) O.** Zur Orientierung über die Entwicklung der Spongien. Von...In: Z. W. Z. xxv, Suppl. p. 127–141, Pl. viii–x.

1875 (β) **Schmidt, (E.) O.** Die Gattung Loxosoma. Von...In: Arch. Mikr. Anat. xii, p. 1–14, Pl. i–iii.

1875 **Schulze, F. E.** Ueber den Bau und die Entwicklung eines Kalk-schwammes, Sycandra raphanus Haeckel. In: Tagebl. 48. Versamml. d. Naturf. u. Aerzte, p. 101–102.

1875 (α) **Schulze, F. E.** Ueber den Bau und die Entwicklung von Sycandra raphanus Haeckel. Von...In: Z. W. Z. xxv, Suppl. p. 247–280, Pl. xviii–xxi.

1875 **Senoner, A.** (Ueber die Badeschwamm- und Korallenfischerei im Adriatischen Meere.) In: Zool. Garten, xvi, p. 149–150.

1875 **Sorby, H. C.** On the Chromatological Relations of Spongilla fluviatilis. By...In: Quart. Journ. Microsc. Sc. (2), xv, p. 47–52.

1875 **Speyer, O.** Die paläontologischen Einschlüsse der Trias, in der Um-gebung Fulda's. In: 2. Ber. Ver. Naturk. Fulda, p. 43–86.

1875 **Stebbing, F. R. R.** On Some new exotic Sessile-eyed Crustaceans. By...In: Ann. and Mag. N. H. (4), xv, p. 184–188, Pl. xv A.

1875 **Sterzel, J. F.** Die fossilen Pflanzen des Rothliegenden von Chemnitz in der Geschichte der Paläontologie. In: 5. Ber. naturw. Ges. Chemnitz, p. 71–243.

1875 **Thomson, C. Wyville.** Report to the Hydrographer of the Admiralty on the Cruise of H.M.S. "Challenger"...By...In: Proc. R. Soc. London, xxiv, p. 33–40.

1875 **Toula, F.** Die Tiefsee-Untersuchungen und ihre wichtigsten Re-sultate. In: Mitth. geogr. Ges. Wien, xviii (2), viii, pp. 49–68, 97–111, 145–165, 1 Pl.

1875 **Willemoes-Suhm, R. von.** Von der Challenger-Expedition. Briefe von...an C. Th. E. v. Siebold. iii. In: Z. W. Z. xxv, p. xxv–xlvi

1875 **Wright, J.** A list of the Cretaceous Microzoa of the North of Ireland. By...In: Annual Rep. and Proc. Belfast Naturalists' Field Club (2), i, App. iii, p. 73–99, Pl. ii–iii.

1876* **Bachmann, I.** (Demonstration von zwei Riesenbecher-Schwämmen, Poterium Poseidonis.) In: Mitth. naturf. Ges. Bern, p. 25.

1876 **Baily, W. H.** Figures of Characteristic British Fossils; with Descriptive Remarks. By...Part v. London.

1876 **Barrois, C.** Mémoire sur l'embryologie de quelques éponges de la Manche. Par...In: Ann. Sc. Nat. (6), iii, Art. no. 11, p. 1–84, Pl. 12–16. Also separately: Thèse présentée à la faculté des sciences de Paris pour obtenir le grade de docteur ès sciences naturelles par Ch. Barrois. 2e Thèse. Embryologie de quelques Éponges de la Manche...Paris. 8º. Original paging and plates.

1876 (α) **Barrois, C.** L'âge de la pierre de Totternhoe. In: Ann. Soc. Géol. Nord, iii, p. 145–149.

1876 (β) **Barrois, C.** Recherches sur le terrain crétacé supérieur de l'Angle-terre et de l'Irlande. In: Mém. Soc. Géol. du Nord, i.

---

* Read 27 Nov. 1875. On the title-page of the volume is printed 1876.

1876 **Benecke, E. W.** Über die Umgebungen von Esino in der Lombardei, von...In: Geognost. Paläont. Beitr. ii, 3, p. 259–317, Pl. xxi–xxiv.

1876 **Blake, J. F.** On Renulina sorbyana. In: Monthly Microsc. Journ. xv, p. 262–264.

1876 **Bowerbank, J. S.** A Monograph of the Siliceo-fibrous Sponges. By... Part vi. In: Proc. Zool. Soc. London, p. 535–540, Pl. lvi–lvii.

1876 (*a*) **Bowerbank, J. S.** Contributions to a General History of the Spongiadae. By...Part viii. In: Proc. Zool. Soc. London, p. 768–775, Pl. lxxviii–lxxxi.

1876 **Bowerbank-Norman.** See Norman.

1876 **Carpenter, W. B.** Remarks on Mr Carter's Paper "On the Polytremata, especially with reference to their Mythical Hybrid Nature." By...In: Ann. and Mag. N. H. (4), xvii, p. 380–387.

1876 **Carter, H. J.** Descriptions and Figures of Deep-Sea Sponges and their Spicules, from the Atlantic Ocean, dredged up on board H.M.S. "Porcupine," chiefly in 1869 (concluded). By...In: Ann. and Mag. N. H. (4), xviii, pp. 226–240, 307–324, 388–410, 458–479, Pl. xii–xvi.

1876 (*a*) **Carter, H. J.** On the Polytremata (Foraminifera) especially with reference to their Mythical Hybrid Nature. By...In: Ann. and Mag. N. H. (4), xvii, p. 185–214, Pl. xii.

1876 **Dawson, G. M.** On some Canadian Species of Spongilla. In: Canad. Naturalist, viii, p. 1–5, Pl. i.

1876 **Dawson, J. W.** Notes on the Occurrence of Eozoon canadense at Côte St Pierre. By...In: Quart. Journ. Geol. Soc. London, xxxii, p. 66–74, Pl. x.

1876 **Duncan, P. M.** On some Unicellular Algae parasitic within Silurian and Tertiary Corals, with a Notice of their Presence in Calceola sandalina and other Fossils. By...In: Quart. Journ. Geol. Soc. London, xxxii, p. 205–211, Pl. xvi.

1876 **Etheridge, R.** ...Further localities for Acanthospongia Smithii, Young, and Estheria Dawsoni, Jones. In: Geol. Mag. (2), iii, p. 576.

1876 **Forel, F. A.** Matériaux pour servir à l'étude de la faune profonde du lac Léman, par...In: Bull. Soc. Vaud. xiv, p. 202–...

1876 **Gastaldi, B.** Sui fossili del calcare dolomitico del Chaberton (Alpi Cozie) studiati da G. Michelotti...In: Atti Real. Accad. Lincei (2), iii, 2, p. 114–121, Pl. i–ii.

1876 **Gosselet, J.** Le calcaire de Givet. In: Ann. Soc. Géol. du Nord, iii, p. 36–75.

1876 **Grimm, O. von.** Kaspiiskoe more i ego fauna...In: Trudi Aralo-Kasp. exped. ii, p. 1–168, Pl. ii–iii (The Caspian Sea and its fauna).

1876 **Haeckel, E. H. P. A.** (Bemerkungen über die Organisation und das System der lebenden Spongien.) In: Zeitschr. d. Geol. Ges. xxviii, p. 632.

1876 **Hayck, G. von.** Handbuch der Zoologie. Von...i...Wien. vi and 437 pp., with woodcuts.

1876 **Henry, J.** L'infralias dans la Franche-Comté, par...In: Mém. Soc. Émul. Doubs (4), x, p. 285–476, Pl. i–v.

1876 **Higgin, T.** On some Sponges recently presented to the Free Museum. In: Proc. Lit. and Phil. Soc. Liverpool, xxx, p. xlix–liii.

1876 **Higgins, H. H.** The Cruise of the Argo. In: Nature, xiv, p. 48.

1876 (**Hyatt, A.** Sponges considered as a distinct Sub-Kingdom of Animals.) In: Proc. Boston Soc. xix, p. 12–17.

1876 **Kayser, E.** Ueber primordiale und untersilurische Fossilien aus der argentinischen Republik, von...In: Palaeontogr. Suppl. iii, 2, 33 pp., 5 Pl.

1876 **Keller, C.** Untersuchungen über die Anatomie und Entwicklungsgeschichte einiger Spongien des Mittelmeeres. Ein Beitrag zur Lösung der Spongienfrage, von...Basel-Genf-Lyon. 4°. 39 pp. and 2 Pl.

1876 **Koch, G. von.** Zur Anatomie von Halisarca Dujardini, Johnston. In: Morph. Jahrb. ii, p. 83–84.

1876 **Koninck, L. G. de.** Recherches sur les fossiles paléozoïques de la Nouvelle-Galles du Sud (Australie). In: Mém. Soc. Sc. Liège (2), vi, p. 1–140, Pl. i–iv.

1876 **Lankester, E. R.** An account of Professor Haeckel's recent additions to the Gastraea-theory. By...In: Quart. Journ. Microsc. Sc. (2), xvi, p. 51–66, Pl. vii–x.

1876 **Linnarsson, J. G. O.** On the Brachiopoda of the Paradoxides beds of Sweden. In: Bihang svensk. akad. handl. iii, No. 12, p. 1–34, Pl. i–iv.

1876 **Lortet, .** Quelques points de l'organisation des éponges fibreuses de Syrie. Extrait. In: Assoc. Franç. Compte-rendu 4. session, p. 815–816.

1876 **Macalister, A.** An Introduction to Animal Morphology and Systematic Zoology. Part i. Invertebrata. By...London.

1876 **Marshall, W.** Ideen ueber die Verwandtschaftsverhältnisse der Hexactinelliden. Von...In: Z. W. Z. xxvii, p. 113–136.

1876 (α) **Marshall, W.** Bemerkung über die Nothwendigkeit, bei den Hornschwämmen die Arten mit hohlen Fasern von denen mit soliden Fasern zu trennen. In: Zeitschr. d. Geol. Ges. xxviii, p. 632.

1876 **Metschnikoff, E.** Beiträge zur Morphologie der Spongien. Von...In: Z. W. Z. xxvii, p. 275–286. Transl. in: Arch. Zool. Expérim. v, p. 357–368.

1876 **Murie, J.** On Steere's sponge, a new genus of the Hexactinellid order of the Spongidae. In: ? Proc. Linn. Soc. London. [See 1877.]

1876 **Norman, A. M.** Crustacea, Tunicata, Polyzoa, Echinodermata, Actinozoa, Foraminifera, Polycystina and Spongida obtained during the cruise of the "Valorous" to Davis' Strait in 1875. In: Proc. R. Soc. London, xxv, p. 202–215.

1876 (a) **Norman, A. M.** On the Polyzoa, Hydrozoa and Spongozoa dredged off the Coast of Durham and North Yorkshire in 1874. In: Rep. 45th Meet. Brit. Assoc. p. 197–199.

1876 **Payer, J.** Die österreichisch-ungarische Nordpol-Expedition in den Jahren 1872–74 nebst einer Skizze der 2. deutschen Nordpol-Expedition 1869–70 und der Polar-Expedition von 1871...Wien. cvi and 696 pp. With 146 illustr. and 3 maps.

1876 **Quenstedt, F. A.** Petrefactenkunde Deutschlands. Der ersten Abtheilung fünfter Band...Die Schwämme, von...Leipzig. 8°. Atlas, Fol. p. 1–96, Pl. ?. Cf. 1878, footnote.

1876 **Roemer, C. F.** Lethaea geognostica...1. Theil. Lethaea palaeozoica, von...Stuttgart. ii and 62 pp. Atlas with 62 Pl.

1876 **Schlüter, C. A.** Verbreitung der Cephalopoden in der oberen Kreide Norddeutschlands. Von...In: Zeitschr. d. geol. Ges. xxviii (3), p. 457–518. Also in: Palaeontogr. xxiv (2), iv, p. 85–134.

1876 **Schmid, E. E.** Der Muschelkalk des östlichen Thüringens...Von... Jena. 20 pp.

1876 **Schmidt, (E.) O.** Nochmals die Gastrula der Kalkschwämme. Von... In: Arch. mikrosk. Anat. xii, p. 551–556.

1876 **Schulze, F. E.** Zur Entwicklungsgeschichte von Sycandra. Von... In: Z. W. Z. xxvii, p. 486–487.

1876 **Smith, S. J. & O. Harger.** Report on the Dredgings in the region of St George's Bank in 1872. In: Trans. Connecticut Acad. iii, 1, p. 1–57, Pl. i–vii.

1876 **Sollas, W. J.** On Eubrochus clausus, a Vitreo-hexactinellid Sponge from the Cambridge "Coprolite" Bed. By...In: Geol. Mag. (2), iii, p. 398–403, Pl. xiv.

1876 **Thomson, C. Wyville.** The "Challenger" Expedition. In: Nature, xiv, p. 492–495.

1876 **Trautschold, H.** Ergänzung zur Fauna des russischen Jura. Von... In: Sapiski S. Petersburgk mineral. Obschtschwa (2), xii, p. 79–111, Pl. iv–ix.

1876 **Tribolet, M. de.** Sur les terrains jurassiques supérieurs de la Haute-Marne comparés à ceux du Jura suisse et français. Par...In: Bull. Soc. Géol. France (3), iv, p. 259–285.

1876 (a) **Tribolet, M. de.** Sur le véritable horizon stratigraphique de l'astartien dans le Jura. Par...In: Mém. Soc. d'Émul. Doubs (4), x, p. 233–244.

1876 **Wichmann, A.** Über Puddingstein. Von...In: N. Jahrb. Mineral. p. 907–918.

1876 **Willemoes-Suhm, R. von.** Von der Challenger-Expedition. Briefe von...an C. Th. v. Siebold. vii. In: Z. W. Z. xxvii, p. xcvii–cviii.

1876 **Williamson, W. C.** Corrections of the Nomenclature of the Objects figured in a Memoir "On some of the minute Objects found in the Mud of the Levant*." In: Mem. Lit. and Phil. Soc. Manchester (3), v, p. 131–137.

1876 **Zittel, K. A.** Ueber Coeloptychium. Ein Beitrag zur Kenntniss der Organisation fossiler Spongien. Von...In: Abh. bayer. Akad. xii (Denkschr. xliv), 3, p. 1–80, Pl. i–vii.

1876 (α) **Zittel, K. A.** (Über sein Handbuch der Paläontologie und das Studium der fossilen Spongien.) In: N. Jahrb. Mineral. p. 286–289.

1876 (β) **Zittel, K. A.** (Über seine neuesten Untersuchungen über fossile Spongien.) In: Zeitschr. d. geol. Ges. xxviii, p. 631–632.

1877 **Blake, J. F. & W. H. Hudleston.** On the Corallian Rocks of England. By...In: Quart. Journ. Geol. Soc. London, xxxiii, p. 260–404, Pl. xii–xvii.

1877 **Bowerbank, J. S.** Description of five new Species of Sponges discovered by A. B. Meyer on the Philippine Islands and New Guinea. By the late...In: Proc. Zool. Soc. London, p. 456–464.

1877 **Carter, H. J.** Remarks on Prof. E. Haeckel's Observations on Wyvillethomsonia Wallichii and Squamulina scopula. By...In: Ann. and Mag. N. H. (4), xx, p. 337–339.

1877 (α) **Carter, H. J.** Arctic and Antarctic Sponges...By...In: Ann. and Mag. N. H. (4), xx, p. 38–42, Pl. i, fig. 1–5.

1877 (β) **(Carter, H. J.** Appendix on Murie's paper, 1877.) Cf. Murie.

1877 (γ) **Carter, H. J.** On two Vitreohexactinellid Sponges. By...In: Ann. and Mag. N. H. (4), xix, p. 121–131, Pl. ix.

1877 (δ) **Carter, H. J.** Sponges Dredged up on board H.M.S. "Porcupine" in 1869–70, returned. By...In: Ann. and Mag. N. H. (4), xix, p. 432–434.

1877 (ε) **Carter, H. J.** On a Melobesian Form of Foraminifera (Gypsina melobesioides, mihi); and further Observations on Carpenteria monticularis. By...In: Ann. and Mag. N. H. (4), xx, p. 172–176.

1877 (ζ) **Carter, H. J.** On a Fossil Species of Sarcohexactinellid Sponge allied to Hyalonema. By...In: Ann. and Mag. N. H. (4), xx, p. 176–178.

1877 **Dybowsky, W.** Ueber die Spongillen (Süsswasserschwämme) der Ostsee-Provinzen. In: Sitz. Ber. naturf. Ges. Dorpat, iv, p. 527–534. [See also ibid. v, p. 10–12.]

1877 **Entz, Géza.** A Gastraeadák jegenleg élö, kepviselöiröl. In: Orvostermészett, értesitö Kolozsvár, ii, p....

* Cf. 1847, Williamson.

1877 **Fullagar, J.** Note on the development of the Spicules of Spongilla fluviatilis. In: Hardwicke's Science Gossip, xiii, June, p. 138. Also in: Monthly Microsc. Journ. xviii, p. 45.

1877 **Grimm, O. von.** Kaspiiskoe more i ego fauna...In: Trudi Aralo-Kasp. exped. p. 1–105, Pl. ix...Continued from 1876.

1877 **Haeckel, E. (H. P. A.).** Die Physemarien (Haliphysema und Gastrophysema), Gastraeaden der Gegenwart. Von...In: Jen. Zeitschr. xi, p. 1–54, Pl. i–vi.

1877 (α) **Haeckel, E. H. P. A.** Nachträge zur Gastraea-Theorie...In: Jen. Zeitschr. xi, (2) iv, p. 55–98.

1877 **Hall, J.** (32 Lithogr. Plates, illustrating a paper in vol. iv, Trans. Albany Inst. p. 195–208...) In: 28th Annual Rep. New York State Museum, p. 99–210, Pl. iv–xxxiv. Cf. 1862, Hall.

1877 **Henrich, C.** Über Spongien oder Meerschwämme.. Von...In: Verh. und Mitth. Siebenbürg. Ver. Hermannstadt, xxvii, p. 29–40.

1877 **Higgin, T. H.** Description of some Sponges obtained during a Cruise of the Steam-Yacht "Argo" in the Caribbean and neighbouring seas. By...In: Ann. and Mag. N. H. (4), xix, p. 291–299, Pl. xiv.

1877 (α) **Higgin, T. (H.).** On the Sponges of the Argo Expedition. In: Proc. Lit. and Phil. Soc. Liverpool, xxxi, p. li–lvi.

1877 **Hull, E.** On the Nature and Origin of the Beds of Chert in the Upper Carboniferous Limestone of Ireland. In: Sc. Trans. Dublin Soc. (2), i, p. 71–84, Pl. iii.

1877 **Huxley, T. H.** A Manual of the Anatomy of Invertebrated Animals. By...London. viii and 698 pp. With 158 woodcuts.

1877 **Hyatt, A.** Revision of the North American Poriferae; with Remarks upon Foreign Species. Part ii. By...In: Mem. Boston Soc. N. H. ii, p. 481–554, Pl. xv–xvii.

1877 **Jukes-Browne, A. J.** Supplementary Notes on the Fauna of the Cambridge Greensand. By...In: Quart. Journ. Geol. Soc. London, xxxiii, p. 485–504, Pl. xxi.

1877 **Kent, W. Saville.** (M. K. A. Zittel on Fossil Hexactinellida.) In: Ann. and Mag. N. H. (4), xx, p. 446–447.

1877 (α) **Kent, W. Saville.** [Prof. Haeckel's Group of the Physamariae (sic).] In: Ann. and Mag. N. H. (4), xx, p. 448. [Editor on 1878 (α).]

1877 **Law, J. K.** Sponges or Spongidae. In: Rep. and Trans. Cardiff Nat. Soc. viii, p. 17–22.

1877 **Marenzeller, E. von.** Die Coelenteraten, Echinodermen und Würmer der K. K. Oesterreichisch-Ungarischen Nordpol-Expedition. Bearbeitet von...In: Denkschr. Mathem.-naturw. Classe K. Akad. Wiss. Wien, xxxv, p. 357–398, Pl. i–iv.

1877 **Marshall, W. & A. B. Meyer.** Ueber einige neue und wenig bekannte Philippinische Hexactinelliden. Von...In: Mitth. Zool. Mus. Dresden, Heft 2, p. 261–279, Pl. xxiv–xxv*.

1877 **Martin, K.** Untersuchungen über die Organisation von Astylospongia, Ferd. Roem., und Bemerkungen über die Natur der Wallsteine, Meyn. Von...In: Arch. Ver. Freunde Naturgesch. Mecklenburg, xxxi, p. 1–32, 1 Pl.

1877 **Metschnikoff, E.** Issljedowaniä o gubkach. (Russian.) (Investigations on Sponges.) In: Sapiski novoross. obschtsch. Odessa, iv, p. 1–10.

1877 **Miers, J. E.** On Species of Crustacea living within the Venus's Flower-basket (Euplectella) and in Meyerina claviformis. By... In: Journ. Linn. Soc. London, xiii, p. 506–512, Pl. xxiv.

1877 **Miller, S. A.** The American Palaeozoic Fossils: a Catalogue of the Genera and Species...Cincinnati. xv and 254 pp.

1877 **Moseley, H. N.** On the Colouring Matters of Various Animals, and especially of Deep-sea Forms dredged by H.M.S. "Challenger." By...In: Quart. Journ. Microsc. Sc. xvii, p. 1–23, Pl. i–ii.

1877 **Munier-Chalmas, E.** Observations sur les Algues calcaires appartenant au groupe des Siphonées verticillées...Note de M....In: C. R. Paris, lxxxv, p. 814–817.

1877 **Murie, J.** On Steere's Sponge, a new Genus of the Hexactinellid Group of the Spongida†. By...In: Trans. Linn. Soc. London (2), i, p. 219–234, Pl. xxxvi–xxxvii. (At the end of Murie's paper there is added an "Appendix by H. J. Carter," p. 231–232.)

1877 **Nicholson, H. A.** The Ancient Life-History of the Earth...By... Edinburgh and London.

1877 **Price, F. G. H.** On the Beds between the Gault and Upper Chalk, near Folkestone. By...In: Quart. Journ. Geol. Soc. London, xxxiii, p. 431–445.

1877 **Quenstedt, F. A.** Petrefactenkunde Deutschlands. Der ersten Abtheilung fünfter Band...Die Schwämme, von...Leipzig. 8°. Atlas, Fol. p. 1–448, Pl. 115–130 (?). Cf. 1878.

1877 **Reinsch, P. F.** Notiz über die mikroskopische Fauna der mittleren und unteren fränkischen Liasschichten. In: N. Jahrb. Mineral. p. 176–178.

1877 **Salensky, W.** Études sur les bryozoaires entoproctes. Par...In: Ann. Sc. Nat. (6), v, p. 1–60, Pl. xii–xv.

1877 **Schmidt, (E.) O.** Das Larvenstadium von Ascetta primordialis und Ascetta clathrus. Von...In: Arch. mikr. Anat. xiv, p. 249–263, Pl. xv–xvi.

---

* This paper is unfortunate, being almost always overlooked. Lendenfeld (1889) states that it appeared in 1879, which cannot be true, since in Nature, xvii (1878), p. 222, there is an abstract of it. Cf. abstracts for 1879 in Zool. Rec. xvi (1881); Spong. pp. 1, 8, 12, 13.   † Read on Jan. 20, 1876.

1877 **Schuffner, O.** Beschreibung einiger neuer Kalkschwämme. Von...
In: Jen. Zeitschr. xi, (2) iv, p. 403–433, Pl. xxiv–xxvi.

1877 **Schulze, F. E.** Untersuchungen über den Bau und die Entwicklung
der Spongien...Die Gattung Halisarca. Von...In: Z. W. Z.
xxviii, p. 1–48, Pl. i–v.

1877 (α) **Schulze, F. E.** Untersuchungen über den Bau und die Entwicklung
der Spongien...Die Familie der Chondrosidae. Von...In:
Z. W. Z. xxix, p. 87–122, Pl. viii–ix.

1877 (β) **Schulze, F. E.** Spongicola fistularis, ein in Spongien wohnendes
Hydrozoon. Von...In: Arch. mikr. Anat. xiii, p. 795–817,
Pl. xlv–xlvii.

1877 **Shaler, N. S.** On the Occurrence of the genus Beatricea in Kentucky.
In: Amer. Naturalist, xi, p. 628.

1877 **Sollas, W. J.** On the Changes produced in the Siliceous Skeletons of
certain Sponges by the Action of Caustic Potash. By...In:
Ann. and Mag. N. H. (4), xx, p. 285–300, Pl. ix.

1877 (α) **Sollas, W. J.** On Pharetrospongia Strahani, Sollas, a Fossil Holo-
rhaphidote Sponge from the Cambridge "Coprolite" Bed. By...
In: Quart. Journ. Geol. Soc. London, xxxiii, p. 242–255, Pl. xi
and woodcuts.

1877 (β) **Sollas, W. J.** On the Structure and Affinities of the Genus Siphonia.
By...In: Quart. Journ. Geol. Soc. xxxiii, p. 790–834, Pl. xxv
and xxvi.

1877 (γ) **Sollas, W. J.** On Stauronema, a new genus of Fossil Hexactinellid
Sponges...By...In: Ann. and Mag. N. H. (4), xix, p. 1–25,
Pl. i–v.

1877 **Steinmann, G.** Über Material aus den Ancyloceras-Mergeln der
unteren Kreide von Schellenberg bei Hallein. In: N. Jahrb.
Mineral. p. 630–631.

1877 **Thomson, C. Wyville.** The Voyage of the "Challenger"...By...
London. 2 vols.

1877 **Toula, F.** Geologische Untersuchungen im westlichen Theile des
Balkan und in den angrenzenden Gebieten. iv...Von...In:
Sitz. ber. Akad. Wien, lxxv, 1, p. 465–549, Pl. i–viii.

1877 **Trautschold, H.** Ueber Kreidefossilien Russlands. In: Bull. Soc. Nat.
Moscou, lii, p. 332–349, Pl. v–vi.

1877 **Waller, J. G.** On a new British Sponge of the genus Microciona. In:
Monthly Microsc. Journ. xviii, p. 261.

1877 **Wright, E. Perceval.** On a New Genus and Species of Sponge. By...
In: Proc. Irish Acad. (2), ii, p. 754–756, Pl. 40.

1877 **Wright, R. R.** The Systematic Position of the Spongiadae. In:
Canad. Journ. xv, p. 417–428.

1877 **Young, J. & J. Young.** On a Carboniferous Hyalonema and other Sponges from Ayrshire. By...In: Ann. and Mag. N. H. (4), xx, p. 425–432, Pl. xiv–xv.

1877 **Zittel, K. A.** Studien über fossile Spongien. Von...i, Hexactinellidae. In: Abhandl. k. bayer. Akad. xiii, p. 1–63. Translated in: Ann. and Mag. N. H. (4), xx, pp. 257, 405, 501*.

1877 (α) **Zittel, K. A.** (Ueber seine Untersuchung fossiler Spongien.) In: N. Jahrb. Mineral. p. 77–78.

1877 (β) **Zittel, K. A.** (Über Schildkrötenreste aus dem lithographischen Schiefern von Eichstädt.) In: N. Jahrb. Mineral. p. 280–281.

1877 (γ) **Zittel, K. A.** Criticism of 1877 Quenstedt, part i. In: N. Jahrb. Mineral. p. 705–709.

1877 (δ) **Zittel, K. A.** (Über Astylospongia.) In: N. Jahrb. Min. p. 709–711.

1878 **Agassiz, A.** ...To C. P. Patterson, Superintendent Coast Survey,... from..., on the Dredging Operations of the United States Coast Survey Steamer "Blake"...In: Bull. Mus. Compar. Zool. Harvard Coll. v, p. 1–9.

1878 **Barrett, S. T.** Descriptions of New Species of Fossils from the Upper Silurian Rocks of Port Jervis, N.Y....In: Ann. N. York Acad. i, p. 121–124.

1878 **Bigsby, J. J.** Thesaurus devonico-carboniferus...By...London... 4°. xiv and 447 pp.

1878 **Blake, J. F.** On the Chalk of Yorkshire In: Proc. Geol. Assoc. v, p. 232–281.

1878 **Brögger, W. C.** Om Paradoxidesskifrene ved Krekling. In: Nyt mag. naturvidensk. xxiv (2), iv, p. 18–88, Pl. i–vi.

1878 **Bütschli, O.** Beiträge zur Kenntniss der Flagellaten und einiger verwandter Organismen. i. Von...In: Z. W. Z. xxx, p. 205–281, Pl. xi–xv.

1878 **Carter, H. J.** Parasites of the Spongida. By...In: Ann. and Mag. N. H. (5), ii, p. 157–172.

1878 (α) **Carter, H. J.** Position of the Sponge-spicule in the Spongida; and Postscript on the Identity of Squamulina scopula with the Sponges. By...In: Ann. and Mag. N. H. (5), i, p. 170–174.

1878 (β) **Carter, H. J.** Note on Tethea muricata, Bowerbank. By... In: Ann. and Mag. N. H. (5), ii, p. 174–177.

1878 (γ) **Carter, H. J.** Mr James Thomson's Fossil Sponges from the Carboniferous System of the South-west of Scotland. By...In: Ann. and Mag. N. H. (5), i, p. 128–143, Pl. ix–x.

1878 (δ) **Carter, H. J.** Emendatory Description of Purisiphonia Clarkei, Bk., a Hexactinellid Fossil Sponge from N.W. Australia. By...In: Ann. and Mag. N. H. (5), i, p. 376–379.

* Zittel's Monograph was reprinted with little alteration from Abh. bayer. Akad. in N. Jahrb. Min., and again separately. See 1878, 1878 (a), and 1879 (β).—G. P. B.

1878 (ε) **Carter, H. J.** On Teichonia, a new Family of Calcareous Sponges, with Descriptions of two Species. By...In: Ann. and Mag. N. H. (5), ii, p. 35–40, Pl. ii.

1878 (ζ) **Carter, H. J.** On Calcareous Hexactinellid Structure in the Devonian Limestone...and further Observations on the Replacement of Silex by Calcite. By...In: Ann. and Mag. N. H. (5), i, p. 412–419.

1878 **Chimmo, W.** Natural History of the Euplectella aspergillum from the Philippine Islands. London. 4°. 22 pp. and 6 Pl.

1878 **Coues, E. & H. C. Yarrow.** Notes on the Natural History of Fort Macon, N. C., and vicinity. In: Proc. Acad. Nat. Sc. Philadelphia, p. 297–315.

1878 **Dybowski, W.** Ueber die Spongillen (Süsswasserschwämme) der Ostseeprovinzen. In: Sitz. ber. naturf. Ges. Dorpat, v, p. 10–12.

1878 (α) **Dybowski, W.** Ueber Spongillen der Ostseeprovinzen. In: Arch. Anat. Phys. p. 11–12. Reprint from 1878, Dybowski [and 1877, id.].

1878 (β) **Dybowski, W.** Mittheilungen über Spongien i, ii...Von...In: Zool. Anz. pp. 30–32, 53–54.

1878 **Etheridge, R.** Palaeontology of the Coast of the Arctic Lands visited by the late British Expedition under Capt. Sir Geo. Nares... In: Quart. Journ. Geol. Soc. London, xxxiv, p. 568–636, Pl. xxv–xxix.

1878 (α) **Etheridge, R.** A Catalogue of Australian Fossils...London. 236 pp.

1878 (β) **Etheridge, R.** Palaeontological Notes. In: Geol. Mag. (2), v, p. 117–119.

1878 **Etheridge, R. & H. A. Nicholson.** On the Genus Palaeacis...By... In: Ann. and Mag. N. H. (5), i, p. 206–227.

1878 **Ford, S. W.** Descriptions of two new species of Primordial Fossils. In: Amer. Journ. Sc. and Arts (3), xv, p. 124–127.

1878 **Frédéricq, L.** Sur la digestion des albuminoïdes chez quelques invertébrés par...In: Bull. Acad. Belgique, xlvi, p. 213–228.

1878 **Ganin, M.** Zur Entwickelung der Spongilla fluviatilis Von...In: Zool. Anz. i, p. 195–199. Transl. in: Journ. Microsc. Soc. London, ii (1879), p. 174–176.

1878 **Heilmann,** . Notice sur les éponges et sur une espèce remarquable des côtes de Provence. In: Mém. Soc. Cannes, vii, p. 127–131, Pl....

1878 **Higgin, T.** On a Fresh-water Sponge from Bahia, Spongilla coralloides, Bow. In: Proc. Lit. and Phil. Soc. Liverpool, xxxii, p. lvi–lviii.

1878 **Hudleston, W. H.** The Yorkshire Oolites, Part ii...In: Proc. Geol. Assoc. v, p. 407–494.

1878 **Hyatt, A.** On a new species of Sponge...By...In: Proc. Acad. Nat. Sc. Philadelphia, ii, p. 163–164, Pl. i.

1878 **James, U. P.** Descriptions of newly discovered species of fossils from the Lower Silurian Formation: Cincinnati Group. In: Palaeontologist, i, p. 1–8.

1878 (α) **James, U. P.** Description of newly discovered species of fossils, and remarks on others, from the Lower and Upper Silurian rocks of Ohio. In: Palaeontologist, ii, p. 9–16.

1878 **Keller, C.** Ueber den Bau von Reniera semitubulosa O. S. Ein Beitrag zur Anatomie der Kieselschwämme. Von...In: Zeitschr. wiss. Zool. xxx, p. 563–586, Pl. xxxvi–xxxvii.

1878 (α) **Keller, C.** Ueber Spermabildung bei Spongilla. Von...In: Zool. Anz. i, p. 314–315. Tr. in: J. R. M. Soc. ii (1879), p. 73–74.

1878 **Kent, W. Saville.** Notes on the Embryology of Sponges. By...In: Ann. and Mag. N. H. (5), ii, p. 139–156, Pl. vi, vii.

1878 (α) **Kent, W. Saville.** Observations upon Professor Ernst Haeckel's Group of the "Physemaria," and on the Affinity of the Sponges. By...In: Ann. and Mag. N. H. (5), i, p. 1–17. [See 1877 (a).]

1878 (β) **Kent, W. Saville.** The Foraminiferal Nature of Haliphysema Tumanowiczii, Bow. (Squamulina scopula, Carter), demonstrated. By...In: Ann. and Mag. N. H. (5), ii, p. 68–78, Pl. iv–v.

1878 (γ) **Kent, W. Saville.** The Origin and Distribution of Organic Colour. In: Nature, xviii, p. 523–524.

1878 (δ) **Kent, W. Saville.** A New Field for the Microscopist. In: Popular Sc. Review, ii, p. 113–..., Pl. iii–iv.

[1878 (ε), (1879?) **Kent, W. Saville.** A Monograph of the Gymnozoidal Discostomatous Flagellata, with a proposed new Scheme of Classification of the Protozoa...Read before Linnean Soc. 21 June 1877.]

1878 **L(ankester), E. R(ay).** The Structure and Development of Sponges... In: Nature, xviii, p. 307–308, and in: J. R. M. Soc. i, p. 223–225.

1878 **Lenz, H.** Die wirbellosen Thiere der Travemünder Bucht...Theil i. Anhang i. zu: Jahresber. 1874–1875 Comm. wiss. Unters. d. Meere.

1878 **Marenzeller, E. v.** Die Aufzucht des Badeschwammes aus Theilstücken. Von...In: Verh. Zool. Bot. Ges. Wien, p. 687–694.

1878 **Marshall, W.** Spongiologische Mittheilungen. In: Amtl. Ber. 50. Vers. deutscher Naturf. p. 187.

1878 **Martin, K.** Niederländische und nordwestdeutsche Sedimentärgeschiebe...Leiden. 106 pp. and 3 Pl.

1878 **Matyasowsky, J. von.** Ein fossiler Spongit aus dem Karpathen-Sandsteine von Kis-Lipnik im Sároser Comitate. In: Termeszetrajzi füzetek, ii, Hungarian text, p. 262–266, transl. p. 297–301, Pl. xii.

1878 **Merejkowsky, C.** Predwaritelny otczet o bjelomorskich gubkach (Preliminary account on sponges of the White Sea)...In: Trudi St Petersburg Obschtsch. ix, p. 249–270.

1878 (α) **Merejkowsky, C.** On Wagnerella, a new Genus of Sponge nearly allied to the Physemaria of Ernst Häckel. By...In: Ann. and Mag. N. H. (5), i, p. 70–77, Pl. vi.

1878 **Mickleborough, J. & A. G. Wetherby.** A classified list of Lower Silurian Fossils, Cincinnati-group. By...In: Journ. Cincinnati Soc. i, p. 61–86.

1878 **Miller, S. A. & C. B. Dyer.** Contributions to Palaeontology. By... In: Journ. Cincinnati Soc. i, p. 24–39, Pl. I–II.

1878 **Murie, J.** (Resemblance of some Stromatoporidae to Hexactinellids.) In: Abstr. Proc. Geol. Soc. London, p.... Also in: Ann. and Mag. N. H. (5), ii, p. 344.

1878 **Nicholson, H. A. & R. Etheridge.** A Monograph of the Silurian Fossils of the Girvan District in Ayrshire, with special reference to those contained in the "Gray Collection." By...Fasciculus i... Edinburgh and London. 135 pp. and 9 Pl.

1878 **Norman, A. M.** On the Genus Haliphysema, with Description of several Forms apparently allied to it. By...In: Ann. and Mag. N. H. (5), i, p. 265–284, Pl. XVI.

1878 **Parfitt, E.** On the Structure of Haliphysema Tumanowiczii. By... In: Ann. and Mag. N. H. (5), ii, p. 88–90.

1878 **Peach, C. W.** Observations on British Polyzoa. By...In: Journ. Linn. Soc. London, xiii, p. 479–486, Pl. XXIII.

1878 **Quenstedt, F. A.** Petrefactenkunde Deutschlands. Der ersten Abtheilung fünfter Band...Die Schwämme, von...Leipzig. 8°. Atlas, Fol. p. 449–612, Pl. 131 (?)–142*.

1878 **Renard, A.** Recherches lithologiques sur les Phthanites du Calcaire carbonifère de Belgique, par...In: Bull. Acad. Belgique, xlvi, p. 471–499, Pl. XLVI.

1878 **Schmidt, (E.) O.** Brehm's Thierleben...Zweite umgearbeitete und vermehrte Auflage...Die niederen Thiere, von...Leipzig.

1878 (α) **Schmidt, E. O.** Bemerkungen zu den Arbeiten über Loxosoma. Von...In: Z. W. Z. xxxi, p. 68–80.

1878 (β) **Schmidt, E. O.** Die Entodermbildung bei den Asconen durch Wanderzellen. In: Amtl. Ber. 50. Vers. d. Naturf. und Aerzte, p. 173.

1878 (γ) **Schmidt, E. O.** Im mexikanischen Meerbusen und bei Barbados gesammelte Spongien. In: Tagebl. 51. Vers. d. Naturf. und Aerzte, p. 86.

1878 (δ) **Schmidt, E. O.** Die Fibrillen der Spongiengattung Filifera, Lkhn. Von...In: Z. W. Z. xxx, p. 661–662.

1878 **Schmidtlein, R.** Beobachtungen über Trächtigkeits- und Eiablage-Perioden verschiedener Seethiere. Januar 1875–Juli 1878. Von ...In: Mitth. Zool. Stat. Neapel, i, p. 124–136.

---

* The book appeared in 5 parts: 1877, Liefer. 1 (Dec. 1876), p. 1–96; Liefer. 2 (Apr. 1877), p. 97–224; Liefer. 3 (Aug. 1877), p. 225–320; Liefer. 4 (Sept. 1877), p. 321–448; 1878, Liefer. 5 (Febr. 1878), p. 449–612. So far as I am aware Pl. 115–130 appeared in 1877, the rest in 1878.

1878 **Schulze, F. E.** Untersuchungen über den Bau und die Entwicklung der Spongien. Vierte Mittheilung. Die Familie der Aplysinidae. Von...In: Z. W. Z. xxx, p. 379–420, Pl. xxi–xxiv.

1878 (α) **Schulze, F. E.** Untersuchungen über den Bau und die Entwicklung der Spongien. Sechste Mittheilung. Die Gattung Spongelia. Von...In: Z. W. Z. xxxii, p. 117–157, Pl. v–viii.

1878 (β) **Schulze, F. E.** Untersuchungen über den Bau und die Entwicklung der Spongien. Fünfte Mittheilung. Die Metamorphose von Sycandra raphanus. Von...In: Zeitschr. f. wiss. Zool. xxxi, p. 262–295, Pl. xviii–xix.

1878 **Simmonds, P. L.** The Commercial Products of the Sea...By... London. viii and 484 pp.*

1878 **Sollas, W. J.** On Two New and Remarkable Species of Cliona. By... In: Ann. and Mag. N. H. (5), i, p. 54–66, Pl. i–ii.

1878 (α) **Sollas, W. J.** On the Structure and Affinities of the Genus Catagma. By...In: Ann. and Mag. N. H. (5), ii, p. 353–364, Pl. xiv, and 3 woodcuts.

1878 (β) **Sollas, W. J.** On the Ventriculite Sponges. By...In: Dixon, Geology of Sussex. New ed. p. 448–455, Pl. xlvi–li.

1878 **Stebbing, T. R. R.** Amphipoda in Sponges. By...In: Ann. and Mag. N. H. (5), ii, p. 427–428.

1878 **Trautschold, H.** Ueber Camerospongia Auerbachi, Eichw. Von... In: Zeitschr. d. geol. Ges. xxx, p. 225–228, Pl. ix.

1878 **Tyler, C.** Biography of James Scott Bowerbank. In: Journ. Roy. Microsc. Soc. i, p. 28–30.

1878 **Verrill, A. E.** Note on Borings of a Sponge in Italian Marble. By... In: Amer. Journ. Sc. and Arts, xvi, p. 406.

1878 **Wallace, S. J.** On the "Geodes" of the Keokuk Formation and the genus Biopalla, with some species. By...In: Amer. Journ. Sc. and Arts (3), xv, p. 366–370.

1878 **Waller, J. G.** On a new British Sponge of the Genus Microciona. By...In: Journ. Quekett Microsc. Club, v, p. 1–5, Pl. i–ii.

1878 (α) **Waller, J. G.** On Variation in Spongilla fluviatilis. In: Journ. Quekett Microsc. Club, v, p. 53–62, Pl. v.

1878 **Young, J. F.** On the Occurrence of a Fresh-water Sponge in the Purbeck Limestone. By...In: Geol. Mag. (2), v, p. 220–221.

1878 **Zarecznego, S.** O średnich warstwach kredowych w Krakowskim okręgu. In: Sprawozd. Komm. Fizyjograf. Akad Umiejętn. Krakow, xii, p. 176–246, Pl. iv–viii.

1878 **Zittel, K. A.** Studien über fossile Spongien. Zweite Abth. Lithistidae. Von...In: Abh. K. Bayr. Akad. xiii, p. 65–154 (1–90), Pl. i–x. [Transl. Ann. and Mag. (5) ii, mostly repr. N. Jahrb. Min. p. 561.]

1878 (α) **Zittel, K. A.** Studien über fossile Spongien. Dritte Abtheilung. Monactinellidae, Tetractinellidae und Calcispongiae. Von...In: A. K. B. A. xiii, p. 91–138 (1–48), Pl. xi–xii. [A. & M. and N. J. M.]

* On the title is printed 1879; the book was published, however, before Nov. 1878.

1878 (β) **Zittel, K. A.** Zur Stammes-Geschichte der Spongien, von... München. 4°. 20 pp.

1878 (γ) **Zittel, K. A.** (Fossile Lithistidae und Hexactinellidae.) In: Amtl. Ber. 50. Vers. d. Naturf. und Aerzte, p. 161.

1878 (δ) **Zit(t)el, K. (A.).** (Criticism of 1877 Quenstedt, part 2, on Jurassic sponges.) In: N. Jahrb. Min. p. 58–62.

1879 **Agassiz, A.** ...(Letter No. 3) To Carlile P. Patterson, Superintendent United States Coast Survey,...from..., on the Dredging Operations carried on...by the United States Coast Survey Steamer "Blake"...In: Bull. Mus. Compar. Anat. Harvard Coll. v, p. 289–302, 2 maps.

1879 **Balfour, F. M.** On the Morphology and Systematic Position of the Spongida. By...In: Quart. Journ. Microsc. Sc. xix, p. 103–109. Transl. in: Arch. Zool. Expérim. viii (1880), p. v–viii.

1879 **Barnard, F.** Notes on Sponge from Northern Territory...In: Quart. Journ. Microsc. Sc. Victoria, i, p. 14–15, Pl. I.

1879 **Carter, H. J.** On a New Genus of Foraminifera (Aphrosina informis), and Spiculation of an unknown Sponge. By...In: Journ. R. Microsc. Soc. London, ii, p. 500–502, Pl. XVII A, fig. 5–12.

1879 (α) **Carter, H. J.** Contributions to our Knowledge of the Spongida. By...In: Ann. and Mag. N. H. (5), iii, pp. 284–304, 343–360, Pl. XXV–XXIX.

1879 (β) **Carter, H. J.** On a New Species of Excavating Sponge (Alectona Millari); and on a New Species of Rhaphidotheca (R. affinis). By...In: Journ. R. Microsc. Soc. London, ii, p. 493–499, Pl. XVII and XVII A, fig. 1–4.

1879 (γ) **Carter, H. J.** On Holasterella, a Fossil Sponge of the Carboniferous Era, and on Hemiasterella, a new Genus of Recent Sponges. By ...In: Ann. and Mag. N. H. (5), iii, p. 141–150, Pl. XXI.

1879 (δ) **Carter, H. J.** Spongiidae*. By...In: Phil. Trans. R. Soc. London, clxviii, p. 286–288.

1879 (ε) **Carter, H. J.** On the Nutritive and Reproductive Processes of Sponges. By...In: Ann. and Mag. N. H. (5), iv, p. 374–386.

1879 (ζ) **Carter, H. J.** Note on the so-called "Farringdon (Coral Rag) Sponges" (Calcispongiae Zittel). By...In: Ann. and Mag. N. H. (5), iv, p. 431–437.

1879 (η) **Carter, H. J.** On the Mode of Growth of Stromatopora including the Commensalism of Caunopora. By...In: Ann. and Mag N. H. (5), iv, p. 101–106.

1879 **Clarke, B.** A new Arrangement of the Classes of Zoology...London. 4°. Ed. ii, 1880.

1879 **Czerniawsky, V.** Pribreschniä gubki Czernago i Kaspiiskago morei. Predvaritelnoe issljedovanie...In: Bull. Soc. nat. Moscou, liii (2), p. 375–379, Pl. v–viii. [Dated 1878.] Cf. 1880, Czerniawsky.

* The article is a chapter of "The Collections from Kerguelen Island."

131

1879 **Dezsö, B.** Die Histiologie und Sprossenentwickelung der Tethyen, besonders der Tethya lyncurium Lieberkühn (autorum). Von... In: Arch. Mikr. Anat. xvi, p. 626–651, Pl. xxx–xxxiii.

1879 (α) **Dezsö, B.** Fortsetzung der Untersuchungen über Tethya lyncurium Autorum. Von... In: Arch. Mikr. Anat. xvii, p. 151–164, Pl. xii.

1879 (β) **Dezsö, B.** A Tethya Lyncurium sarjfeslödésé. In: Orvos-természzett, értesitö Kolozsvár, iv, p. 4–8.

1879 (γ) **Dezsö, B.** Spongiologiai tanulmány. In: Orvos-természzett, értesitö Kolozsvár, iv, p. 164–175.

1879 **Duncan, P. M.** On some spheroidal Lithistid Spongida from the Upper Silurian formation of New Brunswick. By... In: Ann. and Mag. N. H. (5), iv, p. 84–91, Pl. ix.

1879 **Ganin, M. S.** Materialy k posnanju stroeniä i rozwitiä gubok... Warschauw. iv and 88 pp., 7 Pl. (Russian). Materials for the knowledge of the structure and development of Sponges.

1879 **Gerstaecker, A.** (Crustacea.) In: Bronn's Klassen u. Ordn. Thierr. Cf. pp. 445, 446, 538, 554, 563, 570, 571, 574, 575, 577, 578, 580, 581, 585, 586, 588, 778, Pl. vi, fig. 6.

1879 **Haeckel, E. H. (P. A.).** Einstämmiger und vielstämmiger Ursprung. In: Kosmos, ii, p. 360–376.

1879 **Haller, G.** Zur Kenntniss der Mittelmeerfauna der höheren Crustaceen ...Von...In: Zool. Anz. ii, p. 205–207.

1879 **Hyatt, A.** Boston Society of Natural History Guides for Science Teaching, No. 3. Commercial and Other Sponges. Boston. 43 pp.

1879 **Keller, C.** Studien über Organisation und Entwicklung der Chalineen. Von...In: Z. W. Z. xxxiii, p. 317–349, Taf. xviii–xx. Abstr. and rev. in: J. R. M. Soc. London, iii (1880), p. 277–279; Z. J. for 1879 (1880), pp. 200, 212–213; Zool. Rec. xvi, pp. 2, 6, 9–10,14.

1879 (α) **Keller, C.** Etendue de la division des Zoophytes. ii. Par...In: Arch. Sc. Phys. et Nat. ii, p. 711–713.

1879 (β) **Keller, C.** Zur Entwickelungsgeschichte der Chalineen. Vorläufige Mittheilung, von...In: Zool. Anz. ii, p. 302–303.

1879 **Krukenberg, C. F. W.** Vergleichend-physiologische Studien an den Küsten der Adria. Experimentelle Untersuchungen, von...Erste Abtheilung. Heidelberg, Winter. iii and 172 pp.*

1879 (α) **Krukenberg, C. F. W.** Tetronerythrin in Schwämmen. Von... In: Centralbl. medic. Wiss. xvii, p. 705–706.

1879 (β) **Krukenberg, C. F. W.** Über die Enzymbildung in den Geweben und Gefässen der Evertebraten. In: Unters. physiol. Inst. Heidelberg, ii, p. 338–365†.

1879 (**Kusta, J.**) Spongilla jordanensis bei Rakonitz. In: Sitzgsb. Zool.-botan. Ges. Wien, xxix, Versamml. 2. Juli 1879, p. 40.

* On title-page is printed 1880; the above part appeared Oct. 1879.
† On the title-page is printed 1878; the paper appeared, however, according to Zool. Anz. ii, p. 123, in 1879.

9—2

1879 **Landois, H.** Ueber eine Krebsart im Innern von Euplectella aspergillum. In: 7. Jahresber. westfäl. Provinz Vereins, p. 41–42.

1879 **MacIntosh, W. C.** On Budding in the Syllidian Annelids, chiefly with
reference to a branched form procured by H.M.S. Challenger.
By...In: Rep. 49th Meet. Brit. Assoc. p. 372–375.

1879 **Marenzeller, E. von.** Die Aufzucht des Badeschwammes aus Theilstücken. Von...In: Verhandl. Zool.-bot. Ges. Wien, xxviii,
p. 687–694.

1879 **Matthews, J.** (Parasitism of a Coral on a Sponge.) In: Journ. R.
Microsc. Soc. London, ii, p. 110–111.

1879 **Matyasowsky, J. von.** Glenodictyum. In: Földtany Közlöny, ix, p.160.

1879 **Mayer, P.** Wagnerella borealis. Von...In: Zool. Anz. ii, p. 357–358.

1879 **Mazzetti, G. & A. Manzoni.** Le spugne fossili di Montese. In: Atti
Soc. Tosc. Sc. Nat. Pisa, iv, p. 57–65, Pl. viii–ix.

1879 **Merejkowsky, C.** Études sur les Éponges de la Mer Blanche. Par...
In: Mém. Acad. St. Pétersbourg (7), xxvi, p. 1–51, Pl. i–iii.
Transl. (Russian) in: Trudi St. Petersb. Obschtschwa, x, p. 1–82,
Pl. i-iii.

1879 **Metschnikoff, E.** Spongiologische Studien. Von...In: Z. W. Z. xxxii,
p. 349–387, Pl. xx–xxiii.

1879 **Moseley, H. N.** Notes by a Naturalist on the "Challenger"...By...
London. xvi and 620 pp. With a map, 2 Pl. and numerous
woodcuts.

1879 **Noll, F. C.** Einige Beobachtungen im Seewasser-Zimmeraquarium.
Von...In: Zool. Anz. ii, p. 402–405.

1879 (α) **Noll, F. C.** Nachtrag zu der Notiz über Meeresleuchten im Zim-
mer-Aquarium. Von...In: Zool. Anz. ii, p. 455.

1879 **Norman, A. M.** The Mollusca of the Fjords near Bergen, Norway.
In: Journ. Conchol. ii, p. 8–77.

1879 **Reinsch, P. F.** Beobachtungen über entophyte und entozoische
Pflanzen-parasiten...In: Botan. Zeitung, xxxvii, pp. 17–23,
33–43, Pl. i.

1879 **Ryder, J. A.** On the Destructive Nature of the Boring Sponge, with
observations on its gemmules or eggs. In: Amer. Naturalist, xiii,
p. 279–283.

1879 **Schmidt, (E.) O.** Reports on the Dredging, under the Supervision of
Alexander Agassiz, in the Gulf of Mexico...Report on the
Sponges by...First Part...Die Spongien des Meerbusens von
Mexico von...Jena. 4°. p. 1–32, Pl. i–iv.

1879 (α) **Schmidt, (E.) O.** Die Fortsetzung meiner "Spongien des Meerbusens
von Mexico." In: Zool. Anz. p. 379–380.

1879 **Schulze, F. E.** Untersuchungen über den Bau und die Entwicklung
der Spongien. Siebente Mitth. Die Familie der Spongidae.
Von...In: Z. W. Z. xxxii, p. 593–660, Pl. xxxiv–xxxviii.

1879 (α) **Schulze, F. E.** Untersuchungen über den Bau und die Entwicklung der Spongien. Achte Mittheilung. Die Gattung Hircinia Nardo und Oligoceras n. g. Von...In: Z. W. Z. xxxiii, p. 1–38, Pl. I–IV.

1879 (β) **Schulze, F. E.** Ueber die Bildung freischwebender Brutknospen bei einer Spongie, Halisarca lobularis. Von...In: Z. Anz. ii, p. 636–641.

1879 **Selenka, E.** Ueber einen Kieselschwamm von achtstrahligem Bau, und über Entwicklung der Schwammknospen. Von...In: Z. W. Z. xxxiii, p. 467–476, Pl. XXVII–XXVIII.

1879 **Semper, C.** ...Die natürlichen Existenzbedingungen der Thiere. Von ...Leipzig. x and 299 pp., 106 woodcuts.

1879 **Sinzow, J.** On Calcareous Sponges from the Government of Saratow (Russian). (Trans. (Zapiski) New Russian Nat. Hist. Soc. vi, p. 1–40.)

1879 **Sollas, W. J.** On Plocamia plena, a new Species of Echinonematous Sponge. By...In: Ann. and Mag. N. H. (5), iv, p. 44–53 and 4 woodcuts.

1879 (α) **Sollas, W. J.** On Plectronella papillosa, a new genus and species of Echinonematous Sponge. By...In: Ann. and Mag. N. H. (5), iii, p. 17–27, Pl. IV–VII, and woodcuts.

1879 (β) **Sollas, W. J.** On the Genus Catagma (Correction). By...In: Ann. and Mag. N. H. (5), iii, p. 169–170.

1879 (γ) **Sollas, W. J.** Observations on Dactylocalyx pumiceus (Stutchbury) with a Description of a New Variety, Dactylocalyx Stutchburyi. By...In: Journ. R. Microsc. Soc. ii, p. 122–133, Pl. V–VIII.

1879 (δ) **Sollas, W. J.** On Pharetronema zingiberis, a new genus and species of Renierid Sponge. By...In: Ann. and Mag. N. H. (5), iii, p. 404–407, Pl. XXX.

1879 (ε) **Sollas, W. J.** On the replacement of Siliceous Skeletons by Carbonate of Lime. By...In: Rep. 94th Meet. Brit. Assoc. p. 350.

1879 **Studer, T.** Die Fauna von Kerguelensland...In: Arch. f. Naturgesch. xlv, 1, p. 104–141.

1879 **Syen, H. W.** The Structure and Distribution of Sponges. In: Hardwicke's Science Gossip, xv, p. 197–199.

1879 **Toula, Fr.** Neue Ansichten über die systematische Stellung der Dactyloporiden. Von...In: Verh. Geol. Reichsanst. Wien, 1878, p. 301–304. Transl. in: Ann. and Mag. N. H. (5), iii, p. 151–153.

1879 **Trautschold, H.** Die Kalkbrüche von Mjatschkowa...Von...In: Nouv. Mém. soc. natur. Moscou, xiv (xx), p. 1–82, Pl. I–VII.

1879 **Ulrich, E. O.** Descriptions of new genera and species of fossils from the Lower Silurian about Cincinnati. By...In: Journ. Cincinnati Soc. ii, p. 8–30, Pl. VII.

1879 **Verrill, A. E.** Notice on recent additions to the marine Invertebrata, of the North-eastern Coast of America...Part i...By...In: Proc. U. S. Nat. Mus. ii, p. 165–205 (Porifera, p. 204–205).

1879 (α) **Verrill, A. E.** Preliminary Check-list of the Marine Invertebrata of the Atlantic Coast from Cape Cod to the Gulf of St Lawrence ...New Haven.

1879 **Woeckener, H.** Über das Vorkommen von Spongien im Hilssandstein. Von...In: Zeitschr. d. geol. Ges. xxxi, p. 663–664.

1879 **Zittel, K. A.** Handbuch der Palaeontologie, unter Mitwirkung von W. Ph. Schimper...herausgegeben von...i Band, ii Lieferung. München. 8°. N.B. Bd. I, 1, appeared in 1876. The chapter on sponges there begins p. 128 (last page). All the rest of the Sponges is to be found in the second "Lieferung," from p. 129–202.

1879 (α) **Zittel, K. A.** Zusatz zu vorstehendem Aufsatz. Von...In: Zeitschr. d. geol. Ges. xxxi, p. 665–667.

1879 (β) **Zittel, K. A.** Beiträge zur Systematik der fossilen Spongien. With 10 lith. Pl. Stuttgart. 132 pp. Sep. Abdr. N. J. Min. 1877, 1878, 1879. Theilw. abgedr. a. d. Abh. B. Ak. Bd. xiii.

1880 **Balfour, F. M.** A treatise on Comparative Embryology. By... London...Vol. i. xi, 492 and xxii pp., 275 woodcuts. Ed. ii, 1885.

1880 (α) **Balfour, F. M.** Larval Forms: their Nature, Origin and Affinities. By...In: Quart. Journ. Microsc. Sc. (2), xx, p. 381–407, 21 woodcuts.

1880 **Carter, H. J.** Report on Specimens dredged up from the Gulf of Manaar and presented to the Liverpool Free Museum by Capt. W. H. Cawne Warren. By...In: Ann. and Mag. N. H. (5), v, p. 437–457, Pl. xviii–xix, and (5), vi, pp. 35–61, 129–156, Pl. iv–viii.

1880 (α) **Carter, H. J.** On Fossil Sponge-spicules from the carboniferous Strata of Ben Bulben, near Sligo. By...In: Ann. and Mag. N. H. (5), vi, p. 209–214, Pl. xiv B, fig. 1–17.

1880 (β) **Carter, H. J.** Sponges. (Including list of Sponges dredged by the Birmingham Natural History and Microscopical Society, Falmouth Excursion, 1879. Depth 15–20 Fathoms.) By...In: Midland Naturalist, p. 190–195. Also in:...p. 55–60.

1880 ...**Carter, H. J.** See Urban, 1880.

1880 (**Cunningham, R. O.** On Sponges.) In: Proc. Belfast Nat. Hist. Soc. p. 208–209. Abstract.

1880 **Czerniawsky, V.** Spongiae littorales Pontis Euxini et Maris Caspii... In: Bull. Soc. Nat. Moscou, liv, pp. 88–128, 228–320. With 1879, Czerniawsky, separately published in 1880?.

1880 **Dewalque, G.** Prodrome d'une description géologique de la Belgique ...Bruxelles.

1880 **Dezsö, B.** A Magyar tengerpart Szivacsfaunája I....(Sponge-fauna of Hungarian coast.) In: Értek. Termesz., Budapest, x, 13, p. 3–7.

1880 (α) **Dezsö, B.** Új szivacsnem a Magyar tengerböl. In: Termész. Közl. xii, p. 351–352.

1880 **Duncan, P. M.** On a Parasitic Sponge of the Order Calcarea. By... In: Journ. R. Micr. Soc. iii, p. 377–383, Pl. x.

1880 **Dybowski, W.** Studien über die Spongien des russischen Reiches, mit besonderer Berücksichtigung der Spongien-Fauna des Baikal-Sees. Von...In: Mém. Acad. St. Pétersbourg (7), xxvii, 71 pp., Pl. i–iv.

1880 (α) **Dybowski, W.** Einige Bemerkungen über die Veränderlichkeit der Form und Gestalt von Lubomirskia baicalensis und über die Verbreitung der Baikalschwämme im Allgemeinen. Von...In: Bull. Acad. St. Pétersb. xxvii, col. 45–50. Cf. 1881, Dybowski.

1880 **Edwards, A. Milne.** Compte rendu sommaire d'une exploration zoologique faite dans le golfe de Gascogne à bord du navire de l'Etat "le Travailleur"; par...In: C. R. Paris, xci, pp. 311–314, 355–360.

1880 **Fullagar, J.** The Fresh-Water Sponge...In: Hardwicke's Science Gossip, xvi, p. 3–5.

1880 (α) **Fullagar, J.** Development of the Fresh-water Sponge...In: Hardwicke's Science Gossip, xvi, p. 111–112.

1880 **Gümbel, C. W.** Spongien-Nadeln im Flysch. In: Verhandl. K. K. geol. Reichsanst. xiv, p. 213–215.

1880 **Hahn, O.** Die Meteorite (Chondrite) und ihre Organismen. Tübingen ...56 pp., 32 Pl.

1880 **Haswell, W. A.** On some additional new genera and species of Amphi-podous Crustaceans...In: Proc. Linn. Soc. N. S. Wales, iv, p. 319–350, Pl. xviii–xxiv.

1880 **Hinde, G. J.** Fossil Sponge-Spicules from the Upper Chalk. Found in the Interior of a Single Flintstone from Horstead, Norfolk. By...Munich. 8º. (Inaug. Diss.) 83 pp., 5 Pl.

1880 **Jourdan, E.** Recherches zoologiques et histologiques sur les Zo-anthaires du Golfe de Marseille. Par...In: Ann. Sc. Nat. x, p. 1–154, Pl. i–xvii.

1880 **Keller, C.** Neue Coelenteraten aus dem Golf von Neapel. Von...In: Arch. Mikr. Anat. xviii, p. 271–280, Taf. xiii und xiv.

1880 **Kiesow, J.** Über paläozoische Versteinerungen aus dem Diluvium der Umgebung Danzigs. In: Tagebl. 53. Vers. d. Naturf. und Aerzte, p....

1880 **Krukenberg, C. F. W.** Vergleichend-physiologische Studien an den Küsten der Adria. Experimentelle Untersuchungen von... Zweite Abtheilung...Heidelberg. 108 pp.

1880 (α) **Krukenberg, C. F. W.** Vergleichend-physiologische Studien zu Tunis, Mentone und Palermo. Experimentelle Untersuchungen von...Dritte Abtheilung...Heidelberg. 192 pp.

1880 **Lankester, E. Ray.** Nature Series. Degeneration. A Chapter in Darwinism. By...London.

1880 **Lindeman, M.** Die Seefischereien...Von...In: Petermann's Mittheil. Ergänz. xiii, n°. 60.

1880 **MacIntosh, W. C.** On budding in the Syllidian Annelids...In: Rep. Brit. Assoc. 49th Meet. p. 372–375.

1880 **Marshall, W.** Untersuchungen über Dysideiden und Phoriospongien. Von...In: Z. W. Z. xxxv, p. 88–129, Pl. vi–viii. The paper appeared separately as "Habilitations-Schrift," Leipzig.

1880 **Meneghini, G.** Nuovi fossili Siluriani di Sardegna...In: Atti R. Accad. Lincei (3), v, p. 209–220, 1 Pl.

1880 **Merejkowsky, C. de.** Reproduction des Éponges par bourgeonnement extérieur, par...In: Arch. Zool. Expérim. viii, p. 417–432, Pl. xxxi.

1880 **Mills, H.** Fresh-water Sponges. In: Amer. Journ. Microsc. v, p. 125–132.

1880 **Minot, C. S.** A Sketch of Comparative Embryology. (iv, The Embryology of Sponges.)... In: Amer. Naturalist, xiv, p. 479–485.

1880 **Möbius, K.** Beiträge zur Meeresfauna der Insel Mauritius und der Seychellen...Berlin...4°. iv and 352 pp., 1 map and 22 Pl.

1880 **Nebeski, O.** Beiträge zur Kenntniss der Amphipoden der Adria. Von...In: Arb. Z. Inst. Wien, iii, p. 111–162, Pl. x–xiii. (p. 1–52, Pl. i–iv.)

1880 **Nicholson, H. A. & R. Etheridge, jun.** A Monograph of the Silurian Fossils of the Girvan District in Ayrshire...By...Fasc. iii... Edinburgh-London. i–vi and 237–341 pp., Pl. xxi–xxiv.

1880 **Norman, A. M.** Notes on the French exploring Voyage of "Le Travailleur" in the Bay of Biscay. By...In: Ann. and Mag. N. H. (5), vi, p. 430–436.

1880 **Pantanelli, D. S.** I Diaspri della Toscana e i loro fossili. In: Atti R. Accad. Lincei (3), viii, p. 35–66, 1 Pl.

1880 **Potts, E.** (Fresh-water Sponges of Fairmount Park.) In: Proc. Acad. Nat. Sc. Philadelphia, p. 330–331. [Volume dated 1881.]

1880 (α) **Potts, E.** (On Fresh-water Sponges.) In: Proc. Acad. Nat. Sc. Philadelphia, p. 356–357. [Volume dated 1881.]

1880 **Ridley, S. O.** On two Cases of Incorporation by Sponges of Spicules foreign to them. By...In: Journ. Linn. Soc. xv, p. 149–151.

1880 **Roemer, C. F.** Lethaea geognostica. Bd. i, 688 pp. Stuttgart. Pl. A & B.

1880 **Ryder, J. A.** On Camaraphysema, a new type of Sponge. By...In: Proc. U.S. Nation. Mus. p. 269–272. Woodcuts.

1880 **Schlüter, C. A.** Über Lepidospongia rugosa. In: Verhandl. naturh. Ver. pr. Rheinl. und Westphal. xxxvi (4), vi, p. 290.

1880 **Schmidt, (E.) O.** Reports on the Dredging, under the Supervision of Alexander Agassiz, in the Gulf of Mexico...Report on the Sponges by...Die Spongien des Meerbusen von Mexico (und des Caraibischen Meeres) von...Zweites (Schluss-) Heft. Jena. 4º. p. 33–90, Taf. v–x.

1880 (α) **Schmidt, (E.) O.** Zusatz zu obiger Abhandlung (von Keller). Von... In: Arch. Mikr. Anat. xviii, p. 280–282.

1880 (β) **Schmidt, (E.) O.** Die Absonderung und die Auslese im Kampfe um's Dasein. In: Kosmos, iv, p. 329–333.

1880 **Schulze, F. E.** Untersuchungen über den Bau und die Entwicklung der Spongien. Neunte Mitth. Die Plakiniden. Von...In: Z. W. Z. xxxiv, p. 407–451, Pl. xx–xxii. Abstr. in: Journ. Roy. Micr. Soc. (2), i (1881), p. 64–66; Zool. Jahresber. 1880 (1881), pp. 184–186, 189–191.

1880 (α) **Schulze, F. E.** On the Structure and Arrangement of the Soft Parts in Euplectella aspergillum. By...In: Trans. R. Soc. Edinburgh, xxix, p. 661–673, Pl. A. Reprinted in Challenger Report on the Hexactinellida, 1887 (α), p. 23.

1880 **Sollas, W. J.** The Sponge-fauna of Norway; a Report on the Rev. A. M. Norman's Collection of Sponges from the Norwegian Coast. By...In: Ann. and Mag. N. H. (5), v, pp. 130–144, 241–259, 396–409, Pl. vi, vii, x–xii, xvii.

1880 (α) **Sollas, W. J.** On the Flint Nodules of the Trimmingham Chalk. By...In: Ann. and Mag. N. H. (5), vi, pp. 384–395, 437–461, Pl. xix–xx.

1880 (β) **Sollas, W. J.** On the Structure and Affinities of the Genus Protospongia (Salter). By...In: Quart. Journ. Geol. Soc. xxxvi, p. 362–367, 2 woodcuts.

1880 (γ) **Sollas, W. J.** On Sponge-spicules from the Chalk of Trimmingham, Norfolk. By...In: Rep. 50th Meet. Brit. Assoc. p. 586–587.

1880 (δ) **Sollas, W. J.** The Structure and Life-History of a Sponge. By... In: Proc. Bristol Nat. Hist. Soc. (8), i, p. 1–18.

1880 (ε) **Sollas, W. J.** On the Replacement of Siliceous Skeletons by Carbonate of Lime. In: Rep. 49th Meet. Brit. Assoc. p. 350.

1880 **Steinmann, G.** Mikroskopische Thierreste aus dem deutschen Kohlenkalke (Foraminiferen und Spongien). Von...In: Zeitschr. d. geol. Ges. xxxii, p. 394–400, Pl. xix.

1880 **(Stewart, C.** On the Histology of Grantia compressa.) In: Journ. R. Microsc. Soc. (3), i, p. 560.

1880 **Struckmann, C.** Einige ergänzende paläontologische Mittheilungen aus dem oberen Jura von Hannover. In: Zeitschr d. geol. Ges. xxxii, p. 660–663.

138

1880 **Stuxberg, A.** Evertebratfaunan i Sibiriens Ishaf. Förelöpande Studier grundade på de zoologiska undersökningarna under Prof. A. E. Nordenskjöld's Ishafs expedition 1878–79. In: Bihang svensk. vetensk. akad. handl. v, n°. 22, p. 1–76.

1880 **Thomson, C. Wyville.** General Introduction to the Zoological Series of Reports. By...In: Rep. Sc. Res.Voy. Challenger. Zool. i, p. 1–59.

1880 **Urban, W. S. M. D'.** The zoology of Barents Sea. By...In: Ann. and Mag. N. H. (5), vi, p. 253–277. (Sponges by Carter.)

1880 **Vasseur, G.** Reproduction asexuelle de la Leucosolenia botryoides (Ascandra variabilis, H.). In: Arch. Z. Exp. viii, p. 59–66. Woodcuts.

1880 **Vosmaer, G. C. J.** The Sponges of the Leyden Museum. i, The Family of the Desmacidinae. By...In: Notes Leyden Museum, ii, p. 99–164.

1880 (α) **Vosmaer, G. C. J.** Aanteekeningen over Leucandra aspera, H. Bijdrage tot de kennis der Kalksponzen...Leiden. 96 pp., 2 Pl. [See 1881 (γ).]

1880 (β) **Vosmaer, G. C. J.** Eine spongiologische Bibliographie. In: Zool. Anz. iii, p. 478–480.

1880 **Wagner, M.** Über die Entstehung der Arten durch Absonderung. In: Kosmos, iv, pp. 1–10, 89–99, 169–183.

1880 **Waller, J. G.** On an undescribed British Sponge of the Genus Raphiodesma, Bow....In: J. Quekett Micr. Club, vi, p. 97–104, Pl. v.

1880 (**Wallich, G. C.** A Contribution to the Physical History of the Cretaceous Flints. By...) In: Ann. and Mag. N. H. (5), v, p. 183–184.

1880 (α) **Wallich, G. C.** A Contribution to the Physical History of the Cretaceous Flints. By...In: Q. J. G. Soc. London, xxxvi, p. 68–92.

1880 **Zittel, K. A.** Ueber den geologischen Bau der libyschen Wüste... Von...München. 4°. 47 pp.

1881 **Brandt, K.** Ueber das Zusammenleben von Thieren und Algen. In: Arch. Anat. Phys. (Abth. Phys.), p. 570–574. The "Verh. Phys. Ges. Berlin" are published in Arch. Anat. Phys., but, I believe, also separately. Brandt's paper is on p. 22–26*, 3 woodcuts.

1881 **Braun, M.** Ueber die Geschlechtsverhältnisse bei Halisarca lobularis, O. Schm. Von...In: Zool. Anz. iv, p. 232–234.

1881 **Bruder, G.** Zur Kenntniss der Jura-ablagerung von Sternberg bei Zeidler in Böhmen. Von...In: Sitz. Ber. Akad. Wien, lxxxiii, p. 47–99, Pl. i–ii.

1881 **Carrington, J. F. & E. Lovett.** Notes and Observations on British Stalk-eyed Crustacea. In: Zoologist (3), v, pp. 97–101, 137–142, 198–205, 301–307, 358–364, 413–418, 455–461.

* It appeared in: Verh. Phys. Ges. Berlin, "Jahrg. 1881–82, No. 4 u. 5, 2 December 1881"; 9 pp. beginning p. 21 (Sitz. 25 Nov. 1881), p. 22, "Nachtrag zur Sitzung vom 11 November 1881." Colophon: "Abgedruckt im Archiv für Physiologie...Jahrgang 1881."—G. P. B.

1881 **Carter, H. J.** Contributions to our Knowledge of the Spongida...
By...In: Ann. and Mag. N. H. (5), viii, pp. 101–112, 241–259,
Pl. ix. For explanation of the plate see p. 120–121.

1881 (α) **Carter, H. J.** Supplementary Report on Specimens dredged up
from the Gulf of Manaar, together with others from the Sea in
the vicinity of the Basse Rocks and from Bass's Straits respectively, presented to the Liverpool Free Museum by Capt. H.
Cawne Warren. By...In: Ann. and Mag. N. H. (5), vii, p. 361–
385, Pl. xviii; rev. Zool. Jahresber. for 1881 (1882), i, pp. 96,
111, 160.

1881 (β) **Carter, H. J.** History and Classification of the known species of
Spongilla. By...In: Ann. and Mag. N. H. (5), vii, p. 77–107,
Pl. v–vi.

1881 (γ) **Carter, H. J.** On Spongilla cinerea. By...In: Ann. and Mag.
N. H. (5), vii, p. 263.

1881 (δ) **Carter, H. J.** On the Development of the Fibre in the Spongida.
By...In: Ann. and Mag. N. H. (5), viii, p. 113–121, Pl. ix.

1881 (ε) **Carter, H. J.** On Spongiophaga in Spongilla. By...In: Ann. and
Mag. N. H. (5), viii, p. 222.

1881 (ζ) **Carter, H. J.** On Spongiophaga Pottsi, n. sp. By...In: Ann. and
Mag. N. H. (5), viii, p. 354–362, Pl. xvii.

1881 (η) **Carter, H. J.** Addendum to our Knowledge of the Carnosa. By...
In: Ann. and Mag. N. H. (5), viii, p. 450.

1881 (θ) **Carter, H. J.** On the Kunker Formation of the Alluvium in India
compared with the Flint Formation in the Chalk of England.
By...In: Ann. and Mag. N. H. (5), vii, p. 308–312.

1881 **Dawson, J. W.** (On the Structure of a specimen of Uphantaenia,
from the Collection of the American Museum of Natural History,
New York City.) Note by...In Amer. Journ. Sc. p. 132–133.
Also in: Ann. and Mag. N. H. (5), viii, p. 237.

1881 **Duncan, P. M.** On a Lithistid Sponge and on a Form of Aphrocallistes from the Deep Sea off the Coast of Spain. By...In: Journ.
Linn. Soc. xv, p. 320–329, Pl. xxiv–xxv.

1881 (α) **Duncan, P. M.** On some Remarkable Enlargements of the Axial
Canals of Sponge Spicules and their Causes. By...In: Journ.
R. Microsc. Soc. (2), i, p. 557–572, Pl. vii–viii.

1881 (β) **Duncan, P. M.** On a Radiolarian and some Microspongida from
considerable depth in the Atlantic Ocean. By...In: Journ. R.
Microsc. Soc. (2), i, p. 173–179, Pl. iii.

1881 (γ) **Duncan, P. M.** On an Organism which penetrates and excavates
Siliceous sponge-spicula (Spongiophagus Carteri). By...In: Ann.
and Mag. N. H. (5), viii, p. 120–122.

1881 (δ) **Duncan, P. M.** On perforated and excavated Sponge-Spicules. In:
Journ. Microsc. Soc. London (2), i, p. 974.

1881 **Dybowski, W.** Einige Bemerkungen über die Veränderlichkeit der Form und Gestalt von Lubomirskia baicalensis und über die Verbreitung der Baikalschwämme im Allgemeinen. Von...In: Mél. Biol. Bull. Acad. St. Pétersbourg, xi, p. 41–47. Reprint (?) of 1880 (α) Dybowski.

1881 **Edwards, A. Milne.** Compte rendu sommaire d'une exploration zoologique faite dans la Méditerranée, à bord du navire de l'État "le Travailleur." In: C. R. Paris, xciii, p. 876–882.

1881 (α) **Edwards, A. Milne.** Compte rendu sommaire d'une exploration zoologique faite dans l'Atlantique, à bord du navire "le Travailleur," par...In: C. R. Acad. Paris, xciii, p. 931–936.

1881 **Gerster, C.** Die Plänerbildungen um Ortenburg bei Passau. Von... In: Nova Acta acad. nat. curios. xlii, n°. 1, p. 1–59, Pl. i.

1881 **Giard, A.** Fragments biologiques. Par...ii, Deux ennemis de l'ostréiculture. In: Bull. Sc. Départ. Nord, xiii, p. 70–73.

1881 **Giglioli, H. H.** Italian deep-sea exploration in the Mediterranean. In: Nature, xxiv, p. 381–382.

1881 **Horst, R.** Over een kalkspons van de oesterbanken naby Yerseke. In: Tijdschr. Nederl. Dierk. Ver. v, p. cx.

1881 **Jeffreys, J. Gwyn.** The French Deep-Sea Exploration in the Bay of Biscay. In: Rep. 50th Meet. Brit. Assoc. p. 378–387.

1881 (**Johnson, M. Hawkins.** On the burrowing of Cliona.) In: Journ. Quekett Microsc. Club, vi, p. 289–290.

1881 **Krukenberg, C. F. W.** Untersuchung der Fleischextracte verschiedener Fische und Wirbellosen. Von...In: Untersuch. physiol. Inst. Heidelberg, iv, p. 33–63. Separate copy, p. 1–31.

1881 **Leslie, G. & W. A. Herdman.** The Invertebrate Fauna of the Firth of Forth. Part iii. By...In: Proc. Phys. Soc. Edinburgh, vi, p. 268–315. Abstr. and rev. in: Zool. Jahresber. f. 1881 (1882), p. 62.

1881 **Manzoni, A.** Spugne silicee della Molassa miocenica del Bolognese. In: Atti soc. toscan. v, p. 173–176, Pl....

1881 **Mayer, P.** Noch einmal Wagnerella borealis. Von...In: Zool. Anz. iv, p. 592–593.

1881 **Mereschkowsky, C.** Note on Wagnerella borealis, a Protozoan. By... In: Ann. and Mag. N. H. (5), viii, p. 288–290.

1881 **Mills, H.** A variety of Spongilla...In. Amer. Journ. Microsc. vi, p. 30–31.

1881 **Nassonow, N.** Ueber die aushöhlende Kraft und zum feineren Bau der Clione. Von...Vorläufige Mittheilung. In: Zool. Anz. iv, p. 459–460.

1881 **Noll, F. C.** Mein Seewasser Zimmeraquarium...In: Zool. Garten, xxii, pp. 8–19, 32–42, 71–79, 137–147, 168–177, 195–206. With woodcuts.

1881 **Norman, A. M.** Supplementary Paper. By...In: Rep. 50th Meet. Brit. Assoc. p. 387–390.

1881 **Packard, A. S.** Notes on the early larval stages of the Fiddler Crab and of Alpheus. In: Amer. Naturalist, xv, p. 784–789.

1881 **Pavesi, P.** Di una spugna d'acqua dolce nuova per l'Italia. Nota del S. C. prof...In: Rendic. R. Inst. Lomb. (2), xiv, p. 1–6.

1881 **Perrier, E.** Les colonies animales et la formation des organismes par...Paris.

1881 **Potts, E.** (Some new Genera of Fresh-water Sponges.) In: Proc. Acad. Nat. Sc. Philadelphia, p. 149–150. [Vol. dated 1882] Also in: Ann. and Mag. N. H. (5) viii, p. 387–388.

1881 (α) **Potts, E.** (The genus Carterella versus Spongiophaga Pottsi.) In: Proc. Acad. Nat. Sc. Philadelphia, p. 460–463. [Vol. dated 1882.] Also in: Ann. and Mag. N. H. (5), ix (1882), p. 330–333.

1881 (β) **Potts, E.** A New Form of Freshwater Sponge. In: Proc. Acad. Nat. Sc. Philadelphia, p. 176. Also in: Ann. and Mag. N. H. (5), viii, p. 389.

1881 **Priest, B. W.** On the Natural History and Histology of Sponges. By...In: Journ. Quekett Microsc. Club, vi, p. 229–238, Pl. xvii–xviii.

1881 (α) **Priest, B. W.** Further Remarks on the Histology of Sponges. By...In: Journ. Quekett Microsc. Club, vi, p. 269–271. Abstr. and rev. in: Zool. Jahresber. f. 1882 (1883), p. 125.

1881 (β) **Priest, B. W.** The Statoblasts of Fresh-water Sponges. In: Journ. Quekett Microsc. Club, p...

1881 (γ) **Priest, B. W.** On an Undescribed Species of Sponge of the Genus Polymastia, from Honduras...In: Journ. Quekett Microsc. Club, vi, p. 302–304, Pl...

1881 **Rein, J. J.** Japan nach Reisen und Studien im Auftrage der Königlich preussischen Regierung dargestellt von...Bd. i. Leipzig. xiii and 630 pp.

1881 **Ridley, S. O.** Spongida...By...(Chapter xi of: Account of the Zoological Collections made during the Survey of H.M.S. "Alert" in the Straits of Magellan and on the Coast of Patagonia. Communicated by Dr Albert Günther...) In: Proc. Zool. Soc. London, pp. 107–137, 140–141*, Pl. x–xi.

1881 (α) **Ridley, S. O.** On the Genus Plocamia, Schmidt, and on some other sponges of the Order Echinonemata. By...With descriptions of two additional new species of Dirrhopalum by Prof. P. Martin Duncan...In: Journ. Linn. Soc. London, xv, p. 476–497, Pl. xxviii–xxix, Part i.

1881 (β) **Ridley, S. O.** Polyzoa, Coelenterata, and Sponges of Franz-Joseph Land. By...In: Ann. Mag. N. H. (5), vii, p. 442–457.

* In the separate copies the pages of the Explanation of the Plates are not 140–141 but 138–139.

1881 **Schlüter, C. A.** Nadelreste von Astraeospongia aus dem Eifelkalk von Gerolstein. In: Verh. naturh. Ver. pr. Rheinl. und Westphal. xxxvii (4), vii, p. 226.

1881 **Schulze, F. E.** Untersuchungen über den Bau und die Entwickelung der Spongien. Zehnte Mittheilung. Corticium candelabrum, O. Schmidt. Von...In: Zeitschr. wiss. Zool. xxxv, p. 410–430, Pl. xxii.

1881 (α) **Schulze, F. E.** Ueber den Badeschwamm. Von...In: Mitth. Naturw. Ver. Steiermark, p. 48–53.

1881 **Sollas, W. J.** On Astroconia Granti, a new Lyssakine Hexactinellid from the Silurian formation of Canada. By...In: Quart. Journ. Geol. Soc. xxxvii, p. 254–260. Woodcuts.

1881 (α) **Sollas, W. J.** Note on the Occurrence of Sponge spicules in Chert from the Carboniferous Limestone in Ireland. By...In: Ann. and Mag. N. H. (5), vii, p. 141–143.

1881 (β) **Sollas, W. J.** On Astroconia Granti, a new Lyssakine Hexactinellid from the Silurian formation of Canada. By...In: Abstr. Proc. Geol. Soc. London, p. 50–51. Also in: Ann. and Mag. N. H. (5), vii, p. 481–482.

1881 (γ) **Sollas, W. J.** On Sponge-Spicules from the Chalk of Trimmingham, Norfolk. In: Rep. 50th Meet. Brit. Assoc. p. 586–587.

1881 **Steinmann, G.** Ueber Protetraclis Linki, n. f., eine Lithistide des Malms. Von...In: N. Jahrb. Mineral. ii, p. 154–163, Pl. ix.

1881 (**Stewart, Ch.** On the burrowing of Cliona.) In: Journ. Quekett Microsc. Club, vi, p. 289.

1881 **Trautschold, H.** Über devonische Fossilien von Schelonj. In: Bull. soc. natur. Moscou, lvi, 1, p. 432–440, Pl. v.

1881 **Vogt, C.** Sur les prétendus organismes des météorites...In: C. R. Paris, xciii, p. 1166–1168.

1881 **Vosmaer, G. C. J.** Voorloopig berigt omtrent het onderzoek door den ondergeteekende aan de Nederlandsche werktafel in het Zoölogisch Station te Napels verrigt. 20 November 1880– 20 Februarij 1881. In: Nederl. Staats-courant, No. 109.

1881 (α) **Vosmaer, G. C. J.** Versuch einer spongiologischen Stenographie, von...In: Tijdschr. Ned. Dierk. Ver. v, p. 197–206, Pl. vi.

1881 (β) **Vosmaer, G. C. J.** Ueber die Fortpflanzungsverhältnisse bei Spongien. Von...In: Biol. Centralbl. i, p. 103–104.

1881 (γ) **Vosmaer, G. C. J.** Ueber Leucandra aspera, H. Nebst allgemeinen Bemerkungen ueber das Canalsystem der Spongien. Von...In: Tijdschr. Ned. Dierk. Ver. v, p. 144–166, Pl. iii–iv. Abstr. transl. German, from 1880 (α).

1881 **Walcott, C. D.** On the Nature of Cyathophycus; by...In: Amer. Journ. Sc. xxii, p. 394–395. Also in: Ann. and Mag. N. H. (5), viii, p. 459.

1881 **Waller, J. G.** On Cliona celata (Grant), Hymeniacidon celata (Bowerbank). Does the Sponge make the Burrow? By...In: Journ. Quekett Microsc. Club, vi, p. 251–268, Pl. xx–xxi.

1881 **Wallich, G. C.** On Siliceous Sponge-growth in the Cretaceous Ocean. By...In: Ann. and Mag. N. H. (5), vii, p. 261–263.

1881 (α) **Wallich, G. C.** On the Origin and Formation of the Flints of the Upper or White Chalk. By...In: Ann. and Mag. N. H. (5), vii, p. 162–204, Pl. xi.

1881 (β) **Wallich, G. C.** Supplementary Notes on the Flints and the Lithological Identity of the Chalk and Recent Calcareous Deposits in the Ocean. By...In: Ann. and Mag. N. H. (5), viii, p. 46–58.

1881 **Weber, M.** Verslag over de zoölogische onderzoekingen gedurende de vierde reis der "Willem Barents." In: Versl. 4. tocht W. Barents, p. 101–140.

1881 (**White, T. Charters.** Remarks on the burrowing of Cliona.) In: Journ. Quekett Microsc. Club, vi, p. 287–289.

1881 **Whitfield, R. P.** Observations on the Structure of Dictyophyton and its affinities with certain Sponges; by...In: Amer. Journ. Sc. (3), xxii, p. 53–54. Also in: Ann. and Mag. N. H. (5), viii, p. 167–168.

1881 (α) **Whitfield, R. P.** On the nature of Dictyophyton; by...In: Amer. Journ. Sc. (3), xxii, p. 132. Also in: Ann. and Mag. N. H. (5), viii, p. 237.

1881 **Wilson, A.** On the Origin of Colonial Organisms. By...In: Ann. and Mag. N. H. (5), vii, p. 413–416.

1881 **Wright, E. P.** On a New Genus and Species of Sponge with supposed Heteromorphic Zooids. By...(Read, Nov. 10, 1879)...In: Trans. Irish Acad. xxviii, p. 13–20, Pl. i.

1882 **Badock, J.** Red flints in the Chalk. In: Nature, xxv, p. 529.

1882 **Bowerbank, J. S.** A Monograph of the British Spongiadae. By the late...Edited, with additions, by the Rev. **A. M. Norman**... Vol. iv (supplementary). London: printed for the Ray Society, MDCCCLXXXII. 8⁰. xvii and 250 pp., 17 Plates.

1882 **Brandt, K.** Ueber die morphologische und physiologische Bedeutung des Chlorophylls bei Thieren. Von...In: Arch. Anat. Phys. Physiol. Abth. p. 125–151, Pl. i.

1882 **Carter, H. J.** Some Sponges from the West Indies and Acapulco in the Liverpool Free Museum described, with general and classificatory Remarks. By...(Plates xi and xii)...In: Ann. and Mag. N. H. (5), ix, pp. 266–301, 346–368.

1882 (α) **Carter, H. J.** New Sponges, Observations on old ones, and a proposed New Group. By...In: Ann. and Mag. N. H. (5), x, p. 106–125.

1882 (β) **Carter, H. J.** Form and Nature of the Cirrous Appendages on the Statoblasts of Carterella latitenta, Potts, etc., originally designated "Spongiophaga Pottsi." By... In: Ann. and Mag. N. H. (5), ix, p. 390–396, Pl. xiv.

1882 (γ) **Carter, H. J.** Spermatozoa, Polygonal Cell-structure, and the Green Colour in Spongilla, together with a new Species. By... In: Ann. and Mag. N. H. (5), x, p. 362–372, Pl. xvi.

1882 **Costa, A.** Rapporto preliminare e sommario sulle ricerche zoologiche fatte in Sardegna durante la primavera del 1882... In: Rendic. Accad. Sc. Napoli, xxi, p. 189–201.

1882 **Dufour, J.** Notice sur un champignon parasite des Éponges, par... In: Bull. Soc. Vaudoise Sc. Nat. (2), xviii, p. 144–147.

1882 **Dunikowski, E. v.** Die Spongien, Radiolarien und Foraminiferen der unterliassischen Schichten von Schafberg bei Salzburg. Von... In: Denkschr. Akad. Wiss. Wien, xlv, p. 163–194, Pl. i–vi. Sep. copy, p. 1–34, Pl. i–vi.

1882 **Dybowski, W.** Studien über die Süsswasser-Schwämme des Russischen Reiches. Von... In: Mém. Acad. Sc. St. Pétersbourg (7), xxx, p. 1–26, Pl. i–iii.

1882 **Entz, Géza.** Das Konsortialverhältniss von Algen und Tieren. In: Biol. Centralbl. ii, p. 451–464.

1882 **Fuchs, Th.** Welche Ablagerungen haben wir als Tiefsee-bildungen zu betrachten? Von... In: N. Jahrb. Mineral. 2 Beil. Bd. p. 487–584.

1882 (α) **Fuchs, Th.** Was haben wir unter der Tiefsee-Fauna zu verstehen? ... In: Verh. K. K. geol. Reichsanst. xvi, p. 55–68.

1882 (α) **Fullagar, J.** The Fresh-water Sponge. In: Hardwicke's Science Gossip, p. 3–5. With woodcuts.

1882 (α) **Fullagar, J.** Development of the Fresh-water Sponges. In: Hardwicke's Science Gossip, p. 111–112. With woodcuts.

1882 **Geddes, P.** On the Nature and Functions of the "Yellow-Cells" of Radiolarians and Coelenterates. By... In: Proc. R. Soc. Edinburgh, p. 377–396.

1882 **Gerstaecker, A.** Arthropoda in Bronn's Klassen und Ordnungen des Thierreichs, v, 2, p. 97–192, Pl. ix–xvi.

1882 **Graeffe, Ed.** Uebersicht der Seethierfauna des Golfes von Triest nebst Notizen über Vorkommen, Lebensweise, Erscheinungs- und Fortpflanzungszeit der einzelnen Arten, von... ii, Coelenteraten, Spongiariae. In: Arb. Zool. Inst. Wien, iv (1882), p. 313–321.

1882 **Hamann, O.** Zur Entstehung und Entwicklung der grünen Zellen bei Hydra. In: Z. W. Z. xxxvii, p. 457–464, Pl. xxvi.

1882 **Haswell, W. A.** On Australian Fresh-water Sponges. By... In: Proc. Linn. Soc. N. S. Wales, vii, p. 208–210.

1882 **Heider, A. von.** Die Gattung Cladocora Ehrenb., von...In: Sitz. Ber. K. Akad. Wiss. Wien, Math.-naturw. Cl. Bd. lxxxiv, p. 634–667, Taf. ɪ–ɪᴠ.

1882 **Hilgendorf, F.** Über Spongilla fluviatilis, Lbk. var. japonica. In: Sitz. ber. Ges. naturf. Freunde Berlin, p. 26.

1882 **Hinde, G. J.** Notes on Fossil Calcispongiae, with Descriptions of new Species. By...In: Ann. and Mag. N. H. (5), x, p. 185–205, Pl. x–xɪɪ.

1882 **Hyatt, J. D.** The Boring Sponge—Does it Excavate the Burrows in which it is found? By...In: Amer. Monthly Microsc. Journ. iii, p. 81–84. 3 woodcuts.

1882 **Joliet, L.** Observations sur quelques Crustacés de la Méditerranée. In: Arch. Zool. Expérim. x, p. 101–120, Pl. ᴠɪ.

1882 **[Joseph, G.** Ueber einen Grottenschwamm (Spongilla stygia, n. sp.).] In: Jahresber. Schles. Ges. vaterl. Cultur, p. 253.

1882 **Keller, C.** Die Fauna im Suez-Canal und die Diffusion der mediterranen und erythräischen Thierwelt. Eine thiergeographische Untersuchung von...In: N. Denkschr. Schweiz. Ges. Naturw. xxviii, p. 1–39, Pl. ɪ–ɪɪ.

1882 **Kent, W. Saville.** A Manual of the Infusoria: including...an account of the organization and affinities of the Sponges. By...London, 1880–1882. 913 pp., 51 Pl.

1882 **Klebs, G.** Ueber Symbiose ungleichartiger Organismen. Von...In: Biol. Centralbl. ii, p. 289–299.

1882 **Krukenberg, C. Fr. W.** Vergleichend-physiologische Vorträge. ii, Grundzüge einer vergleichende Physiologie der Verdauung. Heidelberg. 82 pp.

1882 (α) **Krukenberg, C. Fr. W.** Vergleichend-physiologische Studien an den Küsten der Adria. 2. Reihe, 3. Abtheil. 154 pp., 9 Pl.

1882 **Lankester, E. Ray.** Dredging in the Norwegian Fjords. In: Nature, xxvi, p. 478–479.

1882 (α) **Lankester, E. Ray.** On the Chlorophyll-corpuscles and Amyloid Deposits of Spongilla and Hydra. By...In: Quart. Journ. Microsc. Sc. xxii, p. 229–254, Pl. xx.

1882 (β) **Lankester, E. Ray.** The chlorophyll-corpuscles of Hydra. In: Nature, xxvii, p. 87–88.

1882 **Lendenfeld, R. von.** Das Hornfaserwachsthum der Aplysinidae. Von... In: Zool. Anz. v, p. 634–636.

1882 **Lenz, H.** Die wirbellosen Thiere der Travemünder Bucht. In: Vierter Ber. Comm. wiss. Unters. d. Meere, ii, p. 169–180.

1882 **Manzoni, A.** La struttura microscopica delle Spugne Silicee del Miocene medio della Provincia di Bologna e di Modena. Per... Bologna. 4°. 24 pp. and 7 Pl.

1882 **Marshall, W.** Die Ontogenie von Reniera filigrana, O. Schm. Von...
In: Z. W. Z. xxxvii, p. 221–246, Pl. xiii–xiv.

1882 **Metschnikoff, E.** Zur Lehre über die intracelluläre Verdauung
niederer Thiere. Von...In: Zool. Anz. v, p. 310-316.

1882 (α) **Metschnikoff, E.** Vergleichend-embryologische Studien. 3. Über
die Gastrula einiger Metazoen. Von...In: Z. W. Z. xxxvii,
p. 286–313, Pl. xix–xx.

1882 **Miller, S. A.** Description of two genera and eight new species of
fossils from the Hudson river Group, with remarks upon others.
By...In: Journ. Cincinnati Soc. Nat. Hist. v, p. 34–44, Pl. i–ii.

1882 **Mills, H.** Notes on the Spongillae of Buffalo. In: Bull. Buffalo Soc.
iv, p. 57–60.

1882 (?) **Mindt, C. S.** Embryology of Sponges. In: Amer. Naturalist, xiv,
p. 479–485. With woodcuts.

1882 **Munier-Chalmas, E.** ...(Barroisia, n. g. des Éponges.) In: Bull. Soc.
Géol. France (3), x, p. 425.

1882 **Nathorst, A. G.** Om aftryk af Medusor i Sveriges Kambriska Lager.
Af...In: Kgl. Svenska Vidensk. Akad. Handlingar, xix, p. 1–33,
Pl. i–vi.

1882 **Neumayer, M.** Ueber den alterthümlichen Character der Tiefsee-
fauna. Von...In: N. Jahrb. Mineral. i, p. 123–131.

1882 **Noll, F. C.** Eau de Javelle als Mittel zum Entfernen der Weichtheile aus
mikroskopischen Präparaten. Von...In: Zool. Anz. v, p. 528–530.

1882 **Norman, A. M.** See 1882, Bowerbank, J. S.

1882 **Počta, P.** Einige Bemerkungen über das Gitterskelet der fossilen
Hexactinelliden...von...In: Sitz. ber. k. Böhm. Ges. Prag,
1882, p. 378–390, 1 Pl.

1882 **Poléjaeff, N.** Ueber das Sperma und die Spermatogenese bei Sycandra
raphanus, Hckl. In: Anz. K. Akad. Wiss. Wien, Math.-nat. Cl.
No. xxv, p. 237–238. Cf. 1883, Poléjaeff.

1882 **Potts, E.** Three more Fresh-water Sponges. In: Proc. Acad. Nat Sc
Philadelphia, 1883, p. 12–14. Also in: Ann. and Mag. N. H. (5), ix
p. 474–476.

1882 (α) **(Potts, E.** Sponges from the neighborhood of Boston.) In: Proc.
Acad. Nat. Sc. Philadelphia, 1883, p. 69–70. Also in: Ann. and
Mag. N. H. (5), x, p. 72–73.

1882 **Rathbun, R.** List of Marine Invertebrates, from the New England
Coast, distributed by the Un. St. National Museum. Series
iii...In: Proc. U. S. Nation. Museum, iv, p. 304–307.

1882 **Schulze, F. E.** Report on the Sponges. By...In: Tizard and Murray,
Exploration of the Faroe Channel...[Proc. R. Soc. Edinburgh,
xi, p. 708 (sep. copy, p. 71).]

1882 (α) **(Schulze, F. E.** Ueber radiäre Symmetrie bei Spongien.) In:
Tagebl. 53. Vers. Naturf. Eisenach, p. 199.

1882 (β) **Schulze, F. E.** Der Badeschwamm. In: Ill. deutsche Monatsch.
(5), iii, p. 188–210. With woodcuts.

1882 (γ) **Schulze, F. E.** Ueber den Badeschwamm. Von...In: Mitth. naturwiss. Ver. Steiermark, p. xlviii–liii.

1882 **Scudder, S. H.** ...Nomenclator zoologicus...Supplemental List, pp.xix and 376 and Universal Index, pp. 340. By...Washington.

1882 **Sollas, W. J.** The Sponge-fauna of Norway; a Report on the Rev. A. M. Norman's Collection of Sponges from the Norwegian Coast. By...In: Ann. and Mag. N. H. (5), ix, pp. 141–165, 426–453, Pl. vi and xvii.

1882 (α) **Sollas, W. J.** (The Group Spongiae.) In: Cassell's Nat. History, vi. London. 4º. p. 312–331. With woodcuts.

1882 (β) **Sollas, W. J.** Spongidae (from the Inferior Oolite). In: Rep. 52nd Meet. Brit. Assoc. p. 534–535.

1882 **Steinmann, G.** Pharetronen-Studien. Von...In: N. Jahrb. Mineral. ii, p. 139–191, Pl. vi–ix.

1882 **Studer, Th.** Beiträge zur Meeresfauna West-Africas. Von...(Schluss.) In: Zool. Anz. v, p. 351–356.

1882 **Tizard & J. Murray.** Exploration of the Faroe Channel during the summer of 1880 in Her Majesty's hired ship "Knight Errant." By...In: Proc. R. Soc. Edinburgh, Session 1881–82, p. 1–83 (?). Cf. 1882, Schulze.

1882 **Vosmaer, G. C. J.** Report on the Sponges dredged up in the Arctic Sea by the "Willem Barents" in the Years 1878 and 1879. By ...In: Niederl. Arch. Zool. Suppl. i, p. 1–58, Pl. i–iv.

1882 (α) **Vosmaer, G. C. J.** ...Porifera. Von...In: Bronn, H. G., Die Klassen und Ordnungen des Thierreichs...ii, p. 1–32, Pl. i, ii, iv.

1882 **Weinland, D. F.** Ueber die in Meteoriten entdeckten Thierreste. Von...Esslingen. 4º. 12 pp. and 2 figs.

1882 **Weltner, W.** Beiträge zur Kenntniss der Spongien. Inaug. Diss. Freiburg. i. Br. 62 pp. and 3 Pl.

1882 **Whitfield, R. P.** Remarks on Dictyophyton, and descriptions of new species of allied forms from the Keokuk beds, at Crawfordsville, Ind. In: Bull. Amer. Mus. N. H. i, p. 10–20. Pl. iii and iv.

1882 **Zittel, K. A.** Notizen über fossile Spongien. In: N. Jahrb. Mineral. ii, p. 203–204.

1883 **Barrois, Ch.** Sur les Dictyospongidae des Psammites du Condros. Par...In: Ann. Soc. Géol. Nord, xi, p. 80–86, Pl. i.

1883 **Bonardi, E. & C. F. Parona.** Ricerche micropaleontologiche sulla argille del bacino lignitico di Leffe in val Gandino. In: Atti soc. italiana, xxvi, p. 182–210, Pl. v.

1883 **Brandt, K.** Ueber die morphologische und physiologische Bedeutung des Chlorophylls bei Thieren...Von...In: Mitth. Zool. Stat. Neapel, iv, p. 191–302, Taf. xix–xx.

1883 (α) **Brandt, K.** Die Fortpflanzung der grünen Körper von Hydra... Von...In: Zool. Anz. vi, p. 438–440.

1883 **Carter, H. J.** On the Presence of Starch-granules in the Ovum of the Marine Sponges, and on the Ovigerous Layer of Suberites domuncula Nardo. By...In: Ann. and Mag. N. H. (5), xii, p. 30–36. With figs. 1–4.

1883 (α) **Carter, H. J.** Contributions to our Knowledge of the Spongida. Pachytragida. By...In: Ann. and Mag. N. H. (5), xi, p. 344–369, Pl. xiv–xv.

1883 (β) **Carter, H. J.** Contributions to our Knowledge of the Spongida. By...In: Ann. and Mag. N. H. (5), xii, p. 308–329, Pl. xi–xiv.

1883 (γ) **Carter, H. J.** Further Observations on the so-called "Farringdon Sponges" (Calcispongiae, Zittel), followed by a Description of an Existing Species of a like kind. By...In: Ann. and Mag. N. H. (5), xi, p. 20–37, Pl. i.

1883 (δ) **Carter, H. J.** New Genus of Sponges. By...In: Ann. and Mag. N. H. (5), xi, p. 369–370, Pl. xv, fig. 10.

1883 (ε) **Carter, H. J.** On the microscopic Structure of thin Slices of Fossil Calcispongiae. By...In: Ann. and Mag. N. H. (5), xii, p. 26–30.

1883 (ζ) **Carter, H. J.** Spicules of Spongilla in the Diluvium of the Altmühl Valley, Bavaria. By...In: Ann. and Mag. N. H. (5), xii, p. 329–333, Pl. xiv.

1883 **Chilton, Ch.** A New-Zealand Freshwater Sponge. In: N. Zealand Journ. Sc. i, p. 383–384.

1883 **Dunikowski, E. von.** Die Pharetronen aus dem Cenoman von Essen und die systematische Stellung der Pharetronen. Von...In: Palaeontographica, xxix [=(3), v], p. 281–324, Pl. xxxvii–xl.

1883 **Edwards, A. Milne.** Rapport préliminaire sur l'expédition du "Talisman" dans l'Océan Atlantique. In: C. R. Acad. Paris, xcvii, p. 1389–1395.

1883 **Faber, G. L.** Sponge Fisheries of the Adriatic. In: The Fisheries of the Adriatic. London. 4°. p. 96.

1883 **Gaudry, A.** Sur quelques-uns des résultats déjà obtenus par les explorations sous-marines faites à bord du Talisman. In: C. R. Acad. Paris, xcvii, p. 140–141.

1883 **Hamann, O.** Die Fortpflanzung der grünen Körper von Hydra. Entgegnung an Herrn Brandt. Von...In: Zool. Anz. vi, p. 367–370.

1883 **Hilgendorf, F.** (Süsswasserschwämme aus Central-Africa.) In: Sitz. Ber. Ges. Nat. Freunde Berlin, p. 87–90. Transl. in: Ann. and Mag. N. H. (5), xii, p. 120–123.

1883 **Hinde, G. J.** Cf. 1884.

1883 **Keeping, W.** The Fossils and Palaeontological Affinities of the Neocomian Deposits of Upware and Brickhill...London, 167 pp and 8 Pl.

1883 **Klemm, D. E.** Ueber alte und neue Ramispongien und andere verwandte Schwammformen aus der Geislinger Gegend. Von...
In: Jahresber. Ver. vaterl. Naturkunde Württemberg, xxxix, p. 243–308.

1883 **Lendenfeld, R. von.** Ueber Coelenteraten der Südsee. Von... Neue Aplysinidae... In: Z. W. Z. xxxviii, p. 234–313, Taf. x–xiii.

1883 (*a*) **Lendenfeld, R. von.** Über eine eigenthümliche Art der Sprossenbildung bei Campanulariden. Von... In: Zool. Anz. vi, p. 42–44.

1883 **Linck, G.** Zwei neue Spongiengattungen. Von... In: N. Jahrb. Mineral. ii, p. 59–62, Pl. ii–iii.

1883 **Margó, T.** Die Classification des Thierreichs... Von... In: Math. Nat. Ber. Ungarn, i, p. 234–260, Pl. v.

1883 **Marion, A. F.** Esquisse d'une topographie zoologique du Golfe de Marseille, par... In: Ann. Mus. Hist. Nat. Marseille, Zool. i (Mém. No. 1), p. 1–108 and 1 map.

1883 (*a*) **Marion, A. F.** Considérations sur les faunes profondes de la Méditerranée d'après les dragages opérés au large des Côtes Méridionales de France par... In: Ann. Mus. Hist. Nat. Marseille, Zool. i (Mém. No. 2), p. 1–50.

1883 **Marshall, W.** Ueber einige neue, von Hrn. Pechüel-Lösche aus dem Congo gesammelte Kieselschwämme. Von... In: Jen. Zeitschr. xvi, p. 553–577, Pl. xxiv.

1883 (*a*) **Marshall, W.** Einige vorläufige Bemerkungen über die Gemmulae der Süsswasserschwämme. Von... In: Zool. Anz. vi, pp. 630–634, 648–652.

1883 **Miller, S. A.** The American Palaeozoic Fossils... Cincinnati. 334 pp.

1883 **Nassonow, N.** Zur Biologie und Anatomie der Clione. Von... In: Z. W. Z. xxxix, p. 295–308, Pl. xviii–xix.

1883 **Počta, P.** Beiträge zur Kenntnis der Spongien der Böhmischen Kreideformation. Von... i. Abtheilung, Hexactinellidae. In: Abh. K. Böhm. Ges. Wiss. Prag, Mathem.-nat. Cl. (6), xii, p. 1–45, Pl. i–ii (but see 1884).

1883 (*a*) **Počta, P.** Ueber isolirte Kieselspongiennadeln aus der Böhm. Kreideformation... Von... In: Sitz. Ber. Böhm. Ges. Wiss. p. 371–384, 1 Pl.

1883 **Poléjaeff, N.** Ueber das Sperma und die Spermatogenese bei Sycandra raphanus, Haeckel. Von... In: Sitz. Ber. Akad. Wiss. Wien, lxxxvi, i. Abth. p. 276–298, Pl. i–ii*.

1883 **Potts, E.** Fresh-water Sponges: what, where, when, and who wants them. By... In: Bull. U. S. Fish. Comm. iii, p. 389–391.

1883 (*a*) **Potts, E.** Our Fresh-water Sponges. In: Amer. Naturalist, xvii, p. 1293–1296. With woodcuts.

* The plates are dated 1882, but "Mittheilungen, Jahrgang 1882," and English and American journals of November, 1882, are acknowledged on p. 275.

1883 **Priest, B. W.** On the Statoblasts of the Freshwater Sponges. By...
In: Journ. Quekett Microsc. Club (2), i, p. 173–181, Pl. VII.

1883 **Rathbun, R.** Sponge-Culture in Florida. In: Science, ii, p. 213

1883 **Retzer, W.** Die deutschen Süsswasserschwämme. Inaug. Diss.
Tübingen. 30 pp., Pl. I and II.

1883 **Ridley, S. O.** Notes on Zoophytes and Sponges obtained by Mr F. Day,
off the east Coast of Scotland. By...In: Journ. Linn. Soc.
London, xvii, p. 106–108.

1883 **Schulze, F. E.** Cancelled.

1883 (α) **Schulze, F. E.** Über radiäre Symmetrie bei Spongien. In: Tagebl.
55. Vers. d. Naturf. und Aerzte, pp. 48, 62, 199.

1883 **Seguenza, G.** Studi geologici e paleontologici sul cretaceo medio
dell' Italia meridionale...In: Atti R. Accad. Lincei (3), xii,
p. 65–214*, Pl. I–XXXI.

1883 **Solger, B.** Ueber einige der anatomischen Untersuchung zugängliche
Lebenserscheinungen der Spongien. In: Biol. Centralbl. iii,
p. 227–235.

1883 **Sollas, W. J.** The Estuaries of the Severn and its Tributaries; an Inquiry
into the Nature and Origin of their Tidal Sediment and Alluvial
Flats. By...In: Quart. Journ. Geol. Soc. xxxix, p. 611–626, 1 Pl.

1883 (α) **Sollas, W. J.** Description of Fossil Sponges from the Inferior
Oolite, with a Notice of some from the Great Oolite. By...In:
Quart. Journ. Geol. Soc. xxxix, p. 541–554, Pl. XX–XXI.

1883 (β) **Sollas, W. J.** Spongidae from the Inferior Oolite. In: Rep. 52nd
Meet. Brit. Assoc. p. 534–535.

1883 **Steinmann, G.** Systematische Stellung der Pharetronen. Von...In:
N. Jahrb. Mineral. i, p. 79.

1883 **Vejdovský, F.** Die Süsswasser-Schwämme Böhmens. Von...In: Abh.
K. Böhm. Ges. Wiss. Math.-nat. Cl. (6), xii, p. 1–43, Pl. I–III.
? Separately as Part i of "Revisio Faunae Bohemicae." Transl. in
Ann. and Mag. N. H. (5), xiii. With remarks by H. J. Carter,
p. 96–102, Pl. VI.

1883 (α) **Vejdovský, F.** Příspěvky k známostem o houbách sladkovodních
...(Freshwater Sponges.) In: Sitz. ber. Böhm. Ges. Wiss.
p. 328–340, 1 Pl.

1883 **Vosmaer, G. C. J.** ...Porifera. Von...In: Bronn, H. G., Die Klassen
und Ordnungen des Thierreichs...ii, p. 33–64, Pl. V–VI.

1883 (α) **Vosmaer, G. C. J.** Studies on Sponges. i, On Velinea gracilis. In.
Mitth. Z. Stat. Neapel, iv, p. 437–447, Pl. XXXI–XXXII.

1883 **Waller, J. G.** On a Newly Discovered British sponge. By...In:
Journ. Quekett Microsc. Club (2), i, p. 216–223, Pl. VIII.

1884 **Barrois, C.** Mémoire sur les Dictyospongidae des Psammites du Condros
(but see 1883). Par...In: Ann. Soc. Géol. du Nord, xi, p. 80–86, Pl. I.

* On title-page is printed 1882; according to Zool. Anz., however, the paper
appeared in 1883.

1884 **Beltrémieux, E.** Faune vivante du département de la Charente-Inférieure...p. 13–14 [fide Topsent 1891 (γ). Cf. Lendenfeld 1889, p. 9: Beltrémieux 1860—G. P. B.].

1884 **Bütschli, O.** Bemerkungen zur Gastraeatheorie. Von...In: Morph. Jahrb. ix, p. 416–427, Pl. xx. Transl. in: Ann. and Mag. N. H. (5), xiii, p. 372–383, Pl. xv.

1884 **Carter, H. J.** Catalogue of Marine Sponges, collected by Mr Jos. Willcox, on the West Coast of Florida. By...In: Proc. Acad. Philadelphia, 1884 (1885), p. 202–209.

1884 (α) **Carter, H. J.** On the Spongia coriacea of Montagu,=Leucosolenia coriacea, Bk., together with a new Variety of Leucosolenia lacunosa, Bk., elucidating the Spicular Structure of some of the Fossil Calcispongiae; followed by Illustrations of the Pin-like Spicules on Verticillites helvetica, De Loriol. By...In: Ann. and Mag. N. H. (5) xiv, p. 17–29, Pl. i.

1884 (β) **Carter, H. J.** Generic Characters of the Sponges described in Mr Carter's "Contributions to our Knowledge of the Spongida."...In: Ann. and Mag. N. H. (5), xiii, p. 129–130.

1884 (γ) **Carter, H. J.** On Grantia ciliata, var. spinispiculum, Crtr. By...In: Ann. and Mag. N. H. (5), xiii, p. 153–163, Pl. viii.

1884 (δ) **Carter, H. J.** The Branched and Unbranched Forms of the Freshwater Sponges considered generally. By...In: Ann. and Mag. N. H. (5), xiii, p. 269–273.

1884 (ε) **Carter, H. J.** Note on the assumed relationship of Parkeria to Stromatopora, and on a microscopic section of Stromatopora mamillata, Fr. Schmidt. By...In: Ann. and Mag. N. H. (5), xiii, p. 353–356.

1884 **Doederlein, L.** Studien an Japanischen Lithistiden. Von...In: Z. W. Z. xl, p. 62–104, Pl. v–vii.

1884 (α) **Doederlein, L.** Ueber Lithistiden aus Japan. In: Amtl. Ber. 56. Vers. d. Naturf. u. Ärzte, Freiburg, p. 110–112.

1884 **Dybowski, W.** Notiz über die aus Süd-Russland stammenden Spongillen, von...In: Sitz. Ber. Dorpater Naturf. Ges. vi, p. 507–515.

1884 (α) **Dybowski, W.** Ein Beitrag zur Kenntnis des Süsswasserschwammes Dosilia Stepanowii. Von...In: Zool. Anz. vii, p. 476–480. Transl. in: Ann. and Mag. N. H. (5), xiv, p. 249–245.

1884 (β) **Dybowski, W.** Mittheilung über einen neuen Fundort des Schwammes Lubomirskia baicalensis. Von...In: Sitz. Ber. Nat. Ges. Dorpat, vii, p. 44–45.

1884 (γ) **Dybowski, W.** Monographie der Spongilla sibirica, Dybow.... von...In: Sitz. Ber. Nat. Ges. Dorpat, vii, p. 64–75, Pl. (i).

1884 (δ) **Dybowski, W.** Samjetka o bodjagach juschnoi Rossii. (Notes on Freshwater Sponges of S. Russia.) In: Trudi Obschtsch. Charkow, xvii, p. 289–295, Pl. vii.

1884 (ε) **Dybowski, W.** (Monographie der Spongilla sibirica. Nachtrag.) In: Sitz. Ber. Nat. Ges. Dorpat, vii, p. 137–139.

1884 (ζ) **Dybowski, W.** Notiz über eine die Entstehung des Baikal-Sees betreffende Hypothese. In: Bull. soc. Imp. nat. Moscou, lix, p. 175–181.

1884 **Fristedt, K.** Om en fossil spongia. Af...In: Öfversigt K. Vetensk. Akad. Förh. p. 55–60, Pl. XII.

1884 **Gardner, J. S.** Chalk and the Origin and Distribution of Deep-Sea Deposits. In: Nature, xxx, p. 192–193.

1884 **Götte, A.** Ueber die Entwicklung der Spongillen. In: Zool. Anz. vii, pp. 676–679, 703–705.

1884 **Hall, J.** (Descript. of the Species of Fossil Reticulate Sponges...) In: 35th Ann. Rep. N. Y. State Mus. p. 465–481, Pl. XVIII–XXI.

1884 **Hinde, G. J.** Catalogue of the Fossil Sponges in the Geological Department of the British Museum...By...London. 4°. viii and 248 pp., with 38 Pl.*

1884 (α) **Hinde, G. J.** On some Fossil Calcisponges from the Well-boring at Richmond, Surrey. By...In: Quart. Journ. Geol. Soc. London, xl, p. 778–783, Pl. XXXV.

1884 (β) **Hinde, G. J.** On the Structure and Affinities of the Family Receptaculitidae, including therein the Genera Ischadites, Murchison (=Tetragonis, Eichwald); Sphaerospongia, Pengelly; Acanthochonia, gen. nov., and Receptaculites, Defrance. By... In: Quart. Journ. Geol. Soc. xl, p. 795–849, Pl. XXXVI–XXXVII.

1884 **Hughes, T. McKenny.** On the so-called Spongia paradoxa, S. Woodw. from the Red and White Chalk of Hunstanton. By...In: Quart. Journ. Geol. Soc. xl, p. 273–279.

1884 **Jeffreys, J. Gwyn.** Chalk and the "Origin and Distribution of Deep-Sea Deposits." In: Nature, xxx, p. 215–216.

1884 **Kiesow, J.** Ueber silurische und devonische Geschiebe Westpreussens. Von...In: Schrift. Nat. Ges. Danzig (2), vi, p. 205–300, Pl. II–IV.

1884 **Koehler, R.** Remarques sur le genre "Caminus," et sur une éponge voisine du "Caminus osculosus," Grube. Par...In: Biblioth. de l'école des hautes études 2. (sc. nat.) xxix, art. No. 4 bis (4 pp.).

1884 **Kraepelin, K.** Zur Biologie und Fauna der Süsswasserbryozoen. Von...In: Zool. Anz. vii, p. 319–321.

1884 **Lendenfeld, R. von.** The Digestion of Sponges effected by Ectoderm or Entoderm? By...In: Proc. Linn. Soc. N. S. W. ix, p. 434–438.

1884 (α) **Lendenfeld, R. von.** A Monograph of the Australian Sponges. By...Parts i and ii. In: Proc. Linn. Soc. N. S. Walos,ix, pp. 121 154, 310–344.

1884 (β) **Lendenfeld, R. von.** Das System der Monactinellidae. Von...In: Zool. Anz. vii, p. 201–206.

1884 (γ) **Lendenfeld, R. von.** On the Occurrence of Flesh-Spicules in Sponges. By...In: Proc. Linn. Soc. N. S. Wales, ix, p. 493–494.

* On the title is printed 1883. The preface is dated Dec. 26th, 1883 (cf. 1884, Poléjaeff) and consequently the work can only be considered to be published in 1884. In Zool. Jahresber. I made a mistake in taking 1883 for the year of publication.

1884 (δ) **Lendenfeld, R. von.** Notes on the Fibres of Certain Australian Hircinidae. By...In: Proc. Linn. Soc. N. S. Wales, ix, p. 641–642.

1884 **MacMunn, C. A.** On the Occurrence of Chlorophyll in Animals. In: Rep. 53rd Meet. Brit. Assoc. p. 532–534.

1884 **Marshall, W.** Agilardiella radiata, eine neue Tetractinellidenform mit radiärem Bau. Von...In: Abh. Akad. Berlin (15 pp., 1 Pl.).

1884 **Mills, H.** Serial Arrangement of Birotulate Spicules in Statoblasts of American Sponges. By...In: Amer. Monthly Microsc. Journ. v, p. 41–42.

1884 **Počta, P.** Beiträge zur Kenntniss der Spongien der Böhmischen Kreideformation. Von...ii. Abtheilung: Lithistidae. In: Abh K. Böhm. Ges. Wiss. Math.-nat. Cl. (6), xii, p. 1–45, Pl. I–II.

1884 (α) **Počta, P.** Ueber isolirte Kieselspongiennadeln aus der Böhm. Kreideformation...Von...In: Sitz. Ber. K. Böhm. Ges. Wiss. Prag, p. 1–14, 1 Pl. Continued from 1883 (α).

1884 (β) **Počta, P.** Ueber Spongiennadeln des Brüsauer Hornsteines... von. In: Sitz. Ber. K. Böhm. Ges. Wiss. Prag, p. 243–254, 2 Pl.

1884 **Poléjaeff, N.** Report on the Calcarea dredged by H.M.S. Challenger during the years 1873–1876. By...In: Rep. Sc. Results voyage Challenger, Zoology, viii, p. 1–76, Pl. I–IX*.

1884 (α) **Poléjaeff, N.** Report on the Keratosa collected by H.M.S. Challenger during the years 1873–1876. In: Rep. Challenger, Zool. xi, p. 1–88, Pl. I–X.

1884 **Potts, E.** Fresh-water Sponges as improbable causes of the pollution of river water. In: Proc. Acad. N. Sc. Philadelphia, p. 28–30.

1884 (α) **Potts, E.** Some Modifications observed in the Form of Sponge Spicules. In: Proc. Acad. N. Sc. Philadelphia, p. 184–185.

1884 (β) **Potts, E.** On the Wide Distribution of some American Sponges. In: Proc. Acad. N. Sc. Philadelphia, p. 215–217.

1884 **Priest, B. W.** Statoblasts of the Fresh-water Sponges. By...In: Amer. Monthly Microsc. Journ. iv, p. 208–213.

1884 **Ridley, S. O.** Spongiida. By...In: Report on the Zoological Collections made in the Indo-Pacific Ocean during the voyage of H.M.S. "Alert" 1881–82, pp. 366–482, 582–630, Pl. xxxix–xliii, liii–liv...London. 8°. Rev. in: Nature, xxx (1884), p. 485–486.

1884 (α) **Ridley, S. O.** Notes on Sponges, with Description of a new Species. By...In: Ann. and Mag. N. H. (5), xiv, p. 183–187.

1884 **Sollas, W. J.** On the Development of Halisarca lobularis (O. Schmidt). By...In: Quart. Journ. Microsc. Sc. xxiv, p. 603–621, Pl. xxxvii, fig. 1–39.

* Title-page dated 1883; Editorial Note dated 28th December, 1883.

1884 (α) **Sollas, W. J.** On the Origin of Freshwater Faunas: A Study in Evolution. By...In: Trans. R. Dublin Soc. (2), iii, p. 87-118.

1884 (β) **Sollas, W. J.** On the Origin of Freshwater Faunas: A Study in Evolution. By...In: Proc. R. Dublin Soc. p. 238-240. Abstract from Sollas, 1884 (α).

1884 **Thoulet, J.** Sur les spicules siliceux d'éponges vivantes. Note de... In: C. R. Paris, xcviii, p. 1000-1001.

1884 **Vejdovsky, F.** Bemerkungen über einige Süsswasserschwämme. In: Sitz. Ber. K. Böhm. Ges. Wiss. p. 55-60, 1 Pl.

1884 **Vosmaer, G. C. J.** ...Porifera. Von...In: Bronn, H. G., Die Klassen und Ordnungen des Thierreichs...ii...p. 65-176, Pl. III, VII-XVIII.

1884 (α) **Vosmaer, G. C. J.** Studies on Sponges. By...ii, On the supposed difference between Leucandra crambessa, H. and aspera (O. S.), H. with an attempt to explain it. iii, On Haeckel's entogastric septa. iv, On the relation between certain Monactinellidae and Ceraospongiae. In: Mitth. Z. Stat. Neapel, v, p. 483-493, Pl. 28-29.

1884 **Wierzejski, A.** O Rozwoju Pąków (Gemmulae) gąbek stodkowodnych, europejskich, tudziéz o gat. Spongilla fragilis, Leidy (S. sibirica, Dyb.). (On the Development of the Gemmules of European Freshwater Sponges and on Spongilla fragilis.) (Polish.) In: Abh. Sitz. Ber. Akad. Krakau, xii, p. 239-279, Pl. IX. Sep. copy, p. 1-43. Transl. in: Arch. Slav. Biol. i (1886), p. 26-47.

1884 **Zittel, K. A.** Ueber Astylospongidae und Anomocladina. Von...In: N. Jahrb. Mineral. ii, p. 75-80, Pl. I-II. Transl. in: Ann. and Mag. N. H. (5), xiv, p. 271-276.

1885 **Aubry,** . Observations géologiques sur le royaume du Choa et les pays Gallas...In: C. R. Acad. Paris, ci, p. 1182-1184.

1885 **Carter, H. J.** Descriptions of Sponges from the Neighbourhood of Port Phillip Heads, South Australia. By...In: Ann. and Mag. N. H. (5), xv, pp. 107-117, 196-222, 301-321, Pl. IV, and (5), xvi, pp. 277-294, 347-368.

1885 (α) **Carter, H. J.** Report on a Collection of Marine Sponges from Japan, made by Dr J. Anderson, F.R.S., Superintendent Indian Museum. Calcutta. By...In: Ann. and Mag. N. H. (5), xv, p. 387-406, Pl. XII-XIV.

1885 (β) **Carter, H. J.** Note on Spongilla fragilis, Leidy, and a new species of Spongilla from Nova Scotia. By...In: Ann. and Mag. N. H. (5), xv, p. 18-20.

1885 (γ) **Carter, H. J.** Mode of Circulation in the Spongida. By...In: Ann. and Mag. N. H. (5), xv, p. 117-122, Pl. IV, figs. 5 a-g, and 7 a-p.

1885 (δ) **Carter, H. J.** On a Variety of the Freshwater Sponge Meyenia fluviatilis. By...In: Ann. and Mag. N. H. (5), xv, p. 453-456.

1885 (ε) **Carter, H. J.** On a Variety of the Freshwater Sponge Meyenia fluviatilis auctt., from Florida. By...In: Ann. and Mag. N. H. (5), xvi, p. 179–181.

1885 **Dunikowski, E.** Ueber Permo-Carbon Schwämme von Spitzbergen ...In: K. Svensk. Vet. Akad. (2), xxi, p. 1–18, Pl. I–II*.

1885 **Dybowski, W.** Mittheilung über einen neuen Fundort des Schwammes Lubomirskia baicalensis von...In: Sitz. Ber. Dorpater Naturf. Ges. vii, p. 44–45.

1885 (α) **Dybowski, W.** Monographie d. Spongilla sibirica, Dybow. von... In: Sitz. Ber. Dorpater Naturf. Ges. vii, pp. 64–75, 137–139, Pl. I.

1885 **Entz, Géza.** Zur näheren Kenntnis der Tintinnoden. Von...In: Mitth. Zool. Stat. Neapel, vi, p. 185–216, Pl. XIII–XIV.

1885 **Erdmann, A.** Ueber einige neue Zoantheen. Ein Beitrag zur anatomischen und systematischen Kenntniss der Actinien. Von... In: Jen. Zeitschr. xix, p. 430–488, Pl. IV–V.

1885 **Fristedt, K.** Bidrag till Kännedomen om de vid Sveriges vestra Kust lefvande Spongiae. Af...In: Kongl. Svenska Vetensk. Akad. Handlingar, xxi, p. 1–56, Pl. I–IV†.

1885 **Goette, A.** Ueber die Entwicklung der Spongillen. Von...In: Zool. Anz. viii, p. 377–380.

1885 **Haacke, W.** Wollkrabben und ihre Mäntel. Von...In: Zool. Garten, xxvi, p. 203–205.

1885 **Hall, J.** On the Fossil Reticulate Sponges constituting the Family Dictyospongidae. By...In: Rep. 54th Meet. Brit. Assoc. p. 725–726.

1885 **Hansen, G. A.** Den Norske Nordhavs-Expedition 1876–1878. Zoologi, Spongiadae, ved...Christiania. 4°. 25 pp., 7 Pl. and 1 Map. Title and letter-press also in English.

1885 **Henderson, J. R.** The Anomura. In: Rep. Challenger Narrative, i, 2, p. 897–901.

1885 **Hinde, G. J.** On Beds of Sponge-remains in the Lower and Upper Greensand of the South of England. Part II. By...In: Phil. Trans. R. Soc. p. 403–453, Pl. XL–XLV.

1885 **Hyatt, A.** Larval Theory of the Origin of Cellular Tissues. By... In: Proc. Boston Soc. N. H. xxiii, p. 45–163.

1885 **Koehler, R.** See 1886.

1885 (α) **Koehler, R.** Recherches sur la Faune marine des îles Anglo-Normandes. Par...In: Bull. Soc. Sc. Nancy, p. 1–70.

1885 **Krukenberg, C. F. W.** Vergleichend-physiologische Vorträge von... iv, Grundzüge einer vergleichenden Physiologie der thierischen Gerüstsubstanzen. Heidelberg, C. Winter'sche Univ. Buchh. (p. 187–269; vide pp. 195, 204–206, 215, 216, 219, 256, 258).

* Query if published in 1884? G. P. B.
† Appeared in 1885. Cf. Zool. Jahresber. f. 1885. In Zool. Rec. and in Zool. Anz. wrongly put in 1888.

1885 (α) **Krukenberg, C. F. W.** Ueber das Zustandekommen der sogenannten Eiweissreactionen. Von...In: Sitz. Ber. Ges. Naturw. Jena, p. 122–132.

1885 (β) **Krukenberg, C. F. W.** Die Beziehungen der Eiweiss-stoffe zu den albuminoiden Substanzen und den Kohlehydraten. Von...In: Sitz. Ber. Ges. Naturw. Jena, p. 133–148.

1885 **Lendenfeld, R. von.** On a Sponge destructive to Oyster Culture in the Clarence River. By...In: Proc. Linn. Soc. N. S. Wales, x, p. 326–329.

1885 (α) **Lendenfeld, R. von.** The Histology and Nervous System of the Calcareous Sponges. By...In: Proc. Linn. Soc. N. S. Wales, ix, p. 977–983.

1885 (β) **Lendenfeld, R. von.** A Monograph of the Australian Sponges. Part iii. By...In: Proc. Linn. Soc. N. S. Wales, ix, p. 1083–1150, Pl. 59–67. Abstr. in: Linn. Soc. N. S. W. Abstr. of Proc. 26 Nov. 1884, p. iii–v.

1885 (γ) **Lendenfeld, R. von.** A Monograph of the Australian Sponges. Parts iv, v, vi. By...In: Proc. Linn. Soc. N. S. Wales, x, pp. 3–22, 282–325, 481–553, Pl. 1–5, 26–35, 36–38.

1885 (δ) **Lendenfeld, R. von.** Die Verwandtschaftsverhältnisse der Myxospongien. Von...In: Zool. Anz. viii, p. 510–515.

1885 (ε) **Lendenfeld, R. von.** Notes to the Australian Sponges recently described by Carter. By...In: Proc. Linn. Soc. N. S. Wales, x, p. 151–156.

1885 (ζ) **Lendenfeld, R. von.** Notes to the Australian Sponges recently described by Carter. By...In: Ann. and Mag. N. H. (5), xvi, p. 20–25. This article is almost, but not quite, a reprint of 1885 (ε).

1885 (η) **Lendenfeld, R. von.** Das Nervensystem der Spongien. Vorläufige Mittheilung. Von...In: Zool. Anz. viii, p. 47–50.

1885 (θ) **Lendenfeld, R. von.** Berichtigung und Ergänzung. Das Nervensystem der Spongien. Von...In: Zool. Anz. viii, p. 448.

1885 (ι) **Lendenfeld, R. von.** Zur Histologie der Spongien. Von...In: Zool. Anz. viii, pp. 466–469, 483–486.

1885 (κ) **Lendenfeld, R. von.** The Homocoela hitherto described from Australia and the new family Homodermidae. By...In: Proc. Linn. Soc. N. S. Wales, ix, p. 896–907.

1885 (λ) **Lendenfeld, R. von.** The Phoriospongiae, Marshall. By...In: Proc. Linn. Soc. N. S. Wales, x, p. 81–84.

1885 (μ) **Lendenfeld, R. von.** Addendum to the Monograph of the Australian Sponges. By...In: Proc. Linn. Soc. N. S. Wales, x, p. 475–476.

1885 (ν) **Lendenfeld, R. von.** Beitrag zur Kenntniss des Nerven- und Muskelsystems der Hornschwämme. Von...In: Sitz. Ber. Akad. Berlin, p. 1015–1020. Transl. in: Ann. and Mag. N. H. (5), xvii (1886), p. 372–377.

157

1885 (ξ) **Lendenfeld, R. von.** Die Verwandtschaftsverhältnisse der Kalk-schwämme. Von...In: Zool. Anz. p. 211–215.

1885 (o) **(Lendenfeld, R. von.** Remarks on the Discovery of Sensitive and Ganglia Cells in Horny Sponges.) In: Proc. Linn. Soc. N. S. Wales, x, p. 246.

1885 **Lockwood, S.** Heteromeyenia Ryderi (a Freshwater Sponge). In: Journ. New York Microsc. Soc. i, p. 37–40.

1885 **MacMunn, C. A.** Further Observations on Enterochlorophyll and Allied Pigments. By...In: Proc. R. Soc. London, xxxviii, p. 319–322.

1885 **Marshall, W.** Bemerkungen über die Coelenteratennatur der Spongien. In: Jen. Zeitschr. xviii, p. 868–880. Transl. in: Ann. and Mag. N. H. (5), xvi, p. 90–100.

1885 (α) **Marshall, W.** Vorläufige Bemerkungen über die Fortpflanzungs-verhältnisse von Spongilla lacustris. In: Sitz. Ber. Naturf. Ges. Leipzig, xi, p. 22–29.

1885 **Petr, F.** Spongilla fragilis (Leidy v. Čechách. (Bohemian with Ger-man abstract.) In: Sitz. Ber. K. Böhm. Ges. Wiss. Prag, p. 298–310 (p. 99–111), 1 Pl.

1885 **Počta, P.** Über zwei neue Spongien aus der Böhmischen Kreide-formation...Von...In: Sitz. Ber. K. Böhm. Ges. Wissensch. Prag, p. 587–592 (p. 1–7), 1 Pl.

1885 (α) **Počta, P.** Beiträge zur Kenntniss der Spongien der böhm. Kreide-formation. iii, Tetractinellidae...Von...In: Abh. K. Böhm. Ges. Wiss. vii, sep. cop. p. 1–40, Pl. I.

1885 **(Poléjaeff, N.** The Calcarea and Keratosa.) In: Rep. Challenger, Narrative, i, p. 639–645.

1885 **Potts, E.** A Freshwater Sponge from Mexico. In: Amer. Naturalist, xix, p. 810–811.

1885 (α) **Potts, E.** A new Freshwater Sponge from Nova-Scotia. In: Proc. Acad. N. Sc. Philadelphia, p. 28–29. Also in: Ann. and Mag. N. H. (5), xv, p. 425–426.

1885 (β) **Potts, E.** Freshwater Sponges from Mexico. In: Proc. U. S. Nation. Mus. viii, p. 587–589.

1885 **Rice, H. J.** Sponges. In: Journ. New York Microsc. Soc. i, p. 116–122.

1885 **Ridley, S. O.** (The Monaxonida.) In: Rep. Challenger, Narrative, i, 2, p. 569–573.

1885 **Ringueberg, E. N. S.** New Fossils from the Four Groups of the Niagara Period of Western New York. In: Proc. Acad. N. Sc. Phila-delphia, 1884, p. 144–150, Pl. II and III.

1885 **Schmidt, O.** Entstehung neuer Arten durch Verfall und Schwund älterer Merkmale. Von...In: Z. W. Z. xlii, p. 639–647.

1885 **Schulze, F. E.** (The Hexactinellida.) In: Rep. Challenger, Narrative, i, 1, p. 437–451.

1885 (α) **Schulze, F. E.** Ueber das Verhältnis der Spongien zu den Choano-flagellaten. Von...In: Sitz. Ber. Akad. Berlin, p. 179–191. Transl. in: Ann. and Mag. N. H. (5), xv, p. 365–377.

158

1885 (β) **Schulze, F. E.** (Lebende Krusten von Oscarella lobularis coerulea mit Brutknospen.) In: Ber. Ges. Naturf. Freunde Berlin, p. 183–184.

1885 **Seely, H. M.** A new Genus of Chazy Sponges, Strephochetus. In: Amer. Journ. Sc. xxx, p. 355–357, fig. 1–3.

1885 **Sollas, W. J.** On the Physical Characters of Calcareous and Siliceous Sponge-Spicules and other Structures. By...In: Sc. Proc. R. Dublin Soc. (2), iv, p. 374–392, Pl. xv.

1885 (α) **Sollas, W. J.** Note on the Structure of the Skeleton in the Anomocladina. By...In: Ann. and Mag. N. H. (5), xv, p. 236–238.

1885 (β) **Sollas, W. J.** (The Tetractinellida.) In: Rep. Challenger, Narrative, i, 1, p. 451–452.

1885 (γ) **Sollas, W. J.** Note on the Artificial Deposition of crystals of calcite on Spicules of a Calci-sponge. By...In: Sc. Proc. R. Dublin Soc. v, p. 73.

1885 (δ) **Sollas, W. J.** On Vetulina stalactites (O. S.) and the Skeleton of the Anomocladina. By...In: Proc. R. Irish Acad. iv, p. 486–492, Pl. iii–iv.

1885 (ε) **Sollas, W. J.** On an Hexactinellid Sponge from the Gault, and a Lithistid from the Lias of England. By...In: Sc. Proc. R. Dublin Soc. iv, p. 443–446, Pl. xxi.

1885 (ζ) **Sollas, W. J.** A Classification of the Sponges. By...In: Sc. Proc. R. Dublin Soc. (2), v, p. 112. Also in: Ann. and Mag. N H. (5), xvi, p. 395.

1885 **Thompson, W. D'Arcy.** A Bibliography of Protozoa, Sponges, Coelenterata, and Worms...Cambridge. viii and 284 pp.

1885 **Vosmaer, G. C. J.** The Sponges of the "Willem Barents" Expedition 1880 and 1881, by...In: Bijdr. Dierk. xii, p. 1–47, Pl. i–v.

1885 (α) **Vosmaer, G. C. J.** ...Porifera. Von...In: Bronn, H. G., Die Klassen und Ordnungen des Thierreichs...ii, p. 177–368, Pl. xix–xxv.

1885 (β) **Vosmaer, G. C. J.** Something about Scudder's Nomenclator Zoologicus. By...In: Zool. Anz. viii, p. 216–219.

1885 **Wierzejski, A.** O Gąbkach stodkowodnych Galicyjskich. (The Freshwater Sponges of Galicia.) In: Osobne odbicie z xix tomu Sprawozdan' Komisyi fizyjogr. Akad. Umiej.

1885 **Wrześniowski, A.** Stanowisko gąbek w systematyce zwierząt Wedtug Marshala. (Position of Sponges in Marshall's System.) In: Wszechswiat. iv, p. 503–504.

1886 **Buccich, G.** Alcune Spugne dell' Adriatico sconosciute e nuove... In: Boll. Soc. Adriat. Sc. nat. Trieste, ix, p. 222–225 (and one lith. Pl.).

1886 **Carter, H. J.** Descriptions of Sponges from the Neighbourhood of Port Phillip Heads, South Australia, continued. By...In: Ann. and Mag. N. H. (5), xvii, pp. 40–53, 112–127, 431–441, 502–516, and (5), xviii, pp. 34–55, 126–149.

1886 (α) **Carter, H. J.** Supplement to the Descriptions of Mr J. Bracebridge Wilson's Australian Sponges. By...In: Ann. and Mag. N. H. (5), xviii, pp. 271–290, 369–379, 445–466, Pl. x.

1886 **Choffat, P.** Note sur la distribution des bancs de Spongiaires à spicules siliceux dans la chaîne du Jura et sur le parallélisme de l'Argovien. In: Bull. Soc. Géol. France (3), xiii, p. 834–841.

1886 **Dendy.** Vide Ridley.

1886 **Duncan, P. M.** On the genus Hindia and its Species. By...In: Ann. and Mag. N. H. (5), xviii, p. 226–228.

1886 **Dybowski, W.** Mittheilung über die aus dem Flusse Niemen stammende Trochospongilla erinaceus, Ehbg. In: Sitz. Ber. Naturf. Ges. Dorpat, vii, p. 295–298.

1886 **Frenzel, J.** Glycerinpraeparate von ganzen Thieren und deren Theilen. Von...In: Zool. Anz. ix, p. 690.

1886 (α) **Frenzel, J.** Verfahren zur Herstellung von Zoologischen und Anatomischen Präparaten mittelst der Glycerindurchtränkung, von...In: Zool. Jahrb. i, p. 216–224.

1886 **Gibson, R. J. H.** On a new species of Sycandra (aspera). In: 1st Rep. Fauna Liverpool Bay, p. 364–367, Pl. i.

1886 **Goette, A.** Untersuchungen zur Entwickelungsgeschichte von Spongilla fluviatilis. Von...In: Abhandl. Entw. Tiere, iii, 64 pp. Woodcuts and 5 Pl.

1886 (α) **Goette, A.** Nachträgliche Bemerkungen zur Entwickelungsgeschichte der Schwämme. Von...In: Zool. Anz. ix, p. 292–295.

1886 **Hansen, G. Armauer.** Bericht über zoologische Untersuchungen, vorgenommen in den Sommern 1884 und 1885 auf Kosten des Museums. Von...In: Bergens Mus. Aarsberetn. for 1885, p. 49–54, Pl. i.

1886 **Heider, K.** Zur Metamorphose der Oscarella lobularis, O. Schm. Von...In: Arb. Zool. Inst. Wien, vi, p. 175–236, Pl. xix–xxi.

1886 **Higgin, T.** Report on the Porifera of the L. M. B. C. District. In: 1st Rep. Fauna Liverpool Bay, p. 72–94.

1886 **Hinde, G. J.** On Beds of Sponge-remains in the Lower and Upper Greensand of the South of England. By...In: Phil. Trans. R. Soc. ii, 1885, p. 403–453, Pl. 40–45.

1886 (α) **Hinde, G. J.** On the Sponge-spicules from the Deposits of St Erth. In: Quart. Journ. Geol. Soc. London, xlvii, p. 214.

1886 **Koehler, R.** Contribution à l'étude de la faune littorale des Iles Anglo-Normandes...Par...In: Ann. Sc. Nat. (6), xx*, art. 4, p. 1–62.

* On the title-page of Vol. xx is printed 1885. According to Zool. Anz. and Zool. Rec. it appeared only in 1886.

1886 **Krukenberg, C. F. W.** Untersuchungen über den chemischen Bau der Eiweissstoffe. Von...In: Jen. Zeitschr. xx (2), xiii, Suppl. Heft 1, p. 39–60.

1886 **Kükenthal, W. & B. Weissenborn.** Ergebnisse eines zoologischen Ausfluges an die Westküste Norwegens....Von...In: Jen. Zeitschr. xix (2), xii, p. 776–789.

1886 **Lackschewitz, P.** Ueber die Kalkschwämme Menorcas. Beitrag zur Spongienfauna des Mittelmeers. Von...In: Zool. Jahrb. i, p. 297–310, Pl. vii.

1886 (a) **Lackschewitz, P.** Ueber die Kalkschwämme Menorcas. In: Sitz. Ber. Naturf. Ges. Dorpat, vii, p. 336–341. Transl. in: Ann. and Mag. N. H. (5), xvii, p. 536–539.

1886 **Lampe, W.** Tetilla japonica, eine neue Tetractinellidenform mit radiärem Bau. Von...In: Arch. Naturgesch. lii, p. 1–18, Pl. i. Also separately as "Inaugural Dissertation."

1886 **Lendenfeld, R. von.** Studies on Sponges...By...i, The vestibule of Dendrilla cavernosa...ii, Raphyrus Hixonii...iii, Halme tingens ...iv, Two cases of mimicry in Sponges...In: Proc. Linn. Soc. N. S. Wales, x, p. 557–574, Pl. 39–44.

1886 (a) **Lendenfeld, R. von.** An Alga forming a pseudomorph of a Siliceous Sponge. By...In: Proc. Linn. Soc. N. S. Wales, x, p. 726–728, Pl. 48, fig. 5.

1886 (β) **Lendenfeld, R. von.** Second addendum to the Monograph of the Australian Sponges. By...In: Proc. Linn. Soc. N. S. Wales, x, p. 845–850.

1886 **Levinsen, G. M. R.** Kara-Havets Svampe (Porifera). Ved...In: Dijmpha-Togtets zool. bot. Udbytte, p. 341–372, Pl. xxix–xxxi.

1886 **Marenzeller, E. von.** Poriferen, Anthozoen, Ctenophoren und Würmer von Jan Mayen...von...In: Internat. Polarforschung 1882–1883, iii, p. 9–24 (sep. 1–16), Pl. i.

1886 **Muller, L.** La pêche des éponges dans l'archipel grec. In: Cosmos, xxxv, p. 360–361.

1886 **Noll, F. C.** Spongilla glomerata, N. Von...In: Zool. Anz. ix, p. 682–684.

1886 **Petr, F.** Dodatky ku fauně českých hub sladkovodních...(Contribution Bohem. Fauna Freshwater sponges.) In: Sitzungsber. Böhm. Ges. Wiss. Prag, p. 147–174 (p. 92–119), 1 Pl.

1886 **Počta, P.** Le développement des Éponges fossiles. In. Arch. Slav. Biol. i, p. 23–25.

1886 (a) **Počta, P.** Ueber einige Spongien aus dem Dogger des Fünfkirchner Gebirges. Von...In: Mitth. Jahrb. K. Ungar. Geol. Anstalt, viii, p. 109–121, Pl. xxiii–xxiv.

1886 **Potts, E.** Freshwater Sponges from Newfoundland: A new Species. In: Proc. Acad. Nat. Sc. Philadelphia, p. 227–230. Also in: Ann. and Mag. N. H. (5), xviii, p. 243–246

1886 **Priest, B. W.** On Spongilla fragilis found in the Thames. By...In: Journ. Quekett Microsc. Club (2), ii, p. 252–254, Pl. xv.

1886 **Rauff, H.** Über die Gattung Hindia Dunc. Von...In: Verh. naturh. Ver. Rheinl. und Westph. xliii (5), iii, p. 163–174. Transl. in: Ann. and Mag. N. H. (5), xviii, p. 169–179.

1886 **Ridley, S. O. & A. Dendy.** Preliminary Report on the Monaxonida collected by H.M.S. "Challenger." By...In: Ann. and Mag. N. H. (5), xviii, pp. 325–351, 470–493.

1886 (α) **Ridley, S. O. & A. Dendy.** On Proteleia Sollasi, a new Genus and Species of Monaxonid Sponges allied to Polymastia. By...In: Ann. and Mag. N. H. (5), xviii, p. 152–159, Pl. v.

1886 **Schulze, F. E.** Ueber den Bau und das System der Hexactinelliden. Von...In: Abh. Akad. Berlin, p. (1–97).

1886 **Seely, H. M.** The Genus Strephochetus: Distribution and Species; by...In: Amer. Journ. Sc. cxxxii (3), xxxii, p. 31–34.

1886 **Senoner, A.** Die Wollkrabben und ihre Mäntel. In: Zool. Garten, xxvii, p. 92.

1886 **Sollas, W. J.** Preliminary Account of the Tetractinellid Sponges dredged by H.M.S. Challenger, 1872–1876. By...Part i. The Choristida...In: Sc. Proc. R. Dublin Soc. v, p. 177–199.

1886 (α) **Sollas, W. J.** Cancelled.

1886 (β) **Sollas, W. J.** Letter on Dr K. Heider's paper on Oscarella lobularis...In: Zool. Anz. ix, p. 518–519.

1886 (γ) **Sollas, W. J.** A Classification of the Sponges. In: Proc. R. Dublin Soc. v, p. 112.

1886 **Vejdovsky, F.** Einiges über "Spongilla glomerata N."...In: Zool. Anz. ix, p. 713–715.

1886 (α) **Vejdovsky, F.** Přehled sladkovodních hub Europských...(Account on Freshwater Sponges of Europe.) In: Sitz. Ber. Böhm. Ges. Wissensch. Math.-naturw. Cl. p. 472–486 (p. 175–189).

1886 **Vosmaer, G. C. J.** ...Porifera...In: Bronn, H. G., Die Klassen und Ordnungen des Thierreichs...ii, p. 369–496, Pl. xxvi–xxxiv. (Title dated 1887.) P. 472–481 transl. English by Dendy in Ann. and Mag. N. H. (5), xix, p. 249–260.

1886 (α) **Vosmaer, G. C. J.** Einige neuere Arbeiten über Schwämme. Kritisch referirt. In: Biol. Centralbl. vi, pp. 181–188, 193–201.

1886 **Walcott, C. D.** Classification of the Cambrian System of North America, by...In: Amer. Journ. Sc. cxxxii (3), xxxii, p. 138–157.

1886 **Walther, J.** I vulcani sottomarini del Golfo di Napoli. Nota del... con la cooperazione di A. Colombo...In: Boll. R. Com. Geol. d' Italia, p. 1–12, Pl. viii.

1886 **Whitfield, R. P.** Notice of a new fossil body, probably a sponge related to Dictyophyton. In: Bull. Am. Mus. New York, i, p. 346–348, Pl. xxxv.

1886 **Wierzejski, A.** Le développement des gemmules des Éponges d'eau douce d'Europe. Par...In: Arch. Slaves Biol. i, p. 26–47.

1886 **Zahálka, Č.** Über Isoraphinia texta, Roem. sp. und Scytalia pertusa, Reuss sp. aus der Umgebung von Raudnitz a.E. Böhmen. In: Sitz. Ber. k. Akad. Wissensch. Wien, xcii, 1 Abth. p. 647–652, Pl. ı, ıı.

1887 **Bell, F. Jeffrey.** The nervous System of Sponges. By...In: Zool. Anz. x, p. 241.

1887 **Carter, H. J.** Report on the Marine Sponges, chiefly from King Island in the Mergui Archipelago, collected for the Trustees of the Indian Museum, Calcutta, by Dr John Anderson...By... In: Journ. Linn. Soc. London, xxi, p. 61–84, Pl. 5–7.

1887 (α) **Carter, H. J.** On the Position of the Ampullaceous Sac and the Function of the Water Canal-system in the Spongida. By... In: Ann. and Mag. N. H. (5), xix, p. 203–212.

1887 (β) **Carter, H. J.** On the Reproductive Elements of the Spongida. By...In: Ann. and Mag. N. H. (5), xix, p. 350–360.

1887 (γ) **Carter, H. J.** Carterius stepanovii, Petr. By...In: Ann. and Mag. N. H. (5), xix, p. 247–248.

1887 (δ) **Carter, H. J.** Description of Chondrosia spurca, n. sp., from the South Coast of Australia. By...In: Ann. and Mag. N. H. (5), xix, p. 286–288.

1887 **Dendy, A.** Report on a Zoological Collection made by the Officers of H.M.S. "Flying Fish," at Christmas Island, Indian Ocean... ix, Porifera, by...In: Proc. Zool. Soc. London, iii, p. 524–526, Pl. xLIV.

1887 (α) **Dendy, A.** The Sponge-fauna of Madras. A Report on a Collection of Sponges obtained in the Neighbourhood of Madras by Edgar Thurston, Esq. By...In: Ann. and Mag. N. H. (5), xx, p. 153–165, Pl. ıx–xıı.

1887 (β) **Dendy, A.** On a remarkable new Species of Cladorhiza obtained by H.M.S. "Challenger." By...In: Ann. and Mag. N. H. (5), xx, p. 279–282, Pl. xv.

1887 (γ) **Dendy, A.** The New System of Chalininae, with some brief Observations upon Zoological Nomenclature. By...In: Ann. and Mag. N. H. (5), xx, p. 326–337.

1887 (δ) **Dendy, A.** (Observations on the West-Indian Chalininae.) In: Proc. Zool. Soc. London, p. 503–507.

1887 **Duncan, P. M.** A Reply to Dr G. J. Hinde's Communication "On the Genus Hindia, Dunc., and the Name of its Typical Species." By...In: Ann. and Mag. N. H. (5), xix, p. 260–264.

1887 **Ebner, V. von.** Ueber den feineren Bau der Skelettheile der Kalkschwämme nebst Bemerkungen über Kalkskelete überhaupt. Von...In: Sitzungsber. K. Akad. Wiss. Abth. 1, xcv, p. 55–149, Taf. ı–ıv.

1887 (α) **Ebner, V. von.** Amphoriscus bucchichii, n. sp. Von...In: Zool. Jahrbüch. ii, p. 981–982.

1887 **Fiedler, K.** Ueber die Entwicklung der Geschlechtsproducte bei Spongilla. Von...In: Zool. Anz. x, p. 631–636. Transl. in: Ann. and Mag. N. H. (5), xx, p. 435–440.

1887 **Fristedt, K.** Meddelanden om Bohuslänska Spongior...In: Öfvers. Kgl. Vet. Akad. Förh. Stockholm, xliv, p. 25–29.

1887 (*a*) **Fristedt, K.** Sponges from the Atlantic and Arctic Oceans and the Behring Sea described by...In: Vega-Expeditionens Vetensk. Iakttagelser (Nordenskiöld), iv, p. 401–471, Pl. 22–31.

1887 **Hardman, E. F.** Note on Professor Hull's Paper. By...In: Proc. R. Soc. London, xlii, p. 308–310.

1887 **Hinde, G. J.** On the Genus Hindia, Duncan, and the Name of its Typical Species. By...In: Ann. and Mag. N. H. (5), xix, p. 67–79.

1887 (*a*) **Hinde, G. J.** A Monograph of the British Fossil Sponges. By... Part i. 92 pp., 8 Pl. London, Palaeontographical Society. 4º.

1887 (*β*) **Hinde, G. J.** On the Organic Origin of the Chert in the Carboniferous Limestone Series of Ireland, and its Similarity to that in the Corresponding Strata in North Wales and Yorkshire. By...In: Geol. Mag. (3), iv, p. 435–446.

1887 **Hull, E.** Note on Dr G. J. Hinde's Paper "On Beds of Sponge-remains in the Lower and Upper Greensand of the South of England"...By...In: Proc. R. Soc. London, xlii, p. 304–308.

1887 **Lavis, H. J. Johnston & G. C. J. Vosmaer.** On Cutting Sections of Sponges and other similar structures with soft and hard tissues. By...In: Journ. R. Microsc. Soc. (2), vii, p. 200–204.

1887 **Lendenfeld, R. von.** On the Systematic Position and Classification of Sponges. By...In: Proc. Zool. Soc. London, 1886, p. 558–662. N.B. The number was published in 1887 (April). Reviews: Nature, xxxv (1887), p. 238; Zool. Anz. x (1887), p. 47; Amer. Naturalist, xxi (1887), p. 935–938; Journ. Microsc. Soc. London, 1887, p. 599–600.

1887 (*a*) **Lendenfeld, R. von.** Der gegenwärtige Stand unsrer Kenntniss der Spongien. Von...In: Zool. Jahrb. Spengel, ii, p. 511–574.

1887 (*β*) **Lendenfeld, R. von.** Mr Dendy on the Chalininae. By...In: Ann. and Mag. N. H. (5), xx, p. 428–432.

1887 (*γ*) **Lendenfeld, R. von.** On the Structure and Life History of Sponges By...In: Zoologist (3), xi, p. 223–232.

1887 (*δ*) **Lendenfeld, R. von.** Synocils, Sinnesorgane der Spongien. Von ...In: Zool. Anz. x, p. 142–145. Transl. in: Amer. Naturalist, xxi, p. 485–486.

1887 (*ε*) **Lendenfeld, R. von.** (The nervous System of Sponges.)...In: Rep. Brit. Assoc. xxiii, p. 710. Also in: Nature, xxxv, p. 538.

1887 (*ζ*) **Lendenfeld, R. von.** Die Chalineen des australischen Gebietes. Von...In: Zool. Jahrbüch. ii, p. 723–828, Pl. XVIII–XXVII.

1887 (η) **Lendenfeld, R. von.** Errata in my paper on the Systematic Position and Classification of Sponges. In: Zool. Anz. x, p. 335–336.

1887 (**Noll, F. C.** Ueber die Silicoblasten der Kieselschwämme.) Kurzer Bericht über die Sitzungen der vereinigten 5. und 9. Sektion für Zoologie und Anatomie der 60. Versamml. d. Naturf. und Aerzte. In: Anat. Anz. ii, p. 764–765, and Biol. Centralbl. vii, p. 767–768.

1887 (α) **Noll, F. C.** (Die Naturgeschichte der Kieselschwämme.) In: Ber. Senckenb. Ges. Frankfurt, p. 69–71.

1887 **Petr, F.** Nové dodatky ku fauně českých hub sladkovōdních...In: Sitz. Ber. Böhm. Ges. Prag, p. 203–214 (p. 121–132), 1 Pl.

1887 **Počta, P.** Über Spongiennadeln in einigen Gesteinen Ungarns. Von ...In: Földtani Közlöny, xvii, p. 107–114, Pl. i.

1887 **Potts, E.** Contributions towards a synopsis of the American forms of Fresh Water Sponges with descriptions of those named by other authors and from all parts of the world. By...In: Proc. Acad. Sc. Philadelphia, pp. i–iv, 157–279, Pl. v–xii. Reprinted and separ. publ. under the title "Fresh Water Sponges." A Monograph... N.B. "paging of the Proc....retained."

1887 **Ridley, S. O. & A. Dendy.** Report on the Monaxonida collected by H.M.S. Challenger during the years 1873–76. By...In: Rep. Scient. Results Challenger. Zoology. Vol. xx, Part lix. 4°. lxviii and 275 pp., Pl. i–li, and 1 map.

1887 **Schlüter, C.** Ueber Scyphia oder Receptaculites cornu-copiae Goldf. sp. und einige verwandte Formen. Von...In: Zeitschr. D. Geol. Ges. xxxix, p. 1–26, Pl. i–ii.

1887 (**Schulze, F. E.** Discussion Vortrag Noll.)...Bericht Sitz. 60. Versammlung Naturf. und Aerzte. In: Anat. Anz. ii, p. 765, and Biol. Centralbl. vii, p. 768.

1887 (α) **Schulze, F. E.** Report on the Hexactinellida collected by H.M.S. Challenger during the years 1873–1876. By...In: Rep. Scient. Results Challenger. Zoology. Vol. xxi, Part liii. 4°. 514 pp., Pl. i–civ, and 1 map.

1887 (β) **Schulze, F. E.** Zur Stammesgeschichte der Hexactinelliden. Von ...In: Abh. K. Preuss. Akad. Wiss. Berlin, Phys.-math. Cl.... p. 1–35.

1887 **S(ollas), W. J.** Art. Spongos, in: Encyclopaedia Britannica 9th Edition...p. 412–429, 26 woodcuts.

1887 **Thomson, J. A.** On the Structure of Suberites domuncula, Olivi (O. S.), together with a Note on Peculiar Capsules found on the Surface of Spongelia...In: Trans. R. Soc. Edinburgh, xxxiii, p. 241–245, Pl. xvi–xvii.

1887 **Topsent, E.** Sur les prétendus prolongements périphériques des Cliones...In: C. R. Paris, cv, p. 1188.

1887 **Weltner, W.** (Über die Spongillen der Spree und des Tegelsee's bei Berlin.) In: Sitz. Ber. Ges. Naturf. Fr. Berlin, p. 152–157*.

1887 **Wierzejski, A.** Bemerkungen über Süsswasser-Schwämme. Von... In: Zool. Anz. x, p. 122–126. Transl. in: Ann. and Mag. N. H. (5), xix, p. 298–302.

1887 (α) **Wierzejski, A.** Les Éponges d'eau douce de Galicie. Par... In: Arch. Slaves Biol. ii, p. 37–40.

1888 **Bidder, G.** Preliminary note on the Physiology of Sponges. By... In: Proc. Cam. Phil. Soc. vi, p. 5. Abstr. and rev. in: Zool. Jahresb. für 1888 (1890), p. 2.

1888 **Braun, M.** Faunistische Untersuchungen der Bucht von Wismar. In: Arch. der Freunde der Naturgesch. in Mecklenburg, Jahrg. 42, 1889, p. 57–84.

1888 **Colombo, A.** La Fauna sottomarina del Golfo di Napoli. Esplorazione sistematica eseguita dal tenente di vascello... In: Rivista Marittima Ott. Dic. 1887 (p. 1–107, Pl. i–vii).

1888 **Dendy, A.** Studies on the Comparative Anatomy of Sponges. i, On the genera Ridleia, n. gen., and Quasillina, Norman. By... In: Quart. Journ. Microsc. Sc. (2), xxviii, p. 513–529, Pl. xlii.

1888 **Fiedler, K. A.** Ueber Ei- und Spermabildung bei Spongilla fluviatilis ...In: Z. W. Z. xlvii, p. 85–128, Taf. xi–xii.

(1888 **Fristedt.** Vide 1885.)

1888 **Geinitz, E.** Receptaculitidae und andere Spongien der mecklenburgischen Silurgeschiebe. Von...In: Zeitschr. D. Geol. Ges. xl, I, p. 17–23.

1888 **Grieg, J. A.** Undersøgelser over dyrelivet i de vestlandske fjorde. Ved...In: Bergens Mus. Aarsber. f. 1887, No. 3, p. 1–13.

1888 **Hinde, G. J.** On some new Species of Urugaya, Carter, with Remarks on the Genus. By...In: Ann. and Mag. N. H. (6), ii, p. 1–12, Pl. iv.

1888 (α) **Hinde, G. J.** On the Chert and Siliceous Schists of the Permo-Carboniferous Strata of Spitzbergen, and on the Characters of the Sponges therefrom, which have been described by Dr E. von Dunikowsky. By...In: Geol. Mag. (3), v, p. 241–251, Pl. viii.

1888 (β) **Hinde, G. J.** A Monograph of the British Fossil Sponges, ii, p. 93–188, Pl. ix.

1888 **Katzer, F.** Spongienschichten im mittelböhmischen Devon (Hercyn), von...In: Sitz. Ber. K. Akad. Wiss. Wien, Math.-nat. Cl. xcvii, p. 300–310, Taf. (i).

1888 **Lang, A.** Ueber den Einfluss der festsitzenden Lebensweise auf die Thiere und über den Ursprung der ungeschlechtlichen Fortpflanzung durch Theilung und Knospung. Von...Jena. 8°. 166 pp.

* The paper was read Dec. 21, 1886, in a private Society; published Jan. 1887.

1888 **Lendenfeld, R. von.** Descriptive Catalogue of the Sponges in the Australian Museum, Sydney. By...(Publication of the Australian Museum.) London. 8°. xvi and 260 pp., xii Pl.

1888 **Lo Biancho, S.** Notizie biologiche riguardanti specialmente il periodo di maturità sessuale degli animali del golfo di Napoli. In: Mitth. Zool. Stat. Neapel, viii, p. 385–440. (Porifera, p. 386.)

1888 **MacMunn, C. A.** On the Chromatology of some British Sponges. By ...In: Journ. Phys. Cam. ix, p. 1–25, Pl. i.

1888 **Mills, H.** A new Freshwater Sponge (Heteromeyenia radiospiculata, n. sp.). By...In: Ann. and Mag. N. H. (6), i, p. 313–314.

1888 **Neumayr, M.** Die Stämme des Thierreiches. Von...Wirbellose Thiere. i. (Porifera, p. 210–237, fig. 35–44.)

1888 **Noll, F. C.** Beiträge zur Naturgeschichte der Kieselschwämme. i, Desmacidon Bosei Noll mit Hinweisen auf Craniella carnosa, Rüppell und Spongilla fragilis, Leidy. Von...In: Abh. Senckenb. naturf. Ges. Frankfurt, xv, Heft 2, p. 1–58, Pl. i–ii.

1888 **Priest, B. W.** On some Remarkable Spicules from the Oamaru Deposit. By...In: J. Quek. Micr. Club (2), iii, p. 254–256, Pl. xix.

1888 (α) **Priest, B. W.** On the Calcarea. By...In: J. Quek. Micr. Cl. (2), iii, p. 99–107, Pl. vii and viii.

1888 **Schneider, R.** Ueber Eisen-Resorption in thierischen Organen und Geweben. Von...In: Abh. Akad. Berlin (p. 1–68, Pl. i–iii).

1888 **Sollas, W. J.** Report on the Tetractinellida collected by H.M.S. Challenger, during the years 1873–1876. By...In: Report, Challenger, Zool., vol. xxv, pp. i–clxvi, 1–458, Pl. i–xliv, 1 map.

1888 (α) **Sollas, W. J.** A contribution to the history of flints. By...In: Proc. R. Dublin Soc. (2), vi, p. 1–5.

1888 **Topsent, E.** Contribution à l'étude des Clionides par...In: Arch. Zool. Expérim. (2), v bis, suppl. p. 1–165, Pl. i–vii*. The paper appeared also as Thèses présentées à la Faculté des Sciences de Paris...soutenues le (12. Juillet) 1888...Poitiers. 8°.

1888 (α) **Topsent, E.** Sur les gemmules de quelques Silicisponges marines ...In: C. R. Acad. Paris, cvi, p. 1298–1300.

1888 **Vosmaer, G. C. J.** Neuere Arbeiten über Schwämme. Von... i, Hyalospongiae; ii, Spiculispongiae u. Cornacuspongiae. In: Biol. Centralblatt, Bd viii, No. 2, pp. 38–45, 220–224.

1888 (**Weltner, W.** Ueber das Fortleben von Spongillen nach der Ausbildung von Schwärmlarven.) In: Sitz. Ber. Ges. Naturf. Fr. Berlin, p. 18–22. Tr. in: Ann. and Mag. N. H. (6), i, p. 172–184.

1888 **Wierzejski, A.** Beitrag zur Kenntniss der Süsswasserschwämme. Von...In: Verh. K. K. Zool. Bot. Ges. Wien, 38, p. 529–536, Pl. xii.

* Appeared 1888 and not 1887.

1889 **Auchenthaler, F.** Ueber den Bau der Rinde von Stelletta grubii, O. S. Von...In: Ann. K. K. naturhist. Hofmus. iv, p. 1–6, Taf. I.

1889 **Bucher, Ch. E.** Brachiospongidae; on a group of Silurian Sponges. New Haven. 28 pp., 6 Pl.

1889 **Carter, H. J.** A Correction in British Spongology. By...In: Ann. and Mag. N. H. (6), iv, p. 249–250.

1889 (α) **Carter, H. J.** Sketch of the History of known Fossil Sponges in Relation to those of the Present Day. By...In: Ann. and Mag. N. H. (6), iv, p. 280–290.

1889 (β) **Carter, H. J. & R. Hope.** On a new British Species of Microciona, Bk., in which the ends of the Tricurvate are Spiniferous, etc. By...In: Ann. and Mag. N. H. (6), iii, p. 99–106, Pl. VI.

1889 **Dawson, J. W. & G. J. Hinde.** On New Species of Fossil Sponges from the Siluro-Cambrian at Little Metis on the Lower St Lawrence. By...In: Trans. R. Soc. Canada, vii, Sect. iv, p. 31–55, Pl. III.

1889 **Dendy, A.** Studies on the Comparative Anatomy of Sponges. ii, On the Anatomy and Histology of Stelospongus flabelliformis, Carter; with notes on the Development. By...In: Quart. Journ. Microsc. Sc. (2), xxix, p. 325–358, Pl. XXX–XXXIII.

1889 (α) **Dendy, A.** An alphabetical list of the Genera and Species of Sponges described by H. J. Carter, Esq., F.R.S., together with a number of his more important references to those of other authors, with an introductory notice. By...In: Proc. R. Soc. Victoria, N. S. i, p. 34–59.

1889 (β) **Dendy, A.** Report on a Second Collection of Sponges from the Gulf of Manaar. By...In: Ann. and Mag. N. H. (6), iii, p. 73–99, Pl. III–V.

1889* **Girod, P.** Les éponges des Eaux douces d'Auvergne. Par...In: Trav. Labor. Zool. Girod, i, p. 1–11, Pl. I.

1889 (α) **Girod, P.** Les Spongilles; leur recherche, leur préparation, leur détermination...In: Rev. Scient. Bourb. ii, p. 16–25, Pl. I.

1889 **Haeckel, E.** Report on the Deep-Sea Keratosa collected by H.M.S. Challenger during the Years 1873–1876. By...In: Rep. Scient. Results Challenger, Zoology, vol. xxxii, Part lxxxii. 4°. 92 pp., 8 Pl.

1889 **Hanitsch, R.** Second Report on the Porifera of the L. M. B. C. District. By...In: Proc. Biol. Soc. Liverpool, iii, p. 155–173, Pl. V–VII.

1889 **Hinde, G. J.** On a true Leuconid Calcisponge from the Middle Lias of Northamptonshire, and on detached Calcisponge Spicules in the Upper Chalk of Surrey. By...In: Ann. and Mag. N. H. (6), iv, p. 352–358, Pl. XVII.

* According to Zool. Anz.

1889 (α) **Hinde, G. J.** On Archaeocyathus, Billings, and on other Genera, allied to or associated with it, from the Cambrian Strata of North America, Spain, Sardinia, and Scotland. By...In: Quart. Journ. Geol. Soc. London, xlv, p. 125–148, Pl. v.

1889 (β) **Hinde, G. J.** On some Fossil Siliceous Sponges from the Quebec Group of Little Métis, Canada...In: Quart. Journ. Geol. Soc. London, xlv, p. 84–85.

1889 (γ) **Hinde, G. J.** On the nature of some fragments of Siliceous Rock from the boulderclay of the "Roode Klif" (Red Cliff) on the southern border of the province of Friesland. By...(Append. to H. v. Cappelle, Les escarpements du "Gaasterland.") In: Bull. Soc. Belge Géol. iii, p. 254–258, Pl. viii.

1889 **Hope, R.** On two new British Species of Sponges, with short notices of an Ovigerous Specimen of Hymeniacidon Dujardinii, Bowk., and of a Fossil Toxite. By...In: Ann. and Mag. N. H. (6), iv, p. 333–342.

1889 **Keller, C.** Die Spongienfauna des rothen Meeres. Von...(1 Hälfte.) In: Zeitschr. wiss. Zool. xlviii, p. 311–405, Pl. xx–xxv. Abstr. and rev. in: Journ. R. Microsc. Soc. London...(1890), p. 192– .

1889 **Kirkpatrick, R.** (Sponges) in W. S. Green, Report of a Deep-sea Trawling Cruise off the S.W. Coast of Ireland. By...In: Ann. and Mag. N. H. (6), iv, p. 446–447.

1889 **Korschelt, E.** Ueber den Bau und das System der Glasschwämme (Hexactinelliden), nebst ihren Beziehungen zu den übrigen Spongien...In: Naturwiss. Wochenschr. iii, p. 171–173.

1889 **Krukenberg, C. F. W.** La rétention de l'urée chez les Sélaciens, avec quelques remarques sur l'accumulation d'autres substances cristalloïdes dans les tissus contractiles de certaines espèces animales. Par...In: Ann. Mus. N. H. Marseille, iii, Mém. 3, p. 1–43, 2 tables.

1889 **Leidy, J.** The Boring Sponge, Cliona. By...In: Proc. Acad. Nat. Sc. Philadelphia, 1889 (1890), p. 70–75.

1889 **Lendenfeld, R. von.** A Monograph of the Horny Sponges. London. 4º. 936 pp., 50 Pl.

1889 (α) **Lendenfeld, R. von.** Das System der Spongien. Von...In: Biol. Centralbl. ix (1889–90), p. 113–127.

1889 (β) **Lendenfeld, R. von.** Fortschritt unsrer Kenntnis der Spongien. Von...In: Zool. Jahrb. iv, Abt. f. Syst. p. 453–484.

1889 (γ) **Lendenfeld, R. von.** Experimentelle Untersuchungen über die Physiologie der Spongien. Von...In: Zeitschr. wiss. Zool. xlviii, p. 406–700, Pl. xxvi–xl.

1889 (δ) **Lendenfeld, R. von.** Notiz über den Bau der Geisselkammern der Spongien. Von...In: Zool. Anz. xii, p. 361–362.

1889 (ε) **Lendenfeld, R. von.** Die Verwandtschaftsverhältnisse der Hornschwämme. Von...In: Zool. Jahrbüch. Syst. Abth. iv, p. 1–93.

1889 **Maas, O.** Zur Metamorphose der Spongillalarve. Von...In: Zool. Anz. xii, p. 483–487, 6 figs.

1889* **Mackay, A. H.** Freshwater-Sponges of Canada and Newfoundland ...By...In: Trans. R. Soc. Canada, vii, Sect. iv, p. 85–95, Pl. IV.

1889 **Maisonneuve, P.** La Faune marine des côtes de Belle-Isle-en-Mer... In: Bull. Biblioth. Scient. de l'Ouest, Année ii, p. 105.

1889 **Marchal, P.** L'acide urique et la fonction rénale chez les Invertébrés. Par...In: Mém. Soc. Zoologique France, iii, p. 31–87.

1889 **Marenzeller, E. von.** Ueber die adriatischen Arten der Schmidt'schen Gattungen Stelletta und Ancorina. Von...In: Ann. K. K. naturhist. Hofmus. Wien, iv, p. 7–20, Pl. II–III.

1889 **Meunier, S.** Sur la Spongeliomorpha Saportai, espèce nouvelle parisienne...In: C. R. Paris, cix, p. 536–537.

1889 **Poléjaeff, N.** Ueber Korotnewia desiderata und die Phylogenie der Hornschwämme. Von...In: Zool. Anz. xii, p. 366–367.

1889 **Potts, E.** Report upon some Fresh-water Sponges collected in Florida by Jos. Willcox, Esq. By...In: Trans. Wagner Free Instit. ii, p. 5–7.

1889 **Richard, J.** Note sur les pêches effectuées par Ch. Rabot dans les lacs Enara, Imandra et dans le Kolozero. Par...In: Bull. Soc. Zool. France, xiv, p. 100–104.

1889 **Schulze, F. E. & R. von Lendenfeld.** Ueber die Bezeichnung der Spongiennadeln. Von...In: Abh. Akad. Wiss. Berlin (p. 1–35).

1889 **Sollas, W. J.** On the Geodine Genera, Synops, Vosm., and Sidonops. A correction. By...In: Proc. R. Dublin Soc. (2), vi, p. 276–277.

1889 **Topsent, E.** Quelques Spongiaires du Banc de Campêche et de la Pointe-à-Pître, par...In: Mém. Soc. Zool. France, ii, p. 30–52. Review: Zool. Jahresber. Neapel für 1889, p. 3.

1889 (α) **Topsent, E.** Cliona celata ou Cliona sulphurea? Par...In: Bull. Soc. Zool. France, xiv, p. 351–353.

1889† (β) **Topsent, E.** Notes spongologiques. Par...In: Arch. Zool. Expérim. (2), vi, p. xxxiii–xliii. Rev. and abstr. in: Zool. Rec. Jahresber.

1889 (γ) **Topsent, E.** Additions à la Faune des Spongiaires de Luc. Par... In: Bull. Soc. Linn. Norm. (4), ii, p. 53–60.

1889 **Vosmaer, G. C. J.** Note on the Metamorphosis of the Sponge-larva. By...In: Tijdschr. Nederl. Dierk. Ver. (2), ii, p. 287–289, Pl. XIV.

1889 (α) **Vosmaer, G. C. J.** Verslag van de werkzaamheden door den onder-geteekende aan de Nederlandsche werktafel in het Zoölogisch Station te Napels verricht. Nov. 1888–Jan. 1889. In: Nederl. Staatscourant.

* "Zool. Record" for 1891 (Minchin) dates publication 1890.—G. P. B.
† According to the "Zool. Anzeiger," xii, p. 349, the 3rd part of Arch. Zool. Expérim. (2), vi, in which the article of Topsent is printed, appeared in 1889.

1889 (β) **Vosmaer, G. C. J.** Neue Arbeiten über Schwämme. In: Biol. Centralbl. ix, p. 405–414.

1889 **Wisniowski, T.** Nowy przyczynek do znajomości górno-jurajskich Monactinellidów i Tetractinellidów...In: Kosmos (Lemberg), xiv, pp. 185–189, 230–237, Taf...

1890 **Beyerinck, M. W.** Culturversuche mit Zoochlorellen, Lichenengonidien und anderen niederen Algen. Von...In: Bot. Zeit. xlviii, pp. 725–739, 741–754, 757–771, 781–785.

1890 **Bouchon-Brandely & A. Berthoule.** Les Pêches maritimes en Algérie et en Tunisie In: Rev. Marit. et Colon.

1890 **Carter, H. J.** (Porifera in: Ridley, Notes on the Zoology of Fernando Noronha.) In: Journ. Linn. Soc. London, xx, p. 564–569.

1890 **Chatin, J.** Contribution à l'étude du noyau chez les Spongiaires... In: C. R. Paris, cxi, p. 889–890.

1890 **Delage, Y.** Sur le développement des Éponges siliceuses et l'homologation des feuillets chez les Spongiaires. Note de M...présentée par M. de Lacaze-Duthiers. In: C. R. Acad. Paris, cx, p. 654–657.

1890 **Dendy, A.** Observations on the West-Indian Chalinine Sponges, with Descriptions of new Species. By...In: Trans. Zool. Soc. London, xii, p. 349–368, Pl. LVIII–LXIII. Abstr. and rev. in: Journ. R. Microsc. Soc. 1890, p. 344; Zool. Rec. xxvii (1892), pp. 2–3, 9; Zool. Jahresber. f. 1890 (1892), pp. 1, 5.

1890 (α) **Dendy, A.** Some old and new Questions concerning Sponges. By ...In: Zool. Anz. xiii, p. 14–17.

1890 (β) **Dendy, A.** On the Pseudogastrula Stage in the Development of Calcareous Sponges. By...In: Proc. R. Soc. Victoria...p. 93–101, Pl. I A.

1890 (γ) **Dendy, A.** Zoology: Invertebrata...In: Handbook for Victoria, Melbourne, p. 74–96.

1890 **Faurot, L.** La pêche des éponges dans le golfe de Gabès...In: Rev. Scient. xlv, p. 428–431.

1890 **Fol, H.** Sur l'anatomie des éponges cornées du genre Hircinia et sur un genre nouveau...In: C. R. Acad. Paris, cx, p. 1209–1211, and in: Ann. and Mag. N. H. (6) vi, p. 172–174.

1890 **Giard, A.** Le laboratoire de Wimereux en 1889 (recherches fauniques), par...In: Bull. Sc. France Belg. xxii, p. 60–87.

1890 **Hanitsch, R.** Third Report on the Porifera of the L. M. B. C. District. By...In: Trans. Biol. Soc. Liverpool, vol. iv, p. 192–238, Pl. x–xv. Review: Z. Jahresber. für 1890 (1892), Porifera, p. 2.

1890 **Hinde, G. J.** On a new Genus of Siliceous Sponges from the Lower Calcareous Grit of Yorkshire. By...In: Quart. Journ. Geol. Soc. London, xlvi, p. 54–61, Pl. VI.

1890 **Korschelt, E. & K. Heider.** Lehrbuch der vergleichenden Ent wickelungsgeschichte der wirbellosen Thiere. i, Porifera, p. 1–18. With 12 woodcuts.

1890 **Lendenfeld, R. von.** Das System der Spongien von...In: Abh. Senckenb. Naturf. Ges. Frankfurt, xvi, p. 361–439.

1890 (α) **Lendenfeld, R. von.** Bemerkung zu dem Schlüssel der Spongiennadeln. Von...In: Biol. Centralbl. x, p. 550–551.

1890 (β) **Lendenfeld, R. von.** Schlüssel zur Bestimmung der Spongiennadeln. Von...In: Biol. Centralbl. x, p. 131–135.

1890 (γ) **Lendenfeld, R. von.** Eine Bemerkung über Synonymie und Nomenclatur. Von...In: Zool. Anz. xiii, p. 115–116.

1890 (δ) **Lendenfeld, R. von.** Experimentelle Untersuchungen über die Physiologie der Spongien. Von...In: Biol. Centralbl. x, pp. 71–81, 102–110.

1890 (ε) **Lendenfeld, R. von.** Die Lebenserscheinungen des Badeschwammes. Von...In: Zool. Garten, xxxi, p. 97–104.

1890 (ζ) **Lendenfeld, R. von & F. E. Schulze.** Die Gattung Stelletta...In: Anh. Abh. K. Pr. Akad. Wiss. Berlin f. 1889, p. 1–75, Pl. i–x (sep. copy).

1890 (η) **Lendenfeld, R. von.** Fortschritt unsrer Kenntniss der Spongien. iii. Von...In: Zool. Jahrb. v, Abth. f. Syst. p. 169–178.

1890 **Maas, O.** Ueber die Entwicklung des Süsswasserschwammes. Von ...In: Zeitschr. wiss. Zool. l, p. 527–554, Pl. xxii–xxiii.

1890 (α) **Maas, O.** Ueber die Entwicklung des Süsswasserschwamms. Inaug. Diss. von...Berlin, p. 1–42.

1890 **Marenzeller, E. von.** Deutsche Benennungen für Poriferen, Coelenteraten, Echinodermen und Würmer. Von...In: Verh. Zool. bot. Ges. Wien, xl, p. 177–184.

1890 **Pérot, F.** Les Spongiaires fossiles de l'Allier et du bassin de la Loire. In: Rev. Sc. Bourbonnais, ii, p. 263–267, Pl. vii.

1890 **Počta, P.** Ueber einige Spongien aus dem Cuvieri-Pläner von Paderborn. Von...In: Zeitschr. D. Geol. Ges. xlii, p. 217–232, Pl. vi–viii.

1890 **Ridley & Carter.** See Carter.

1890 **Schulze, F. E.** (and Lendenfeld). Vide 1890 (ζ), Lendenfeld.

1890 **Sovinsky, W.** Sur les Spongillidae du Dnièpre. In: Rev. Sc. Nat. Pétersbourg, i, p. 199.

1890 **Topsent, E.** Éponges de la Manche, par...In: Mém. Soc. Zool. France, iii, p. 195–205.

1890 (α) **Topsent, E.** Notice préliminaire sur les Spongiaires recueillis durant les campagnes de l'Hirondelle...Par...In: Bull. Soc. Zool. France, xv, pp. 26–32, 65–71.

1890 (β) **Topsent, E.** Études de Spongiaires, par...In: Rev. Biol. Nord France, ii, p. 289–298.

1890 **Ulrich, E. O.** American Palaeozoic Sponges. By...In: Geol. Survey, Illinois, viii. Part ii, Palaeontology of Illinois, Sect. iii, p. 209–241, 10 woodcuts.

1890 (α) **Ulrich, E. O.** Sponges of the Devonian and Carboniferous Systems. By...In: Geol. Survey, Illinois, viii. Part ii, Palaeontology of Illinois, Sect. iv, p. 243–251, Pl. vi, fig. 2–3; Pl. iii, fig. 2; Pl. vii, fig. 4.

1890 **Ulrich, E. O. & Olivier Everett.** Descriptions of Lower Silurian Sponges. By...In: Geol. Survey, Illinois, viii. Part ii, Palaeontology of Illinois, Sect. v, p. 253–281, Pl. i–viii.

1890 **Vosmaer, G. C. J.** Notes on some Species of Stelletta and other Genera allied to it. By...In: Tijdschr. Ned. Dierk. Ver. (2), iii, p. 35–37.

1890 **Weber, M.** Spongillidae des Indischen Archipels. Von...In: Zool Ergebn. Reise Niederl. Ost-Indien, p. 30–47, Pl. iv.

1890 **Weber, M. & Mm. A. Weber-van Bosse.** Quelques nouveaux cas de symbiose. Par...In: Zool. Ergebn. Reise Niederl. Ost-Indien, p. 48–72, Pl. v.

1890 **Zykow, W.** Notice sur les Spongillides des environs de Moscou. Par...In: Bull. Soc. Imp. Natural. Moscou (2), iv, p. 170–172.

1891 **Bidder, G. (P.).** (Review of 1891, Dendy.) In: Quart. Journ. Microsc. Sc. (2), xxxii, p. 625–632*.

1891 **Delage, Y.** Sur le développement des éponges (Spongilla fluviatilis). In: C. R. Acad. Paris, cxiii, p. 267–269.

1891 **Dendy, A.** A Monograph of the Victorian Sponges, by...Part i, The Organisation and Classification of the Calcarea Homocoela, with Descriptions of the Victorian Species...In: Trans. R. Soc. Victoria, iii, p. 1–81, Pl. 1–11. Abstr. and rev. in: Journ. R. Microsc. Soc. London, p. 610–611; Quart. Journ. Microsc. Sc. (2), xxxii, p. 625–632.

1891 (α) **Dendy, A.** Studies on the Comparative Anatomy of Sponges. iii, On the Anatomy of Grantia labyrinthica, Carter, and the so-called Family Teichonidae. By...In: Quart. Journ. Microsc. Sc. (2), xxxii, p. 1–40, Pl. i–iv.

1891 (β) **Dendy, A.** Studies on the Comparative Anatomy of Sponges. iv, On the Flagellated Chambers and Ova of Halichondria panicea. By...In: Quart. Journ. Microsc. Sc. (2), xxxii, p. 41–48, Pl. v.

1891 (γ) **Dendy, A.** Preliminary Account of Synute pulchella, a New Genus and Species of Calcareous Sponges. By...In: Proc. R. Soc. Victoria, p. 1–6.

1891 **Famintzin, A.** Beitrag zur Symbiose von Algen und Thieren. Von... In: Mém. Acad. Pétersbourg (7), xxxviii (p. 1–16, 1 Pl.).

* Reprinted under the title: Notes on Calcareous Sponges, p. 1–8.

1891 **Grentzenberg, M.** Die Spongienfauna der Ostsee. Inaugural-Dissertation zur Erlangung der Doktorwürde der philosophischen Fakultät der Christian-Albrechts-Universität zu Kiel vorgelegt von...Kiel. 8⁰. 54 pp. and 1 Pl.

1891 **G(roult), P.** Les Éponges...In: Le Naturaliste, xiii, p. 290–292.

1891 **Hanitsch, R.** Notes on Some Sponges collected by Professor Herdman off the West Coast of Ireland from the "Argo"...In: Trans. Biol. Soc. Liverpool, v, p. 213–222, Pl. xi–xii.

1891 **Harvey, A.** Pelotechten balanoides. By...In: Trans. Canad. Inst. i, p. 213–215.

1891 **Jennings, A. V.** On a Variety of Alectona Millari (Carter). By... In: Journ. Linn. Soc. London, xxiii, p. 531–539, Pl. xiii.

1891 **Keller, C.** Das Spongin und seine mechanische Leistung im Spongienorganismus. Von...In: Festschr. 50. jähr. Doctor-Jub. Nägeli u. Kölliker, p. 151–160, 1 Pl. Also separ. Zürich, Müller. Fol. 12 pp., 1 Pl.

1891 (α) **Keller, C.** Die Spongienfauna des rothen Meeres. Von...(2. Hälfte)...In: Z. W. Z. lii, p. 294–368, Taf. xvi–xx.

1891 (**Leidy, J.**) Note on the Boring Sponge of the Oyster. In: Proc. Acad. Nat. Sc. Philadelphia, 1891*, p. 122.

1891 **Lendenfeld, R. von.** Die Spongien der Adria. i, Die Kalkschwämme. Von...In: Zeitschr. wiss. Zool. liii, pp. 185–321, 361–433, Pl. viii–xv.

1891 (α) **Lendenfeld, R. von.** Das System der Kalkschwämme. (Vorläufige Mittheilung.) von...In: Sitzungsber. K. Akad. Wiss. Wien, Math.-naturw. Cl. c, p. 4–19.

1891 (β) **Lendenfeld, R. von.** Ueber die Kieselnadeln von Geodia...In: Zool. Anz. xiv, p. 407–409.

1891 (γ) **Lendenfeld, R. von.** Bemerkungen über die Spongien im Kanal von Lesina. Von...In: Zool. Garten, xxxii, p. 263–265.

1891 **Murray, J. & R. Irvine.** On Silicia and the Siliceous Remains of Organisms in Modern Seas. By...In: Proc. R. Soc. Edinburgh, xviii, p. 229–250.

1891 **Rauff, H.** (Ueber den Bau des Stützskeletes bei den Anomocladinen, sowie einiger Tetracladinen.) In: Verh. naturh. Ver. preuss. Rheinl. xlviii, p. 33–37.

1891 (α) **Rauff, H.** Vorläufige Mittheilung über das Skelet der Anomocladinen, sowie über eine eigenthümliche Gruppe fossiler Kalkschwämme (Polysteganinae), die nach dem Syconen-Typus gebaut sind. Von...In: N. Jahrb. Mineral. i, p. 278–284.

1891 (β) **Rauff, H.** Ueber Palaeospongia prisca, Bornem., Eophyton z. Th., Chondrites antiquus, Haliserites z. Th. und ähnliche Gebilde. Von...In: N. Jahrb. Mineral. ii, p. 92–104.

* On the title-page is printed 1892; but it is stated in the book that p. 1–440 were published in 1891.

1891 (γ) **Rauff, H.** [Ueber eine eigenthümliche Gruppe fossiler Kalk-schwämme (Polysteganinae).] In: Verh. naturh. Ver. preuss. Rheinl. xlviii, p. 45–50.

1891 **Schimkewitsch, W.** Versuch einer Klassifikation des Thierreichs. Von...In: Biol. Centralbl. xi, p. 291–295.

1891 **Thiele, J.** Die Stammesverwandtschaft der Mollusken. Ein Beitrag zur Phylogenie der Thiere. Von...In: Jen. Zeitschr. xxv, p. 480–543.

1891 **Topsent, E.** Essai sur la Faune des Spongiaires de Roscoff par...In: Arch. Zool. Exp. (2), ix, p. 523–554, Pl. xxii, fig. 1–8.

1891 (α) **Topsent, E.** Deuxième Contribution à l'Étude des Clionides par ...In: Arch. Zool. Expérim. (2), ix, p. 555–592, Pl. xxii, fig. 9–17.

1891 (β) **Topsent, E.** Voyage de la Goëlette "Melita" aux Canaries et au Sénégal...Spongiaires. Par...In: Mém. Soc. Zool. France, iv, p. 11–15, Pl. ii.

1891 (γ) **Topsent, E.** Spongiaires des Côtes Océaniques de France. Par... In: Bull. Soc. Zool. France, xvi, p. 125–129.

1891 (δ) **Topsent, E.** Sur la distribution géographique de quelques Micro-sclerophora, par...In: Bull. Soc. Zool. France, xv, p. 231–233.

1891 **Vaughan-Jennings.** See Jennings.

1891 **Weltner, W.** Die Süsswasserschwämme...In: Zacharias, Die Thier-und Pflanzenwelt des Süsswassers, i, p. 185–236. With woodcuts 36–46.

1891 **Wilson, H. V.** Notes on the Development of some Sponges...In: Journ. Morphol. Boston, v, p. 511–519. Abstr. and rev. in: Journ. R. Microsc. Soc. London, 1892, p. 217; Zool. Jahresber. f. 1891 (1893), pp. 2, 5.

1892 **Appellöf, A.** Om Bergensfjordenes faunistiske praeg. Ved...In: Bergens Mus. Aarsberetn. f. 1891, No. 2, p. 1–14.

1892 **Bidder, G. (P.).** Note on Excretion in Sponges. By...In: Proc. R. Soc. London, li, p. 474–485. With 4 woodcuts.

1892 (α) **Bidder, G. (P.).** On the Flask-shaped Ectoderm and Spongoblasts in one of the Keratosa. By...In: Proc. R. Soc. lii, p. 134–139. With 3 woodcuts.

1892 **Buck, E.** Einiges über den Bodenseeschwamm...In: D. Fischerei Zeit. xv, p. 366–367.

1892 **Chworostansky, O.** Ueber die Zonen des Küstenstriches der Solowezki-Inseln...Von...In: Zool. Anz. xv, p. 214–215.

1892 **Delage, Y.** Embryogénie des Éponges. Développement post-larvaire des Éponges siliceuses et fibreuses marines et d'eau douce...par...In: Arch. Zool. Expérim. (2), x, p. 345–498, Pl. xiv–xxi*.

* According to Maas (1892 (α), p. 332) this paper is published "Juni 1892"; accord-ing to Delage (1899, p. 62, footnote) "au commencement de 1893."

1892 **Dendy, A.** The Discovery of the true Nature of the so-called family Teichonidae. By...In: Zool. Anz. xv, p. 245–246.

1892 (α) **Dendy, A.** Synopsis of the Australian Calcarea Heterocoela; with a proposed Classification of the Group and Descriptions of some New Genera and Species. By...In: Proc. R. Soc. Victoria, p. 69–116.

1892 (**Doederlein, L.** Ueber Petrostroma schulzei, n. g., n. sp. der Kalkschwämme.) In: Verh. deutsch. Zool. Gesellsch. ii, p. 143–145.

1892 **Gourret, P.** Notes zoologiques sur l'étang des Eaux-Blanches (Cette). ...In: Ann. Mus. H. N. Marseille, iv, mém. 2, 26 pp.

1892 **Griffiths, A. B.** The Physiology of the Invertebrata. In: Nature, xlvi, p. 414–415.

1892 **Hanitsch, R.** Foraminifer or Sponge?...In: Nature, xlvii, p. 439.

1892 **Hinde, G. J. & W. M. Holmes.** On the Sponge-Remains in the Lower Tertiary Strata near Oamaru, Otago, New Zealand. By...In: Journ. Linn. Soc. London, xxiv, p. 177–262, Pl. 7–15*.

1892 **Hornell, J. A.** A Strange Commensalism-Sponge and Annelid. In: Nature, xlvii, p. 78.

1892 **Lendenfeld, R. von.** Bemerkung über die Homodermidae. Von... In: Zool. Anz. xv, p. 109.

1892 (α) **Lendenfeld, R. von.** Ueber Minchin's Angaben betreffs der Histologie der Kalkschwämme. Von...In: Zool. Anz. xv, p. 277–279.

1892 (β) **Lendenfeld, R. von.** Berichtigung. Von...In: Zool. Anz. xv, p. 370.

1892 (γ) **Lendenfeld, R. von.** Note on Mr Minchin's Paper on Ascetta. By...In: Ann. and Mag. N. H. (6), ix, p. 337.

1892 (δ) **Lendenfeld, R. von.** Bemerkungen über die neuerlich von Dendy beschriebenen Kalkschwämme. Von...In: Biol. Centralbl. xii, p. 58–60.

1892 (ε) **Lendenfeld, R. von.** Die Spongien der Adria. ii, Die Hexaceratina. Von...In: Z. W. Z. liv, p. 275–315, Pl. XIII.

1892 (ζ) **Lendenfeld, R. von.** Note on Dr Hinde's Tertiary Sponge-spicules. By...In: Ann. and Mag. N. H. (6), x, p. 268.

1892 **Maas, O.** Die Metamorphose von Esperia lorenzi, O. S., nebst Beobachtungen an andern Schwammlarven. Von...In: Mitth. Zool. Stat. Neapel, x, p. 408–440, Pl. 27–28. Abstr. and rev. in: Zool. Jahresb. für 1892 (1893), pp. 1, 5.

1892 (α) **Maas, O.** Die Auffassung des Spongienkörpers und einige neuere Arbeiten über Schwämme. Von...In: Biol. Centralbl. xii, p. 566–572. Transl. in: Ann. and Mag. N. H. (6), x, p. 399–405.

1892 **Marshall, W.** Spongiologische Beiträge von...In: Festschrift 70. Wiederk. Geburtst. Leuckart. Also separ. Leipzig, Winter. 4°. 36 pp. and 8 Pl.

* The article is read Febr. 1891; published 1892; on the title-page of Vol. xxiv is printed 1894 (cf. p. ii: "Dates of Publication...").

176

1892 **Minchin, E. A.** Note on a Sieve-like Membrane across the Oscula of a Species of Leucosolenia, with some Observations on the Histology of the Sponge. By...In: Quart. Journ. Microsc. Sc. (2), xxxiii, p. 251–272, Pl. x–xi.

1892 (a) **Minchin, E. A.** The Oscula and Anatomy of Leucosolenia clathrus, O.S. By...In: Quart.J.Micr.Sc.(2), xxxiii, p. 477–495, Pl.xxix.

1892 (β) **Minchin, E. A.** Some Points in the Histology of Leucosolenia (Ascetta) clathrus, O. S. By...In: Zool. Anz. xv, p. 180–184. With 3 figs.

1892 (γ) **Minchin, E. A.** Dr. von Lendenfeld als Kritiker. Von...In: Zool. Anz. xv, p. 415–417.

1892 (δ) **Minchin, E. A.** Dr. von Lendenfeld on the Central Cavity in Euplectella. By...In: Ann. and Mag. N. H. (6), ix, p. 408–409.

1892 (**Norman, A. M.**) Museum Normanianum, or a Catalogue of the Invertebrata of Europe, and the Arctic and North Atlantic Oceans, which are contained in the collection of...(Printed for private distribution.) Durham. 21 pp.

1892 **Ott, Ch.** La Pêche de l'Éponge. In: Bull. Soc. Sc. Basse-Alsace. Tome xxvi, p. 225–228.

1892 **Pearcey, F. G.** Foraminifer or Sponge?...In: Nature, xlvii, p. 390.

1892 **Počta, P.** Ueber Spongien aus der oberen Kreide Frankreichs in dem Kgl. mineralogischen Museum in Dresden. Mit Vorwort von H. B. Geinitz...In: Mitth. K. miner. Geol. Mus. Dresden, xi. 26 pp., Pl. i–iv. Also separ. Cassel, Fischer. 4°.

1892 **Rauff, H.** Changed to 1891 (γ).

1892 (a) **Rauff, H.** (Ueber den Fossilisationsprocess derjenigen verkieselten Spongien, deren Hauptmasse jetzt aus Chalcedon gebildet wird, während das Skelet darin aus Kalkspath besteht, obschon das lebende Thier ein rein kieseliges Gerüst besass.) In: Verh. Nat. Ver. Bonn, xlix, Korresp.-bl. p. 51–57.

1892 (β) **Rauff, H.** Untersuchungen über die Organisation und systematische Stellung der Receptaculitiden. Von...In: Abh. Akad. München, xvii, p. 645–722. 12 woodcuts and 7 Pl.

1892 **Topsent, E.** Contribution à l'étude des Spongiaires de l'Atlantique Nord...In: Résultats des Campagnes scientifiques accomplies sur son Yacht par Albert Ier Prince Souverain de Monaco... Fasc. ii, avec onze planches. 4°. 165 pp.

1892 (a) **Topsent, E.** Diagnoses d'éponges nouvelles de la Méditerranée et plus particulièrement de Banyuls; par...In: Arch. Zool. Expér. (2), x, Notes et revue, p. xvii–xxviii. Review: Zool. Jahresber. f. 1892 (1893), Porifera, p. 4.

1892 (β) **Topsent, E.** Éponges de la Mer Rouge, par...In: Mém. Soc. Zool. France, v, p. 21–29, Pl. i.

1892 (γ) **Topsent, E.** Notes histologiques au sujet de Leucosolenia coriacea (Mont.), Bwk. par...In: Bull. Soc. Zool. France, xvii, p. 125–129. With 5 woodcuts.

1892 (δ) **Topsent, E.** Sur une Éponge du Lac de Tibériade Potamolepis Barroisi, n. sp. Par...In: Rev. Biol. Nord France, Lille, v, p. 85–91, Pl. ii.

1892 (ε) **Topsent, E.** Exposé des Principes actuels de la Classification des Spongiaires. Par...In: Rev. Biol. Nord France, iv, pp. 281–300, 411–414, 457–462, Pl. xi–xii.

1892 **Vosmaer, G. C. J.** On the Canal System of the Homocoela and on the Morphological value of the terms Osculum and Pore in Sponges. By...In: Tijdschr. Ned. Dierk. Ver. (2), iii, p. 235–242. Abstr. and rev. in: Journ. R. Microsc. Soc. London, 1893, p. 195–196.

1892 **Wierzejski, A.** Ueber das Vorkommen von Carterius Stepanovii, Petr, und Heteromeyenia repens, Potts, in Galizien. Von...In: Biol. Centralbl. xii, p. 142–145.

1892 **Zykoff, W.** Entwicklungsgeschichte von Ephydatia Mülleri, Liebk. aus den Gemmulae. Eine biologisch-embryologische Skizze. Von...In: Biol. Centralbl. xii, p. 713–716.

1892 (α) **Zykoff, W.** Die Entwicklung der Gemmulae der Ephydatia fluviatilis, Auct. (Vorläufige Mittheilung.) Von...In: Zool. Anz. xv, p. 95–96. Cf. 1892 (β).

1892 (β) **Zykoff, W.** Die Entwickelung der Gemmulae bei Ephydatia fluviatilis, Auct. Von...In: Bull. Soc. Natur. Moscou (2), vi, p. 1–16, Pl. i–ii.

1893 **Berghaus,** . Korallen- und Schwammfischerei in Italien im Jahre 1891. Von...In: Mitth. D. Fisch. Ver. Sect. Küst. Hochsee Fisch. p. 47–49.

1893 **Celesia, P.** Della Suberites domuncula e della sua simbiosi coi paguri ...In: Atti Soc. Ligustica Sc. Nat. Genova, iv, Tav. i–iv. Also separ.: Genova, Ciminago. 63 pp., 4 Tavv.

1893 **Chopin, A.** Something about Sponges. By...In: Amer. Monthly Microsc. Journ. xiv, p. 342–347.

1893 **Delage, Y.** Note additionnelle sur l'embryogénie des Éponges. Par... In: Arch. Zool. Expérim. (3), i, Notes et Revue, p. iii–vi.

1893 **Dendy, A.** Studies on the Comparative Anatomy of Sponges. v, Observations on the Structure and Classification of the Calcarea Heterocoela. By...In: Quart. Journ. Microsc. Sc. (2), xxxv, p. 159–257, Pl. 10–14. Abstr. and rev. in: Journ. R. Microsc. Soc. London, p. 745–746.

1893 (α) **Dendy, A.** On a New Species of Leucosolenia from the neighbourhood of Port Phillip Heads. By...In: Proc. R. Soc. Victoria, p. 178–180.

1893 (β) **Dendy, A.** Note on the History of the so-called family Teichonidae By...In: Zool. Anz. xvi, p. 43–44. Also in: Ann. and Mag. N. H. (6), xi, p. 50–52.

1893 **Granger, A.** L'Éponge maisonnette, Suberites domuncula...In: Naturaliste, xv, p. 80–81.

1893 **Hanitsch, R.** Foraminifer or Sponge? In: Nature, xlvii, p. 365.

1893 (α) **Hanitsch, R.** Foraminifer or Sponge? In: Nature, xlvii, p. 439.

1893 **Heider, K.** Poriferen. In: E. Korschelt und K. Heider's Lehrbuch der vergl. Entwicklungsgeschichte der wirbellosen Thiere. Jena. 8°. Vide p. 1–18, fig. 1–12.

1893 **Hinde, G. J.** On Palaeosaccus Dawsoni, Hinde, a New Genus and Species of Hexactinellid Sponge from the Quebec Group (Ordovician) at Little Métis, Quebec, Canada. By...In: Geol. Mag. (3), x, p. 56–58, Pl. iv.

[1893 (α) **Hinde, G. J.** A Monograph of the British Fossil Sponges. By... Part iii, p. 189–254, Pl. x–xix.]

1893 **Kent, W. Saville.** The Great Barrier Reef of Australia; its products and potentialities. By...London. 4°. xvii and 387 pp., 48 and 16 Pl.

1893 **Lambe, L. M.** On some Sponges from the Pacific Coast of Canada and Behring Sea. By...In: Proc. and Trans. R. Soc. Canada, x (Section iv), p. 67–78, Pl. iii–vi.

1893 **Lendenfeld** is changed to 1892 (ε).

1893 (α) **Lendenfeld** is changed to 1892 (ζ).

1893 (β) **Lendenfeld, R. von.** Bemerkung über das Entoderm der Spongien. Von...In: Zool. Anzeig. xvi, p. 12–13.

1893 **Levinsen, G. M. R.** Annulata, Hydroidae, Anthozoa, Porifera. Ved ...In: Det videnskabelige udbytte af Kanonbaaden "Hauchs" togter, i, 1883–86, p. 321–425, Pl. i–iii.

1893 **Maas, O.** Die Embryonal-Entwicklung und Metamorphose der Cornacuspongien. Von...In: Zool. Jahrb. Abth. Morph. vii, p. 331–448, Pl. 19–23.

1893 (α) **(Maas, O.)** Ueber die erste Differenzierung von Generations- und Somazellen bei Spongien. In: Verh. D. Zool. Gesellsch. iii, p. 27–35. With 6 woodcuts.

1893 **Norman, A. M.** A month on the Trondhjem Fjord. By...In: Ann. and Mag. N. H. (6), xii, pp. 341–367, 441–452, Pl. xvi and xix. Vide pp. 347, 349.

1893 **Pearcey, F. G.** Foraminifer or Sponge?...In: Nature, xlvii, p. 390.

1893 **Pekelharing.** See Vosmaer & Pekelharing.

1893 **Rauff, H.** Palaeospongiologie. Von...In: Palaeontogr. xl, p. 1–232, fig. 1–48*.

1893 (α) **Rauff, H.** Ueber angebliche Spongien aus dem Archaicum. Von ...In: N. Jahrb. Mineral. p. 57–67, 3 figs.

* I am not quite sure whether p. 121–232 appeared in 1893 or 1894; probably in 1893. The pages in question are mentioned neither in Zool. Jahresber. nor in Zool. Anz

1893 **Schulze, F. E.** Ueber die Ableitung der Hexactinelliden-Nadeln vom regulären Hexactine. Von...In: Sitz. ber. K. Preuss. Akad. Wiss. Berlin, xlvi, p. 991–997. Abstr. and rev. in: Naturw. Rundschau, ix (1894), p. 129–130.

1893 (*a*) **Schulze, F. E.** Revision des Systemes der Hyalonematiden. Von ...In: Sitz. ber. K. Preuss. Akad. Wiss. Berlin, xxx, p. 541–589.

1893 **Stedman, J. M.** Has the Fresh-Water Sponge a Nervous System? By...In: Amer. Monthly Microsc. Journ. xiv, p. 48.

1893 **Topsent, E.** Mission scientifique de M. Ch. Alluaud aux Iles Séchelles (Mars-Mai 1892), Spongiaires, par...In: Bull. Soc. Zool. France, xviii, p. 172–175.

1893 (*a*) **Topsent, E.** Note sur la faune des Spongillides de France, par... In: Bull. Soc. Zool. France, xviii, p. 176.

1893 (*β*) **Topsent, E.** Note sur quelques éponges du Golfe de Tadjoura recueillies par M. le Dr L. Faurot, par...In: Bull. Soc. Zool. France, xviii, p. 177–182.

1893 (*γ*) **Topsent, E.** is changed to 1890 (*β*).

1893 (*δ*) **Topsent, E.** is changed to 1892 (*ε*).

1893 (*ε*) **Topsent, E.** Contribution à l'histologie des Spongiaires (par)... In: C. R. Paris, cxvii, p. 444–446.

1893 (*ζ*) **Topsent, E.** is changed to 1892 (*δ*).

1893 (*η*) **Topsent, E.** Sur une Ephydatie (E. fluviatilis, autt.) du Lac de Houleh (Syrie). Par...In: Rev. Biol. N. France, v, p. 326–327.

1893 **Valle, A. della.** Gammarini del Golfo di Napoli. Monografia di...In: Fauna und Flora des Golfes von Neapel. 948 pp. and 61 Pl.

1893 **Vosmaer, G. C. J. & C. A. Pekelharing.** On Sollas's Membrane in Sponges. By...In: Tijdschrift Ned. Dierk. Ver. (2), iv, p. 38–56, Pl. ii. Also in: Onderzoek. Physiol. Lab. Utrecht (4), iii, p. 185–207, Pl. iv (1894).

1893 (**Weltner, W.** Ueber Metallausgüsse des Canalsystemes eines Süsswasserschwammes.) In: Sitz. Ber. Ges. Nat. Freunde, Berlin, p. 5–6.

1893 (*a*) **Weltner, W.** Spongillidenstudien. Von...(i–ii.) In: Arch. f. Naturgesch. lix, p. 209–284, Pl. viii–ix.

1893 (*β*) **Weltner, W.** Bemerkungen über den Bau und die Entwicklung der Gemmulae der Spongilliden. Von...In: Biol. Centralbl. xiii, p. 119–126.

1893 (*γ*) (**Weltner, W.** Ueber die Autorenbezeichnung von Spongilla erinaceus.) In: Sitz. Ber. Ges. Nat. Freunde, Berlin, p. 7–13.

1893 (*δ*) **Weltner, W.** Bemerkungen über die Gattung Ceratella s. Solanderia...In: Sitz. Ber. Ges. Nat. Freunde, Berlin, p. 13–18.

1893 **Wilson, H. V.** Remarks on the general morphology of Sponges... In: Journ. Elisha Mitchell Sc. Soc. Raleigh, ix, 1892, p. 31–48.

1894 **Bidder, G. (P.).** The collar-cells of sponges. By...In: Zool. Anz. xvii, p. 167–168.

1894 **Dendy, A.** Studies on the Comparative Anatomy of Sponges. vi, On the Anatomy and Relationships of Lelapia australis, a Living Representative of the Fossil Pharetrones. By...In: Quart. Journ. Microsc. Sc. xxxvi, p. 127–142, Pl. xiii.

1894 **Garbini, A.** Contributo allo studio delle Spongille italiane...In: Accad. Agric., Arti e Comm. Verona (3), lxx (23 pp., 3 figs.). Also separately (?).

1894 **Girod, P.** Les Éponges d'eau douce...In: Naturaliste, xvi, p. 180–182.

1894 **Graëlls, M.** L'Exploitation des Éponges à Batabanó...par...In: Rev. Sc. Nat. Appliq. p. (1–10).

1894 **Hanitsch, R.** Revision of the Generic Nomenclature and Classification in Bowerbank's "British Spongiadae." By...In: Trans. Liverpool Biol. Soc. viii, p. 173–206.

1894 (α) **Hanitsch, R.** Amphiute, eine neue Gattung heterocoeler Kalkschwämme. Von...In: Zool. Anz. xvii, p. 433.

1894 **Heider, K.** Berichtigung. Von...In: Zool. Anz. xvii, p. 392–395.

1894 **Herdman, W.** The Seventh Annual Report of the Liverpool Marine Biology Committee...In: Trans.Liverpool Biol.Soc. viii,p. 3–55, Pl. i–v.

1894 **Ijima, I.** Notice of new Hexactinellida from Sagami Bay...In: Zool. Anz. xvii, p. 365–369.

1894 **Kirk, H. B.** Contribution to a Knowledge of the New Zealand Sponges. By...In: Trans. N. Zealand Inst. xxvi, p. 175–179, Pl. xxii.

1894 **Lambe, L. M.** Sponges from the Pacific Coast of Canada. By...In: Proc. and Trans. R. Soc. Canada, xi (Section iv), p. 25–43, Pl. ii–iv.

1894 **Lendenfeld, R. von.** Die Tetractinelliden der Adria (mit einem Anhange über die Lithistiden). Von...In: Denkschr. Mathem.-naturw. Cl. K. Akad. Wiss. Wien, lxi, p. 91–204, Pl. i–viii.

1894 (α) **Lendenfeld, R. von.** Eine neue Pachastrella. Von...In: Sitz. Ber. K. Akad. Wiss. Wien, Math.-naturw. Cl. ciii, p. 439–442, Pl. i.

1894 (β) **Lendenfeld, R. von.** Ergebnisse neuerer Untersuchungen über Spongienepithelien. In: Zool. Centralbl. i, p. 506–510.

1894 (γ) **Lendenfeld, R. von.** Tethranthella, eine neue Lithistide. Von...In: Zool. Anz. xvii, p. 49–51.

1894 (δ) **Lendenfeld, R. von.** Die systematische Stellung von Placospongia. Von...In: Biol. Centralbl. xiv, p. 114–115.

1894 (ζ) **Lendenfeld, R. von.** Tethranthella oder Crambe. Von...In: Zool. Anz. xvii, p. 243–246. (1894 (ε) is changed to 1893.)

1894 (η) **Lendenfeld, R. von.** Bemerkungen über Tinctionsmittel für Spongien. Von...In: Zeitschr. wiss. Mikrosk. xi, p. 22–24.

1894 **Letellier, A.** Une action purement mécanique suffit aux Cliones pour creuser leurs galeries dans les valves des Huîtres. Note de... In: C. R. cxviii, p. 986–989. Cf. 1895.

1894 **Levander, K. M.** Ephydatia fluviatilis in den Esboer Skären bei Helsingfors...In: Meddel. Soc. Fauna und Flora Fenn. p. 9–11.

1894 **Levinsen, G. M. R.** Studier over Svampe-Spicula: Cheler og Ankere. Ved...In: Vidensk. Medd. naturh. Foren. Kjøbenhavn, 1893*, p. 1–21, Pl. I.

1894 **Masterman, A. T.** On the Nutritive and Excretory Processes in Porifera. By...In: Ann. and Mag. N. H. (6), xiii, p. 485–496. With 6 woodcuts.

1894 (α) **Masterman, A. T.** On the Nutritive and Excretory Processes of Porifera. By...In: Ann. and Mag. N. H. (6), xiv, p. 48–49.

1894 **Moller, A. F.** Esponjas de S. Thomé...In: Annaes Sc. Nat. Porto, i, p. 202–203.

1894 **Nöldeke, B.** Die Metamorphose des Süsswasserschwammes...Von... In: Zool. Jahrb. Abth. Anat. und Ontog. Bd. viii, p. 153–189, Pl. 8–9. Also separately as "Inaugur. Dissert." p. 1–42, Pl. I–II.

1894 **Pekelharing.** See Vosmaer & Pekelharing.

1894 **Petr, F.** Evropské houby sladkovodních. Napsal...Chrudimi. 32 pp., 2 Pl. European Freshwater Sponges. By...Chrudim. [See 1895.]

1894 **Rauff, H.†** (Palaeospongiologie.) In: Palaeontographica, xl, p. 233–346, fig. 49–75, Pl. I–XVII.

1894 **Schulze, F. E.** Aus Hexactinelliden hergestellte Artefacte von der Philippinen-Insel Cebu...In: Sitz. Ber. Ges. Nat. Freunde, Berlin, p. 137–141. With 2 figs.

1894 **Topsent, E.** Application de la taxonomie actuelle à une collection de Spongiaires du Banc de Campêche et de la Guadeloupe décrite précédemment...par...In: Mém. Soc. Zool. France, vii, p. 27–36.

1894 (α) **Topsent, E.** Campagne de la Melita, 1892. Éponges du Golfe de Gabès, par...In: Mém. Soc. Zool. France, vii, p. 37–44, Pl. I.

1894 (β) **Topsent, E.** Étude sur la faune des Spongiaires du Pas-de-Calais suivie d'une Application de la Nomenclature actuelle à la Monographie de Bowerbank. Par...In: Rev. Biol. Nord de la France, vii, p. 6–28.

1894 (γ) **Topsent, E.** Une Réforme dans la Classification des Halichondrina, par...In: Mém. Soc. Zool. France, vii, p. 5–26.

1894 (δ) **Topsent, E.** A Propos de Tetranthella fruticosa (Schm.), Lend. par...In: Rev. Biol. Nord de la France, vi, p. 313–314.

1894 (ε) **Topsent, E.** Nouvelle série de diagnoses d'éponges de Roscoff et de Banyuls. Par...In: Arch. Zool. Expérim. (3), i, p. xxxiii–xliii‡.

---

* According to Zool. Anz. xvii, p. 146, published in 1894.   † Cf. 1893.
‡ According to Zool. Anz. xvii, p. 146, and Zool. Jahresber. für 1894, p. 2, this part of the Arch. appeared in 1894.

1894 (ζ) **Topsent, E.** Étude monographique des Spongiaires de France.
i, Tetractinellida, par...In: Arch. Zool. Expérim. (3), ii, p. 259–
400, Pl. xi–xvi.

1894 (η) **Topsent, E.** Sur le mécanisme de la perforation des Cliones.
Par...In: Arch. Zool. Expérim. (3), ii, p. x–xiii. Abstr. and
rev. in: Zool. Jahresber. für 1894 (1895), pp. 2, 11.

1894 **Traxler, L.** Spongilliden der Umgebung von Jaransk. Von...In:
Zool. Anz. xvii, p. 363–364.

1894 **Vosmaer, G. C. J.** Preliminary Notes on some Tetractinellids of the
Bay of Naples. By...In: Tijdschr. Nederl. Dierk. Ver. (2), iv,
p. 269–286.

1894 (α) **Vosmaer, G. C. J.** Note on Suberites fruticosus and Suberites
crambe of Oscar Schmidt. By...In: Tijdschr. Nederl. Dierk.
Ver. (2), iv, p. 287–288.

1894 **Weltner, W.** Spongien. Von...In: Wissensch. Meeresunters.
herausgeg. v. d. Kommission 2. Unters. d. M. (2), i, p. 325–328.

1894 (α) **Weltner, W.** Anleitung zum Sammeln von Süsswasserschwämmen
nebst Bemerkungen über die in ihnen lebenden Insektenlarven.
Von...In: Entom. Nachr. (Karsch), xx, p. 145–151, 10 figs.

1894 (β) **Weltner, W.** (Über zwei neue Cirripedien aus dem indischen
Ocean...) In: Sitz. Ber. Ges. Nat. Freunde, Berlin, p. 80–87,
6 figs.

1894 **Wilson, H. V.** Observations on the Gemmule and Egg Development
of Marine Sponges. By...In: Journ. Morph. ix, p. 277–406,
Pl. xiv–xxv.

1894 (β) **Wilson, H. V.** Embryology of Sponges...In: Amer. Naturalist,
xxviii, p. 73–76.

1895 **Bidder, G.** The Collar-cells of Heterocoela. By...In: Quart. Journ.
Microsc. Sc. xxxviii, p. 9–43, 4 figs, Pl. 2.

1895 **Buck, E.** Beobachtungen an Schwämmen des Bodensees und ihre
Züchtung im Aquarium...In: Ber. Offenbach Ver. f. Naturk.
p. 25–70.

1895 **Cayeux, L.** De l'existence de nombreux débris de Spongiaires dans
le Précambrien de Bretagne (première note)...In: Ann. Soc.
Geol. Nord, xxiii, p. 52–65, Pl. i–ii.

1895 (α) **Cayeux, L.** De l'existence de nombreux débris de Spongiaires dans
les phtanites du Précambrien de Bretagne. Note de...In: C. R.
Paris, cxx, p. 279 282.

1895 **Deecke, W.** Eocäne Kieselschwämme als Diluvialgeschiebe in Vor-
pommern und Mecklenburg...In: Mittheil. Naturw. Ver. Greifs-
wald, 26 Jahrg., p. 166–170, Taf. i.

1895 **Dendy, A.** Catalogue of Non-Calcareous Sponges collected by
J. Bracebridge Wilson, Esq., M.A., in the neighbourhood of Port
Phillip Heads. Part i. By...In: Proc. R. Soc. Victoria (2), vii,
p. 232–260. Abstr. and rev. in: Zool. Jahresber. für 1895 (1896), p. 1.

1895 **Desmazières, O.** Note bibliographique sur les Spongiaires d'Anjou et de la Touraine. In: Bull. Soc. étud. scient. Angers, 1894, p. 147–153.

1895 **Hanitsch, R.** Notes on a Collection of Sponges from the West Coast of Portugal. By...In: Trans. Liverpool Biol. Soc. ix, p. 205–219, Pl. xii–xiii. Abstr. and rev. in: Journ. R. Microsc. Soc. London, 1896, p. 77.

1895 (α) **Hanitsch, R.** American Fresh-water Sponges in Ireland. In: Nature, li, p. 511.

1895 (β) **Hanitsch, R.** The Fresh-water Sponges of Ireland, with Remarks on the general distribution of the group. By...In: Irish Naturalist, iv, p. 122–131, Pl. 4.

1895 **Heider, A. von.** Liste der Schmidt'schen Spongien in der zoologischen Abtheilung des steiermärkischen Landes-Museums. Von...In: Mitth. naturw. Ver. Steiermark, 1894, p. 276–285.

1895 **Hundeshagen, Fr.** Ueber jodhaltige Spongien und Jodospongin... In: Zeitschr. angew. Chemie, p. 473–476.

1895 **Ijima, I.** On Two New Hexactinellida from Sagami Bay. By...In: Zool. Mag. Tokyo, vii, p. 93–96. Abstr. and rev. in: Journ. R. Microsc. Soc. London, p. 541.

1895 **James, J. F.** Sponges: recent and fossil...In: Amer. Naturalist, xxix, p. 536–545, 7 figs.

1895 **Jennings, A. Vaughan.** On the True Nature of "Möbiusispongia parasitica," Duncan. By...In: Journ. Linn. Soc. xxv, p. 317–319.

1895 **Kirk, H. B.** Further Contribution to a Knowledge of the New Zealand Sponges. By...In: Trans. N. Zealand Institute, xxvii, p. 287–292, Pl. xxiv–xxv.

1895* **Lambe, L. M.** Sponges from the Western Coast of North America. By...In: Proc. and Trans. R. Soc. Canada, xii, Section iv, p. 113–138, Pl. ii–iv.

1895 **Lendenfeld, R. von.** Papillina, Osculina und ihre Beziehungen unter einander und zu Bohrschwämmen. Von...In: Zool. Anzeig. xviii, p. 149–151.

1895 (α) **Lendenfeld, R. von.** Entwicklung und Nahrungsaufnahme der Oscarella: Kritische Bemerkungen. Von...In: Zool. Anz. xviii, p. 17–19.

1895 **Letellier, A.** Une action purement mécanique permet d'expliquer comment les Cliones creusent leurs galeries dans les valves des huîtres...In: Bull. Soc. Linn. Normand. (4), viii (3), p. 149–166. Cf. prelim. account in 1894.

---

* According to Zool. Anz. xix (1896) the paper appeared in 1894. On the title of the xiith Volume of Trans. Soc. Canada is, however, printed 1895. A specimen I saw in the Utrecht Univers. Library bears a note that it arrived in Oct. 1895. Therefore I feel inclined rather to put Lambe's article in 1895.

1895 **Malfatti, P.** Silicospongie plioceniche...In: Atti R. Accad. Lincei (5), iv (Rendic.), p. 116–121.

1895 **Merrill, J. A.** Fossil Sponges of Flint Nodules in the Lower Cretaceous of Texas...In: Bull. Mus. Comp. Zool. Harvard Coll. xxviii, p. 1–26.

1895 **Minchin, E. A.** On the Origin of the Triradiate Spicules of Leucosolenia. By...In: Proc. R. Soc. London, lviii, p. 204–205. Also in: Ann. and Mag. N. H. (6), xvi, p. 427–428.

1895 **Pérez, J.** Sur l'homologie des feuillets blastodermiques des Éponges ...In: Act. Soc. Linn. Bordeaux, xlvii [(5), vii], 1894, p. 322–327.

1895 **Petr, Fr.** (Süsswasserschwämme Europa's.) Bohemian (1894), summarised Zool. Centralbl. ii, p. 749.

1895 **Rauff, H.** Palaeospongiologie. Von...In: Palaeontographica, xli, p. 223–272, fig. 76–124, Pl. xx–xxvi*.

1895 **Reeker, R.** Sammeln von Süsswasserschwämmen. In: Jahresber. Westfäl. Prov. Ver. 1895, p. 53–56.

1895 **Schulze, F. E.** Hexactinelliden des Indischen Oceanes. i. Theil. Die Hyalonematiden. Von...In: Abh. K. Preuss. Akad. Wiss. Berlin, 1894†, p. 1–60, Pl. i–ix.

1895 **Topsent, E.** Campagnes du Yacht Princesse Alice. Notice sur les Spongiaires recueillis en 1894 et 1895, par...In: Bull. Soc. Zoo. France, xx, p. 213–216.

1895 **Traxler, L.** Beiträge zur Kenntnis der Süsswasserschwämme. Von... In: Földtani Közlöny, xxv, p. 241–242. Hungarian, p. 181–185.

1895 (α) **Traxler, L.** Spikule von Süsswasserschwämmen aus Brasilien. Von...In: Földtani Közlöny, xxv, p. 238–240, Pl. iii. Hungarian, p. 178–180.

1895 (β) **Traxler, L.** Die Schwammspikule des Schlammes im See Héviz. Von...In Földtani Közlöny, xxv, p. 142–145, Taf. ii. Hungarian, p. 109–112.

1895 (α) **Weltner, W.** Spongillidenstudien. iii, Katalog und Verbreitung der bekannten Süsswasserschwämme...In: Arch. f. Naturgesch. lxvi, p. 114–144. (1895 Weltner is cancelled.)

1896 **Allen, E. J.** Report on the Sponge Fishery of Florida and the Artificial Culture of Sponges. By...In: Journ. Mar. Biol. Ass. (2), iv, p. 188–194.

1896 (α) **Allen, E. J.** Supplement to Report on the Sponge Fishery of Florida and the Artificial Culture of Sponges. By...In: Journ. Mar. Biol. Ass. (2), iv, p. 289–292.

---

* The pages are sometimes (e.g. Zool. Jahresber.) quoted from separate copies, which seem to be sold.

† "Ausgegeben am 16. Juli 1895."

1896 **Bidder, G. (P.).** Note on Projects for the Improvement of Sponge-Fisheries. By...In: Journ. Mar. Biol. Ass. (2), iv, p. 195–202.

1896 **Breitfuss, L.** Kalkschwämme der Bremer Expedition nach Ost-Spitzbergen im Jahre 1889 (Prof. W. Kükenthal und Dr A. Walter). Vorläufige Mittheilung. Von...In: Zool. Anz. xix, p. 426–432.

1896 (α) **Breitfuss, L.** Kalkschwämme von Ternate (Molukken), nach den Sammlungen Prof. W. Kükenthal's. (Vorläufige Mittheilung.) Von...In: Zool. Anz. xix, p. 433–435.

1896 (β) **Breitfuss, L.** Amphoriscus Semoni, eine neue Art heterocoeler Kalkschwämme. (Vorläufige Mittheilung.) Von...In: Zool. Anz. xix, p. 435–436.

1896 **Dendy, A.** Catalogue of Non-Calcareous Sponges collected by J. Bracebridge Wilson, Esq., M.A., in the neighbourhood of Port Phillip Heads. Part ii. By...In: Proc. R. Soc. Victoria (2), viii, p. 14–51.

1896 **Haeckel, E. (H. P. A.).** Systematische Phylogenie. Entwurf eines Natürlichen Systems der Organismen auf Grund ihrer Stammesgeschichte, von...Zweiter Theil: Systematische Phylogenie der wirbellosen Thiere (Invertebrata). Berlin, pp. 720. 8°.

1896 **Hinde, G. Jennings.** Descriptions of new Fossils from the Carboniferous Limestone. i, On Pemmatites constipatus, sp. nov., a Lithistid Sponge...By...In: Quart. Journ. Geol. Soc. lii, pp. 438–440, 450, Pl. xxii, fig. 1, 1 a–1 m.

1896 **Ijima, I.** Notice of New Hexactinellida from Sagami Bay...In: Zool. Anz. xix, p. 249–254.

1896 (α) **Ijima, I.** Long-Lines as Zoological Collecting Apparatus. By... In: Zool. Magazine, viii, pp. 13–17, 19–23, 39–46.

1896 **Jennings, A. Vaughan.** On the True Nature of "Möbiusispongia parasitica"...By...In: Journ. Linn. Soc. London, xxv, p. 317–319.

1896 **Kieschnick, O.** Silicispongiae von Ternate nach den Sammlungen von Herrn Prof. Dr W. Kükenthal. Von...In: Zool. Anz. xix, p. 526–534. Preliminary account.

1896 **Kirk, H. B.** New Zealand Sponges; Third Paper. By...In: Trans. N. Zealand Inst. xxviii, p. 205–210, Pl. iii–iv.

1896 **Lambe, L. M.** Sponges from the Atlantic Coast of Canada...In: Trans. R. Soc. Canada (2), ii, p. 181–211, Pl. i–iii.

1896 **Lendenfeld, R. von.** Die Berechtigung des Gattungsnamens Homandra. Von...In: Zool. Anz. xix, p. 495–496.

1896 **Leonardi, C.** La pesca delle spugne nelle acque di Lampedusa. In: Boll. Natur. Coll. xvi, no. vi, p. 73–74.

1896 **Maas, O.** Erledigte und strittige Fragen der Schwammentwicklung. Von...In: Biol. Centralbl. xvi, p. 231–239.

1896 **Minchin, E. A.** Note on the Larva and the Post-larval Development of Leucosolenia variabilis, H. sp., with Remarks on the Development of other Asconidae. By...In: Proc. R. Soc. London, lx, p. 42–52, 7 zincos.

1896 (α) **Minchin, E. A.** Suggestions for a Natural Classification of the Asconidae. By...In: Ann. and Mag. N. H. (6), xviii, p. 349–362.

1896 **Rothpletz, A.** Ueber Phymatoderma, ein Diatomeen einschliessender Hornschwamm. Von...In: Zeitschr. D. Geol. Ges. xlviii, p. 905–909.

1896 **Schulze, F. E.** Ueber diplodale Spongienkammern. Von...In: Sitzungsber. K. Pr. Akad. Wissensch. Berlin, xxxix, p. 891–897, Pl. v. Abstr. and rev. in: Journ. R. Micr. Soc. London, 1896, p. 636.

1896 (α) **Schulze, F. E.** Hexactinelliden des Indischen Oceanes. ii. Theil. Die Hexasterophora. Von...In: Abh. Akad. Wissensch. Berlin, 92 pp., 8 Pl. Also separ.* 4°, p. 1–92, Pl. i–viii.

1896 **S(o)ukatschoff, B.** Quelques nouvelles formes d'éponges, recueillies dans le lac de Baïkal (Sibérie orientale) par...In: Trav. Soc. St. Pétersb. xxv (2), p. 1–11 (Russian), p. 11–19 (French transl.), Pl. i.

1896 **Topsent, E.** Matériaux pour servir à l'étude de la Faune des Spongiaires de France, par...In: Mém. Soc. Zool. France, ix, p. 113–133.

1896 (α) **Topsent, E.** Campagnes du Yacht Princesse Alice. Sur deux curieuses Espérellines des Açores, par...In: Bull. Soc. Zool. France, xxi, p. 147–150.

1896 (β) **Topsent, E.** Éponges...In: Koehler, Résult. Sc. Campagne "Caudan," Lyon, p. 273–279, Pl. viii.

1896 (γ) **Topsent, E.** Étude monographique des Spongiaires de France. ii, Carnosa par...In: Arch. Zool. Expérim. (3), iii, p. 493†–590, Pl. xxi–xxiii. Abstr. and rev. in: Zool. Jahresber. für 1896, p. 2.

1896 **Traxler, L.** Ueber einen neuen Süsswasserschwamm aus Neu-Seeland. Von...In: Termész. Füzetek, xix, p. 102–105, Pl. ii.

1896 **Weltner, W.** Der Bau des Süsswasserschwammes. Von...In: Blätter f. Aquar. u. Terrar. Freunde, vii, p. 277–285.

1896 (α) **Weltner, W.** Die Colenteraten und Schwämme des süssen Wassers Ost-Afrikas...In: Ost-Afrika, iv, p. 1–8.

1897 **Breitfuss, L. L.** Ascandra hermesi, ein neuer homocöler Kalkschwamm aus der Adria. Von...In: Z. W. Z. lxiii, p. 39–42, 2 figs. Abstr. and rev. in: Zool. Centralbl. iv (1898), p. 913.

---

* On the title is erroneously printed 1895; it is stated to have been "ausgegeben 7. April 1896."

† P. 493–496 published in 1895.

1897 **Dendy, A.** Catalogue of Non-Calcareous Sponges collected by J. Bracebridge Wilson, Esq., M.A., in the neighbourhood of Port Phillip Heads. Part iii. By...In: Proc. R. Soc. Victoria (2), ix, p. 230–259. Abstr. and rev. in: Journ. R. Microsc. Soc. London...p. 541.

1897 **Doederlein, L.** Ueber die Lithonina, eine neue Gruppe von Kalkschwämmen. Von...In: Zool. Jahrb. Abth. System. x, p. 15–32, Taf. 2–6. Abstr. and rev. in: Zool. Centralbl. iv, p. 515–516; Journ. R. Micr. Soc. London, 1897, p. 299.

1897 **Dun, W. S.** Note on the Occurrence of Sponge Remains in the Lower Silurian of New South Wales...In: Proc. Linn. Soc. N. S. Wales, xxii, p. 436–437.

1897 **Garbini, A.** Due Spongille del Lago di Garda nuove per l'Europa. Von (sic)...In: Zool. Anz. xx, p. 477–478. Abstr. and rev. in: Journ. R. Microsc. Soc. London, 1898, p. 202.

1897 **Ijima, I.** Revision of Hexactinellids with Discoctasters, with Descriptions of Five New Species. By...In: Annot. Zool. Japon. i, p. 43–59.

1897 **Kieschnick, O.** Berichtigung...Von...In: Zool. Anz. xx, p. 28.

1897 **Lendenfeld, R. von.** Die Clavulina der Adria, von...In: Nova Acta Acad. Leop. Carol. lxix, p. 1–251, Pl. i–xii. Also separ. Leipzig, Engelmann. 4°.

1897 (α) **Lendenfeld, R. von.** Der Tierstamm der Spongien. Von...In: Zool. Garten, xxxviii, pp. 6–13, 44–51, 71–80, figs. 1–36.

1897 (β) **Lendenfeld, R. von.** Note on some Sponges from the Auckland Islands. By...In: Ann. and Mag. N. H. (6), xix, p. 124.

1897 (γ) **Lendenfeld, R. von.** (Review of recent spongological literature.) In: Zool. Centralbl. iv, pp. 515–516, 609–612.

1897 (δ) **Lendenfeld, R. von.** (Review of Minchin's paper on the position of sponges.) In: Zool. Centralbl. iv, p. 910–912.

1897 (ε) **Lendenfeld, R. von.** Spongien von Sansibar. Von...In: Abhandl Senckenb. Naturf. Ges. xxi, p. 93–133, Pl. ix–x. Also separ. Frankfurt. 4°.

1897 (ζ) **Lendenfeld, R. von.** On the Spongida...(Notes on Rockall Island, etc.) In: Trans. R. Irish Acad. xxxi, p. 82–88, 4 figs.

1897 **Lindgren, N. G.** Beitrag zur Kenntniss der Spongienfauna des Malaiischen Archipels und der Chinesischen Meere. Von...In: Zool. Anz. xx, p. 480–487. Preliminary account of the paper, which appeared 1898. N.B. Some corrections are to be found in Z. A. xxi, p. 40, "Berichtigung."

1897 **Loisel, G.** Contribution à la physiologie et à l'histologie des Éponges... In: C. R. Soc. Biol. Paris (10), iv, p. 934–935; v, pp. 68–69, 351–354.

1897 **Minchin, E. A.** The Position of Sponges in the Animal Kingdom. In: Science Progress (2), i, p. 426–460 (p. 1–35 in separ. copy). Abstr. and rev. in: Journ. R. Micr. Soc. London, 1897, p. 393.

1897 (α) **Minchin, E. A.** Ascandra or Homandra? A Test-Case for the Rules of Zoological Nomenclature. By...In: Zool. Anz. xx, p. 49–50.

1897 **Oppliger, F.** Die Juraspongien von Baden...In: Abhandl. Schweiz. palaeont. Ges. xxiv, p. 1–58, Pl. I–XI. Also separ. Genf-Berlin. 4º.

1897 **Rousseau, E.** Eine neue Methode zur Entkalkung und Entkieselung der Schwämme. Vorläufige Mittheilungen. Von...In: Zeitschr. wiss. Mikrosk. xiv, p. 205–209.

1897 **Schulz, E.** Anatomische und histologische Untersuchung der von den Professoren Dr Semon und Dr Kükenthal während der Jahre 1893 und 1894 im Molukkenarchipel gewonnenen Hornspongien. Inaug. Diss....von...Leipzig. 62 pp., 7 Pl.

1897 **Schulze, F. E.** Revision des Systemes der Asconematiden und Rosselliden. Von...In: Sitz. Ber. Akad. Wissensch. Berlin, Jahrg. 1897, p. 520–558. Abstr. and rev. in: Journ. R. Microsc. Soc. London, p. 541.

1897 (α) **Schulze, F. E.** Ueber einige Symmetrieverhältnisse bei Hexactinelliden Nadeln. In: Verh. D. Zool. Gesellsch. vii, p. 35–37.

1897 **Tempère, J.** Spicules d'Éponges. Par...In: Micrographe Préparateur, v, p. 17–19, Pl. 1–2, 5–6.

1897 **Topsent, E.** Spongiaires de la Baie d'Amboine, par...In: Rev. suisse de Zoologie, iv, p. 421–487, Pl. XVIII–XXI.

1897 (α) **Topsent, E.** Sur le genre Halicnemia Bowerbank, par...In: Mém. Soc. Zool. France, x, p. 235–251, 2 figs.

1897* **Whitelegge, Th.** The Sponges of Funafuti. By...In: Mem. Austral. Mus. iii, p. 321–332, Pl. XVIII.

1897 **Zeise, O.** Die Spongien der Stramberger Schichten...In: Palaeontogr. Suppl. ii, pp. i–iv, 289–342, Pl. XIX–XXI.

1898 **Bidder, G. P.** The Skeleton and Classification of Calcareous Sponges. By...In: Proc. R. Soc. London, lxiv, p. 61–76, 10 figs.

1898 **Breitfuss, L. L.** Kalkschwammfauna des Weissen Meeres und der Eismeerküsten des Europäischen Russlands...Von...In: Mém. Acad. Imp. Sc. Pétersbourg (8), vi, 41 pp., 4 Pl. Also separ. Petersburg-Leipzig. 4º. p. 1–40, Pl. I–IV.

1898 (α) **Breitfuss, L. L.** Die Kalkschwämme der Sammlung Plate. Von ...In: Zool. Jahrbüch. Suppl. iv, p. 155 170, Taf. 27.

1898 (β) **Breitfuss, L. L.** Die arktische Kalkschwammfauna. Von...In: Arch. Naturg. Jahrg. 1898, Bd. i, p. 277–316.

1898 (γ) **Breitfuss, L. L.** Kalkschwammfauna der Westküste Portugals. Von...In: Zool. Jahrb. Abth. Syst. xi, p. 89–102, Taf. 11. Abstr. and rev. in: Journ. R. Soc. Microsc. London, p. 434.

---

* According to Bibliogr. Zool. (Zool. Anz.), iii, p. 43, published in 1905, which is certainly a misprint. Cf. Zool. Jahresber. f. 1897.

1898 (δ) **Breitfuss, L. L.** Die Kalkschwammfauna von Spitzbergen. Nach den Sammlungen der Bremer Expedition nach Ost-Spitzbergen im Jahre 1889 (Prof. W. Kükenthal und Dr A. Walter). Von ...In: Zool. Jahrb. Abth. Syst. xi, p. 103–120, Taf. 12–13.

1898 (ε) **Breitfuss, L. L.** Kalkschwämme von Ternate von...In: Abh. Senckenb. naturf. Ges. xxiv, p. 169–177. Also separ. Frankfurt. 4°.

1898 (ζ) **Breitfuss, L. L.** Amphoriscus semoni, ein neuer heterocoeler Kalkschwamm. Von...In: Denkschr. medic. naturw. Ges. Jena, viii (Semon, Zool. Forschungsreis. Austr. v), p. 381–384.

1898 (η) **Breitfuss, L. L.** Catalog der Calcarea der Zoologischen Sammlung des Königlichen Museums für Naturkunde zu Berlin. Von... In: Arch. f. Naturgesch. lxiii, p. 205–226, 2 figs.

1898 (θ) **Breitfuss, L. L.** Note sur la faune des Calcaires de l'Océan Arctique ...In: Ann. Mus. Zool. Acad. Imp. St Pétersbourg...p. 12–38.

1898 **Delage, Y.** Sur la place des Spongiaires dans la classification. Note de M.... présentée par M. de Lacaze-Duthiers. In: C. R. Acad. Paris, cxxvi, p. 545–548. Abstr. and rev. in: Rev. Scient. (4), ix, p. 276; Journ. R. Microsc. Soc. London, p. 434–435.

1898 (α) **Delage, Y.** Les larves des Spongiaires et l'homologation des feuillets. Note de M....présentée par M. H. de Lacaze-Duthiers. In: C. R. Acad. Paris, cxxvi, p. 767–769. Abstr. and rev. in: Rev. Scient. (4), ix, p. 370.

1898 (β) **Delage, Y.** L'état actuel de la biologie et de l'industrie des éponges. Prem. part. Par...In: Rev. Gén. des Sc. Paris, ix, p. 733–749.

1898 **Dendy, A.** On the Sponges described in Dieffenbach's "New Zealand." By...In: Trans. New Zealand Inst. xxx, p. 316–320, Pl. xxxiii–xxxiv.

1898 **Harnack, E.** Ueber das Iodospongin, die jodhaltige eiweissartige Substanz aus dem Badeschwamm...In: Zeitschr. Phys. Chemie, xxiv, p. 412–424.

1898 **Hesse, O.** (Die Schwammfischerei bei der 'Insel Lampedusa.) In: Zool. Garten, xxxix, p. 323.

1898 **Ijima, I.** The Genera and Species of Rossellidae. Preliminary Notice ...In: Annot. Zool. Japon. ii, p. 41–55.

1898 **Kieschnick, O.** Die Kieselschwämme von Amboina. Inaug. Diss.... von...Jena. 66 pp.

1898 **Kirk, H. B.** ...Notes on New Zealand Sponges: Fourth Paper. By ...In: Trans. New Zealand Inst. xxx, p. 313–316, Pl. xxxi–xxxii.

1898 **Lacaze-Duthiers, H.** Les Éponges sont-elles des Coelentérés? Par... In: Arch. Zool. Expérim. (3), vi, Notes, p. ii–vi.

1898 **Lendenfeld, R. von.** (Review of his own papers, published in 1897.) In: Zool. Centralbl. v, p. 205–219.

1898 (α) **Lendenfeld, R. von.** Neuere Arbeiten über Spongien. Von...In: Zool. Centralbl. v, p. 689–698.

1898 **Lindgren, N. G.** Beitrag zur Kenntniss der Spongienfauna des Malayischen Archipels und der chinesischen Meere. Inaug. Diss., 96 pp. 4 Pl. Von...In: Zool. Jahrbüch. xi (Abth. Syst.), p. 283–378, Pl. 17–20. Cf. 1897, Lindgren.

1898 **Loisel, G.** Contribution à l'histo-physiologie des Éponges...Par... In: Journ. Anat. et Phys. xxxiv, pp. 1–43, 187–234, Pl. ɪ et v.

1898 (*a*) **Loisel, G.** Contribution à l'histo-physiologie des Éponges...In: C. R. Soc. Biol. Paris (10), v, pp. 68–69, 351–354.

1898 **Maas, O.** Die Keimblätter der Spongien und die Metamorphose von Oscarella (Halisarca). Von...In: Zeitschr. wiss. Zool. lxiii, p. 665–679, Pl. xli.

1898 (*a*) **Maas, O.** Die Entwicklung der Spongien. Eine Zusammenstellung der Thatsachen und Folgerungen auf Grund neuerer Arbeiten. Von...In: Zool. Centralbl. v, p. 581–599, 12 figs.

1898 (*β*) **(Maas, O.)** Die Ausbildung des Canalsystems und des Kalkskelets bei jungen Syconen. Von...In: Verh. D. Zool. Gesellsch. viii Jahresvers. p. 132–140 (Annex: "Discussion," p. 140–141), 3 figs.

1898 **Minchin, E. A.** Materials for a Monograph of the Ascons. i, On the Origin and Growth of the Triradiate and Quadriradiate Spicules in the Family Clathrinidae. By...In: Quart. Journ. Microsc. Sc. xl, p. 469–587, Pl. 38–42.

1898 **Moore, J. P.** Hyalodendron navalium, a new genus and species of Euplectellid Sponge...In: Proc. Acad. Nat. Sc. Philadelphia... p. 430–434, Pl. xix.

1898 **(Pekelharing, C. A.** Mededeeling aangaande een onderzoek door Dr G. C. J. Vosmaer en hemzelven verricht: "Over het opnemen van voedsel bij sponsen.") In: Versl. Gew. Verg. K. Akad. Wetensch. Amsterdam, Afd. Wis- en Natuurk. p. 494–498. Preliminary account. Cf. 1898 (*β*), Vosmaer.

1898 **Perrier, E.** Sur la place des Éponges dans la classification et sur la signification attribuée aux feuillets embryonnaires. Note de M...In: C. R. Acad. Paris, cxxvi, p. 579–583.

1898 (*a*) **Perrier, E.** Les larves des Spongiaires et l'homologation des feuillets. Note de M....In: C. R. Acad. Paris, cxxvi, p. 802–805. Transl. in: Ann. and Mag. N. H. (7), i, p. 408–410.

1898 **Petr, F.** Ueber die Bedeutung der Parenchymnadeln bei den Süsswasserschwämmen. Von...In: Zool. Anz. xxi, p. 226–227.

1898 **Schulz, E.** Hornschwämme von Ternate. In: Abhandl. Senckenb Naturf. Ges. xxiv, p. 185 188. Also sopar. Frankfurt. 1°.

1898 **Smith, H. M.** The Florida Commercial Sponges. By...In: Bull. U. S. Fish Comm. xvii, p. 225–240, Pl. 12–31.

1898 **Spangenberg, G.** (*a*) Demonstration von Spongia Ottoi, Geinitz, einer Hexaktinellide; (*b*) Zusammenvorkommen von Kalkschwamm und Lithistide. Von...In: Tagbl. Ges. Deutsch. Naturf. Abth. f. Geol. u. Min. 1, xix, p. 199–201. [Not seen, G.P.B.: Z. R. gives ʟxix.]

1898 **Thiele, J.** Studien über pazifische Spongien. Von...i. In: Zoologica, Heft 24, p. 1–72, Pl. I–VIII.

1898 **Topsent, E.** Sur les Hadromerina de l'Adriatique. Par...In: Bull. Soc. Scient. Ouest, p. 117–130.

1898 (α) **Topsent, E.** De la digestion chez les Éponges. Par...In: Arch. Zool. Expérim. (3), vi, Notes, p. 26–31.

1898 (β) **Topsent, E.** Introduction à l'étude monographique des Monaxonides de France. Classification des Hadromerina par...In: Arch. Zool. Expérim. (3), vi, p. 91–113.

1898 (γ) **Topsent, E.** Éponges nouvelles des Açores. (Première série.) Par...In: Mém. Soc. Zool. France, xi, p. 225–255, 2 figs.

1898 (ε) **Topsent, E.** Sur quelques Éponges de la Calle...Par...In: Arch. Zool. Expérim. (3), vi, Notes, p. 33–37.

1898 **Traxler, L.** Beiträge zur Kenntnis der Spongilla novae terrae, Potts ...Von...In: Termész. Füzetek, xxi, p. 314–324, Pl. XIV. German, pp. 319–323, 324.

1898 **Tschernychew, Th.** (Note sur les Éponges artinskiennes [=Lower Permian, G. P. B.] et carbonifères de l'Oural et du Timan...) (Russian.) In: Bull. Acad. Pétersbourg (5), ix, p. 1–36, Pl. I–V.

1898 **Vosmaer, G. C. J. & C. A. Pekelharing.** Ueber die Nahrungsaufnahme bei Schwämmen. Von...In: Arch. Anat. und Physiol., Physiol. Abth. p. 168–186. German translation of the first chapter of 1898 (α).

1898 (α) **Vosmaer, G. C. J. & C. A. Pekelharing.** Observations on Sponges. By...In: Verh. K. Akad. Wetensch. Amsterdam, Tweede Sectie, vi, No. 3, p. 1–51, Pl. I–IV. Cf. 1898. The first chapter also in: Onderz. Phys. Lab. Utrecht (5), i, p. 1–30 (1899).

1898 (β) **(Vosmaer & Pekelharing.** Over het opnemen van veodsel bij sponsen.) In: K. Akad. Wetensch. Amsterdam, Versl. vergad. p. 494–498.

1898 **Weltner, W.** Die Gattung Damiria. Von...In: Zool. Anz. xxi, p. 429–431.

1898 (α) **Weltner, W.** Viaggio del Dr A. Borelli nel Chaco Boliviano e nella Repubblica Argentina. xiv, Ephydatia Ramsayi (Haswell) forma talaensis aus Argentinien. Von...In: Boll. Mus. Zool. Anat. Comp. Torino, xiii, 3 pp.

1898 (β) **Weltner, W.** Ostafrikanische Süsswasserschwämme, gesammelt von Dr F. Stuhlmann 1888 und 1889...In: Mittheil. Naturhist. Mus. Hamburg, xv, p. 119–131, 1 Pl.

1898 (γ) **Weltner, W.** (Über Formolconservirung von Süsswasserthieren.) Von...In: Sitz. Ber. Ges. naturf. Freunde, Berlin, p. 57–63.

1898 **Wilson, H. V.** On the feasibility of raising Sponges from the Egg. By...In: Bull. U. S. Fish. Comm. xvii, p. 241–245.

1899 **Delage, Y.** Spongiaires. In: Delage et Hérouard, Traité de zoologie concrète, Tome ii, 1re Partie, p. 49–244, Pl. 4–15, text-figures.

1899 (α) **(Delage, Y.** On the Position of Sponges in the Animal Kingdom.) In: Proc. Fourth Internat. Congr. Zool. (Cambridge), p. 57–62.

1899 **Evans, R.** A Description of Two New Species of Spongilla from Lake Tanganyika. By...In: Quart. Journ. Microsc. Sc. xli, p. 471–488, Pl. 37–38.

1899 (*a*) **Evans, R.** The Structure and Metamorphosis of the Larva of Spongilla lacustris. By...In: Quart. Journ. Microsc. Sc. xlii, p. 363–476, Pl. 35–41.

1899 **Girod, P.** Sur la position systématique de Carterius bohemicus (Ephydatia bohemica, Petr)...In: Bull. Soc. Zool. France, xxiv, p. 54–56.

1899 (*a*) **Girod, P.** Considérations sur la distribution géographique des Spongilles d'Europe...In: Bull. Soc. Zool. France, xxiv, p. 51–53.

1899 **Gravier, Ch.** Sur une nouvelle espèce d'Éponge d'eau-douce du genre Parmula Carter et sur la biologie des éponges de ce genre. Par ...In: Bull. Mus. H. N. Paris, p. 126–129.

1899 (*a*) **Gravier, Ch.** Sur une collection d'Éponges (Hexactinellides) du Japon. Par...In: Bull. Mus. H. N. Paris, v, p. 419–423.

1899 **Hall, T. S.** Two new Victorian Palaeozoic Sponges. By...In: Proc. R. Soc. Victoria, Melbourne (2), xi, p. 152–155, Pl. xiv.

1899 **Johnson, J. Y.** Notes on some Sponges belonging to the Clionidae obtained at Madeira. By...In: Journ. R. Microsc. Soc. London, p. 461–463, Pl. vi, fig. 1–6.

1899 (**Kent, Saville.** Discussion on the position of Sponges.) In: Proc. Fourth Internat. Congr. Zool. (Cambridge), p. 66–68. Cf. 1899 (*a*), Delage.

1899 **Lindgren, N. G.** Einige Bemerkungen zu meinem Aufsatz "Beitrag zur Kenntniss der Spongienfauna des Malayischen Archipels und der Chinesischen Meere." Von...In: Zool. Anz. xxii, p. 87–89.

1899 **Maas, O.** Ueber Reifung und Befruchtung bei Spongien. Von...In: Anat. Anz. xvi, p. 290–298, 12 figs.

1899 (**Minchin, E. A.** Discussion on the position of Sponges.) In: Proc. Fourth Internat. Congr. Zool. (Cambridge), p. 62–65. Cf. 1899 (*a*), Delage.

1899 **Petr, F.** Studie o houbách sladkovodních. Část 1. O vývoji a významu jehlic parenchymových. Napsal...In: Rozpravy české akad. cis. Frant. Jos. viii, p. 1–35, Pl. i–ii. (Studies on the Freshwater Sponges. Part i, On the development and the significance of the spicules of the parenchyma By...In: Trans. Bohemian Acad. Franz-Joseph, Prag.) [Abstr. in: Zool. Anz. xxi, p. 226–227.]

1899 **Ravn, J. P. L.** Et Par danske Kridtspongier. In: Medd. dansk. geol. Foren. v, p. 23–32, 1 Pl.

1899 **Rousseau, L.** Quelques mots à propos de la technique microscopique dans l'étude de Spongiaires...In: Ann. Soc. Belge Microsc. xxiv, p. 51–56.

1899 **Schrammen, A.** Beitrag zur Kenntnis der obersenonen Tetractinelliden...In: Mitth. Römer-Mus. Hildesheim, no. 10, 9 pp., 3 Pl.

1899 **Schulze, F. E.** Zur Histologie der Hexactinelliden. Von...In: Sitz. Ber. K. Pr. Akad. Wiss. Berlin, xiv, p. 198-209, 3 figs.

1899 (α) **Schulze, F. E.** Amerikanische Hexactinelliden nach dem Materiale der Albatross-Expedition bearbeitet von...Mit einem Atlas von 19 Tafeln...Jena. 4°. 126 pp., Pl. I–XIX.

1899 (β) **(Schulze, F. E.** Ueber Hyalonema affine W. Marshall.) In: Sitz. Ber. Ges. Naturf. Fr. Berlin...p. 112–129, 1 fig.

1899 (γ) **(Schulze, F. E.** Discussion über die System-Stellung der Spongien.) In: Proc. Fourth Internat. Congr. Zool. (Cambridge), p. 68. Cf. 1899 (α), Delage.

1899 **Sukatschoff, B.** Über den feineren Bau einiger Cuticulae und der Spongienfasern. Von...In: Z. W. Z. lxvi, p. 377–406, Pl. XXIV–XXVI.

1899 **Thiele, J.** Studien über pazifische Spongien, II. Von...In: Zoologica, Heft 24, p. 1–33, Pl. I–V.

1899 (α) **Thiele, J.** Ueber Crambe crambe (O. Schmidt)...In: Arch. f. Naturgesch. lxv, p. 87–94, Pl. VII.

1899 **Topsent, E.** Documents sur la faune des Spongiaires des côtes de Belgique...In: Arch. Biol. xvi, p. 105–115.

1899 **Tschernyschew, Th.** Ueber die Artinsk- und Carbon-Schwämme vom Ural und vom Timan...In: Verh. Russ. K. Mineral. Ges. St Petersburg (2), xxxvi, p. 1–54, Pl. I–V. [Translation of 1898?—G. P. B.]

1899 **(Vosmaer, G. C. J.** Discussion on the position of Sponges.) In: Proc. Fourth Internat. Congr. Zool. (Cambridge), p. 65–66. Cf. 1899 (α), Delage.

1899 **Vosmaer, G. C. J. & C. A. Pekelharing.** Over het opnemen van voedsel bij Sponzen. Door...In: Onderz. Phys. Labor. Utrecht (5), i, p. 1–30. Transl. of 1898 (α).

1899 (α) **Vosmaer, G. C. J. & C. A. Pekelharing.** De Kraagcellen van Sponzen. Door...In: Onderz. Phys. Labor. Utrecht (5), i, p. 31–39. Transl. of 1898 (α).

1900 **Delage, Y.** Développement d'une éponge siliceuse. Par...In: Zoologie descriptive (redact. L. Boutan). Paris, Oct. Doin. i, p. 175–189.

1900 **Evans, R.** A Description of Ephydatia blembingia, with an Account of the Formation and Structure of the Gemmulae...In: Quart. Journ. Microsc. Sc. (2), xliv, p. 71–109, Pl. I–IV.

1900 **Hempel, E.** Süsswasserschwämme (Spongillen) bei Chemnitz...In: Ber. Nat. Ges. Chemnitz, xiv, p. lxi–lxiii.

1900 **Hinde, G. J.** On some Remarkable Calcisponges from the Eocene Strata of Victoria (Australia). By...In: Quart. Journ. Geol. Soc. lvi, p. 50–66, Pl. III–V.

1900 **Kieschnick, O.** Kieselschwämme von Amboina. Von...In: Denkschr. Medic. Naturw. Ges. Jena, viii (Zool. Erforsch. Austr. Semon), p. 545–582, Taf. XLIV–XLV.

1900 **Kirkpatrick, R.** Description of Sponges from Funafuti. By...In: Ann. and Mag. N. H. (7), vi, p. 345–362, Pl. xiii–xv.

1900 (α) **Kirkpatrick, R.*** On the Sponges of Christmas Island. In: Proc. Zool. Soc. London, p. 127–141, Pl. xii–xiii.

1900 **Lambe, L. M.** Catalogue of the Recent Marine Sponges of Canada and Alaska. By...In: Ottawa Naturalist, xiv, p. 153–172.

1900 (α) **Lambe, L. M.** Sponges from the Coasts of North-eastern Canada and Greenland. By...In: Trans. R. Soc. Canada (2), vi, p. 19–49, Pl. i–vi.

1900 **Lister, J. J.** Astrosclera Willeyana, the type of a new family of Sponges. By...In: Willey's Zoological Results, iv, p. 459–482, Pl. xlv–xlviii.

1900 (α) **Lister, J. J.** The Skeleton of Astrosclera compared with that of the Pharetronid Sponges...In: Proc. Cambridge Phil. Soc. p. 189–190.

1900 **Maas, O.** Die Weiterentwicklung der Syconen nach der Metamorphose. Von...In: Z. W. Z. lxvii, p. 215–240, Pl. ix-xii.

1900 (α) **(Maas, O.** Ueber die sogen. Biokrystalle und die Skeletbildungen niederer Thiere. Auszug aus dem Vortrag, gehalten am 19. Juni 1900.) In: Sitz. ber. Ges. Morphol. und Physiol. München, xvi, p. 42–45.

1900 (β) **Maas, O.** Eine neue zusammenfassende Darstellung der Schwämme. The Porifera by E. A. Minchin...In: Biol. Centralbl. xx, p. 789–793.

1900 **MacKay, A. H.** A Fresh Water Sponge from Sable Island...In: Trans. Nova Scotia Inst. Sc. Halifax, x, p. 319–322, figs.

1900 **MacMunn, C. A.** On Spongioporphyrin: the Pigment of Suberites Wilsoni. By...In: Quart. Journ. Microsc. Sc. xliii, p. 337–349, Pl. 16.

1900 **(Minchin, E. A.** Porifera.) In: E. Ray Lankester, A Treatise on Zoology...Part ii, The Porifera and Coelentera...London, Adam and Charles Black, 1900, p. 1–178, and 97 woodcuts.

1900 (α) **Minchin, E. A.** Éponges calcaires. Par...In: Zoologie descriptive des Invertébrés (redact. L. Boutan). Paris, Oct. Doin. i, p. 107–147.

1900 **Orueta, D. de.** Descripción de unas Esponjas del Cantábrico...In: Act. Soc. Españ. H. N. p. 103–107, 4 figs.

1900 **Rothpletz, A.** Ueber einen neuen jurassischen Hornschwamm und die darin eingeschlossenen Diatomeen. Von...In. Zeitschr. D. Geol. Ges. lii, p. 154–160.

1900 (α) **Rothpletz, A.** Nachtrag zu meinem Aufsatz über einen neuen jurassischen Hornschwamm und die darin eingeschlossenen Diatomeen. Von...In: Zeitschr. D. Geol. Ges. lii, p. 388–389.

* Kirkpatrick's paper forms the fourth chapter of: "On the Marine Fauna of Christmas Island (Indian Ocean)," by C. W. Andrews, E. A. Smith, H. M. Bernard, R. Kirkpatrick and F. C. Chapman (Proc. Z. S. p. 115–141).

195

1900 **Schulz, E.** Die Hornschwämme von Thursday Island und Amboina.
Von...In: Denkschr. Medic. Naturw. Ges. Jena, viii (Zool.
Forsch. Austr. Semon), p. 525–544, Taf. XLIII.

1900 **Schulze, F. E.** Die Hexactinelliden. Von...In: Fauna Arctica
(Römer und Schaudinn). Jena, Fischer. 4º. p. 87–108, Pl. I–IV.

1900 (α) **Schulze, F. E.** Hexactinelliden des Indischen Oceanes. iii Theil.
Von...In: Abh. K. Pr. Akad. Wiss. 46 pp., Pl. I–VII.

1900 (β) **(Schulze, F. E.** Ueber Corbitella speciosa, Quoy et Gaimard und
Corbitella corbicula, Bowerbank.) In: Sitz. Ber. Ges. Naturf.
Freunde, Berlin, p. 156–165.

1900 (γ) **Schulze, F. E.** Mittelmeer-Hexactinelliden. Von...In: Denkschr.
Math.-naturw. Cl. Kais. Akad. Wiss. Wien (Ber. Comm. Erforsch.
östl. Mittelm. xxiii; Zool. Ergebn. xii), Bd. lxix, p. 1–8, Pl. I.
Separ. copy.

1900 (δ) **Schulze, F. E.** Hexactinelliden des Rothen Meeres. Von...In:
Denkschr. Math.-naturw. Cl. Kais. Akad. Wiss. Wien (Ber. Comm. f.
oceanogr. Forsch. Zool. Ergebn. xvi), Bd. lxix, p. 311–324, Pl. I–III.

1900 **Thiele, J.** Kieselschwämme von Ternate. i. Von...In: Abhandl.
Senckenb. Naturf. Ges. xxv, p. 19–80, Pl. II–III.

1900 **Topsent, E.** Étude monographique des Spongiaires de France.
iii, Monaxonida (Hadromerina). Par...In: Arch. Zool. Expérim.
(3), viii, p. 1–331, Pl. I–VIII.

1900 (α) **Topsent, E.** Éponges siliceuses. Par...In: Zoologie descriptive des
Invertébrés (redact. L. Boutan), Paris, Oct. Doin. i, p. 148–174.

1900 **Waller, J. G.** On an Undescribed British Sponge of the genus Rhaphio-
desma, Bowerbank. By...In: Journ. Quekett Microsc. Club (2),
vii, p. 253–256, Pl. 14.

1900 **Weltner, W.** Süsswasser-Schwämme. Bearbeitet von...In: Denkschr.
Medic. Naturw. Ges. Jena, viii (Zool. Forschungsreise Austr.
Semon), p. 517–524, Taf. XLII.

1900 **Zemlitschka, F.** Ueber die Aufnahme fester Theilchen durch die
Kragenzellen von Sycandra. Von...In: Z. W. Z. lxvii, p. 241–
246, 2 figs.

1901 **Arnesen, E.** (1) Spongier fra den norske Kyst. i, Calcarea. Systematisk
katalog med bemerkninger og bestemmelsestabel...Ved...In:
Bergens Museums Aarborg 1900, v, p. 1–46, Pl. I.

1901 (α) **Arnesen, E.** (2) Calcarea...In: Meeresfauna von Bergen, redig.
v. Dr A. Appellöf, p. 65–72.

1901 **Blasius, W.** Süsswasser-Schwämme, Spongilliden, bei Braunschweig
Von...In: Jahresber. Ver. Naturw. Braunschweig, xii, p. 43–45.

1901 **Bütschli, O.** Einige Beobachtungen über Kiesel- und Kalknadeln von
Spongien. Von...In: Z. W. Z. lxix, p. 235–286, Pl. XIX–XX and
2 woodcuts.

1901 **Cotte, J.** Notes biologiques sur le Suberites domuncula (Spongiaires)
par...Paris, L. Boyer. 128 pp.

13—2

1901 (α) **Cotte, J.** Note sur les diastases du Suberites domuncula (Spongiaires). Par...In: C. R. Soc. Biol. Paris, liii, p. 95–97.

1901 **Evans, R.** A Description of Ephydatia blembingia, with an Account of the Formation and Structure of the Gemmule. By...In: Quart. Journ. Microsc. Sc. xliv, p. 71–109, Pl. 1–4.

1901 **Fielder, W.** An Australian Collection of Sponges...In: Amer. Monthly Microsc. Journ. xxi, p. 327–342 and 1 Pl.*

1901 **Ijima, I.** Studies on the Hexactinellida. Contribution i. (Euplectellidae.) By...In: Journ. Coll. Science Imp. Univ. Tokyo, Japan, xv, p. 1–299, Pl. i–xiv.

1901 **Kirkpatrick, R.** Description of a new Hexactinellid Sponge from South Africa. By...In: Ann. and Mag. N. H. (7), vii, p. 457–460, Pl. viii.

1901 **Koorders, S. H.** Notiz über Symbiose einer Cladophora mit Ephydatia fluviatilis, in einem Gebirgssee in Java. Von...In: Ann. Jardin Botan. Buitenzorg (2), iii, p. 8–16, Pl. i–ii.

1901 **Lendenfeld, R. von.** (Review of Ijima, 1901) Z. Centralbl. 8, p. 562–566.

1901 **Levander, K. M.** Anteckningar till Finnlands Spongillidfauna af... In: Meddel. Soc. Fauna Flora Fenn. xxvii, p. 56–60.

1901 **Maas, O.** Ueber Entstehung und Wachstum der Kieselgebilde bei Spongien. Von...In: Sitz. Ber. mathem.-phys. Cl. K. bayer. Akademie Wiss. xxx, p. 553–569, Pl. v.

1901 (α) **Maas, O.** Die Knospenentwicklung der Tethya und ihr Vergleich mit der geschlechtlichen Fortpflanzung der Schwämme. Von... In: Z. W. Z. lxx, p. 263–288, Pl. xiii–xiv.

1901 **Malfatti, P.** Contributo alla Spongiofauna del Cenozoico italiano. In: Palaeontogr. Ital. vi, p. 267–302, 6 Pl.

1901 **Orueta, D. de.** Descripción de algunas Esponjas del Cantábrico...In: Bol. Soc. Españ. H. N. i, p. 331–335, 4 figs., 2 Pl.

1901 **Schrammen, A.** Neue Kieselschwämme aus der oberen Kreide der Umgebung von Hannover und Hildesheim...In: Mitth. Römer. Mus. Hildesheim, xiv, 26 pp., 5 Pl.

1901 **Smith, H. M.** Notes on the Florida Sponge Fishery in 1899...In: Bull. U. S. Fish. Comm. xix, p. 149–151.

1901 **Topsent, E.** Notice préliminaire sur les éponges recueillies par l'expédition antarctique belge, par...In: Arch. Zool. Expérim. (3), ix, Notes, p. 5–16.

1901 (α) **Topsent, E.** Les Spongiaires de l'expédition antarctique belge et la bipolarité des faunes. Note de...In: C. R. Paris, cxxxii, p. 168–169.

1901 **Weltner, W.** Süsswasserspongien von Celebes (Spongillidenstudien iv). Von...In: Arch. Naturgesch. lxvii, Beiheft, p. 187–204, Pl. vi–vii.

* The paper was read in Dec. 1900, and, therefore, probably not "published" before Jan. 1901.

1901* **Whitelegge, Th.** Report on Sponges from the Coastal Beaches of New South Wales. By...In: Rec. Australian Museum, iv, p. 55–118, Pl. x–xv.

1901 **Whitfield, R. P.** Notice of a New Sponge from Bermuda and of some other forms from the Bahamas. By...In: Bull. Amer. Mus. Nat. Hist. xiv, p. 47–50, Pl. i–v.

1902 **Allen, E. J. & R. A. Todd.** The Fauna of the Exe Estuary. By...In: Journ. Mar. Biol. Assoc. New Ser. vi, p. 295–335 and 1 chart.

1902 **Bidder, G.** Notes on Plymouth Sponges. By...In: Journ. Mar. Biol. Assoc. New Ser. vi, p. 376–382.

1902 **Cotte, J.** Note sur la nature des produits de désassimilation chez les Spongiaires...In: C. R. Soc. Biol. Paris, liv, p. 1317-1318.

1902 (α) **Cotte, J.** Observations sur les gemmules de Suberites domuncula ...In: C. R. Soc. Biol. Paris, liv, p. 1493–1495.

1902 (β) **Cotte, J.** Comment les choanocytes de Sycandra raphanus absorbent-ils les particules alimentaires, par...In: C. R. Soc. Biol. Paris, liv, p. 1315–1317.

1902 (γ) **Cotte, J.** Note sur le mode de perforation des Cliones...In: C. R. Soc. Biol. Paris, liv, p. 636–637.

1902 **Hutchinson, A.** Appendix (on the mineralogical character of the skeleton of Astrosclera)...In: Zool. Res. Willey, vi, p. 735.

1902 **Ijima, I.** Studies on the Hexactinellida. Contribution ii. (The genera Corbitella and Heterotella.) By...In: Journ. Coll. Sc. Univ. Tokyo, xvii, p. 1–34, Pl. (i).

1902 (α) **Ijima, I.** Note on Walteria leuckarti, Ij. By...In: Annot. Zool. Japonenses, iv, p. 119–122.

1902 (β) **Ijima, I.** Über die von mir in der Sagami-See gesammelten Hexactinelliden...In: Verh. 5. intern. Z. Congress, p. 689–692.

1902 **Kirkpatrick, R.** Descriptions of South African Sponges. By...In: Mar. Investig. S. Africa, Departm. Agric. (i), p. 219–232, Pl. i–iii. In: Z. Anz. xxvii, p. 190, mentioned under the same title, with add.: Rep. Govern. Biol. Cape Good Hope 1901, p. 54–67. (Reprint ?)

1902 (α) **Kirkpatrick, R.** Porifera. In: Rep. Coll. Nat. Hist. Southern Cross, p. 317–318.

1902 **Korschelt, E. & K. Heider.** Lehrbuch der vergleichenden Entwicklungsgeschichte der wirbellosen Thiere, von...(Allgem. Theil; erste und zweite Aufl.) Jena, p. 1–538.

1902 **Lameere, A.** De l'origine des éponges, par...In: Ann. Soc. malac. Belg. xxxvi, p. vii–viii.

1902 **Lauterborn, R.** Ein für Deutschland neuer Süsswasserschwamm (Carterius Stepanowi, Dyb.)...Von...In: Biol. Centralbl. xxii, p. 519–535, 5 figs.

---

* This paper appeared in 1901 according to Zool. Jahresber. Neapel.

1902 **Lundbeck, W.** Porifera. (Part i.) Homorrhaphidae and Heteror rhaphidae. By...In: The Danish Ingolf-Expedition, vi, p. 1–108, Pl. I–XIX and 1 map.

1902 **Schrammen, A.** Neue Hexactinelliden aus der oberen Kreide. In: Mitth. Roemer-Mus. Hildesheim, No. 15, 26 pp., 4 Pl.

1902 **Schulze, F. E.** An Account of the Indian Triaxonia collected by the R. Indian Mar. Surv. Ship Investigator. By...(transl. by R. v. Lendenfeld). Calcutta. 4°. 113 pp., 23 Pl.

1902 **Sollas, Igerna B. J.** On the Sponges collected during the "Skeat Expedition" to the Malay Peninsula, 1899–1900. By...In: Proc. Zool. Soc. London, ii, Part 1, p. 210–221, Pl. XIV–XV.

1902
1902 (α) **Swartschewsky, B.** (Baikal Fresh-water sponges) Russian. In: Zapiski Kiev. Obshch. xvii, 2 (1901), p. ix–xiv, 1 Pl., and p. 329–352, Pl. III–VI.

1902 **Topsent, E.** Les Asterostreptidae par...In: Bull. Sc. et Médic. de l'Ouest...xi (p. 1–18).

1902 (α) **Topsent, E.** Sur l'orientation des Crinorhiza. Par...In: C. R. Acad. Sc. Paris, cxxxiv, p. 58–60.

1902 (β) **Topsent, E.** Considérations sur la faune des Spongiaires des Côtes d'Algérie. Éponges de la Calle. Par...In: Arch. Zool. Expérim. (3), ix, p. 327–370, Pl. XIII–XIV.

1902 (γ) **Topsent, E.** Éponges nouvelles des Açores. (Deuxième série.) Par...In: Mém. Soc. Zool. France, xiv, p. 448–466.

1902 (δ)* **Topsent, E** Spongiaires par...In: Expéd. antarct. Belge. Résultats Voyage S. Y. Belgica 1897–1899...54 pp., Pl. I–VI.

1902 **Urban, F.** Rhabdodermella nuttingi, nov. gen. et nov. spec. Von... In: Z. W. Z. lxxi, p. 268-275, Pl. XIV.

1902 **Vosmaer, G. C. J.** On the Shape of some Siliceous Spicules of Sponges. By...In: Kon. Akad. Wetensch. Amsterdam, Proceedings Meeting...p. 104–114. English transl. from 1902 (β).

1902 (α) **Vosmaer, G. C. J. & J. H. Vernhout.** The Porifera of the Siboga-Expedition. i. The Genus Placospongia. By...In: Siboga-Expeditie, vi A, p. 1–17, Pl. I–V.

1902 (β) **Vosmaer, G. C. J.** Over den vorm van sommige kiezel-spicula by Sponzen...In: Kon. Akad. Wetensch. Amsterdam, Versl. gew. verg. wis- en natuurk. afd. p. 167–178.

1902 **Whitelegge, Th.** Supplementary Notes to the Report on Sponges from the Coastal Beaches of New South Wales. By...In: Rec. Austral. Mus. iv, p. 211–216.

1902 (α) **Whitelegge, Th.** Notes on Lendenfeld's Types described in the Catalogue of Sponges in the Australian Museum. By...In: Rec. Austral. Mus. iv, p. 274–288.

1902 **Wilson, H. V.** On the Asexual Origin of the ciliated Sponge Larva. By...In: Amer. Naturalist, xxxvi, p. 451–459.

* On title is printed 1901. According to Zool. Anzeig. it appeared, however, only in 1902.

199

1902 (α) **Wilson, H. V.** The Sponges collected in Porto Rico in 1899 by the U. S. Fish. Commission Steamer Fish Hawk. By...In: Bull. U. S. Fish. Comm. 1900, ii, p. 375–411, 30 figs.

1903 **Albert I.**, Prince de Monaco. Sur la quatrième campagne de la Princesse Alice II. In: C. R. Paris, cxxxvi, p. 211–215. With 1 woodcut.

1903 (**Anonymous.**) Die Schwammfischerei im Golf von Gabes an der tunesischen Küste im Jahre 1901. In: Mitth. deutsch. Seefischerei-Ver. xix, p. 34.

1903 (α) (**Anonymous.**) Perlen-, Korallen-, Schwamm- und Austern-fischerei in Columbien. In: Mitth. d. Seefisch. Ver. xix, p. 84.

1903 (β) (**Anonymous.**) Schwammfischerei in Key-West. In: Mitth. d. Seefisch. Ver. xix, p. 446.

1903 **Arnesen, Emily.** Spongien von der norwegischen Küste. ii, Monaxonida: Halichondrina. Von...In: Bergens Mus. Aarbog, p. 1–30, Pl. i–vii.

1903 **B.** Die italienische Korallen- und Schwammfischerei im Jahre 1900 und 1901. In: Mitth. d. Seefisch. Ver. xix, p. 222–223.

1903 **Baar, R.** Hornschwämme aus dem Pacific. Ergebnisse einer Reise nach dem Pacific, Schauinsland 1896–97. Von...In: Zool. Jahrb. Abth. Syst. xix, p. 27–36, 3 figs.

1903 **Bertrand, G.** Recherches sur l'existence normale de l'arsenic dans l'organisme. In: Résult. Camp. Sc. Albert I. Fasc. 24.

1903 **Cotte, J.** Contribution à l'étude de la nutrition chez les Spongiaires par...In: Bull. Sc. France et Belgique, xxxviii*, p. 420–573, 10 woodcuts.

1903 (α) **Cotte, J.** Les éponges élaborent-elles de l'amidon?...In: C. R. Soc. Biol.,Paris, lv, p. 674–676.

1903 (β) **Cotte, J.** Sur la présence du manganèse et du fer chez les Éponges ...In: C. R. Soc. Biol. Paris, lv, p. 139–141.

1903 (γ) **Cotte, J.** Sur la présence de la tyrosinase chez Suberites domuncula...In: C. R. Soc. Biol. Paris, lv, p. 137–139.

1903 (δ) **Cotte, J.** Sur la nature des lipochromes...In: C. R. Soc. Biol. Paris, lv, p. 812–813.

1903 (ε) **Cotte, J.** Sur quelques phénomènes dégéneratifs, observés chez Sycandra raphanus...In: C. R. Assoc. Franç. Avanc. Sc. xxxi, p. 733–739, 5 figs.

1903 **Duerden, J. E.** West Indian Sponge-Incrusting Actinians. In: Bull. Amer. Mus. N. H. xix, p. 495–503, Pl. xliv–xlvii.

1903 **Görich, W.** Zur Kenntniss der Spermatogenese bei den Poriferen und Coelenteraten. Von...In: Z. Anz. xxvii, p. 64–70. With 3 woodcuts.

* Also published as: Thèses présentées à la faculté des Sciences de Paris, Sér. A, No. 448. No. d'ordre 1130. Lille.

1903 (α) **Görich, W.** Weiteres ueber die Spermatogenese bei den Poriferen und Coelenteraten. Von...In: Z. Anz. xxvii, p. 172–174.

1903 **Henking, H.** Schwammfischerei der Bahama-Inseln. In: Mitth. d. Seefisch. Ver. xix, p. 305.

1903 (α) **Henking, H.** Schwammfischerei Haitis. In: Mitth. d. Seefisch. Ver. xix, p. 305.

1903 (β) **(Henking, H.)** Beaufsichtigung der Schwammfischerei mit Taucherapparaten. In: Mitth. d. Seefisch. Ver. xix, p. 445–446.

1903 **Ijima, I.** Studies on the Hexactinellida. Contribution iii...By... In: Journ. Coll. Sc. Imp. Univ. Tokyo, xviii, p. 1–124, Pl. i–viii.

1903* **Kirkpatrick, R.** Descriptions of South African Sponges. Part ii. By...In: Marine Invest. South Africa, Dep. Agric. ii, p. 171–180, Pl. iv.

1903 (α) **Kirkpatrick, R.** Descriptions of South African Sponges. Part iii. By...In: Marine Invest. South Africa, ii, p. 233–264, Pl. v–vi.

1903 **Lendenfeld, R. von.** Tetraxonia bearbeitet von...In: Das Tierreich, Lfg. 19, p. 1–168, 44 woodcuts.

1903 (α) **Lendenfeld, R. von.** Eine biologische Notiz über Spongilla fragilis, Leidy. Von...In: Arch. Naturg. lxix, p. 181–182, Pl. x.

1903 **Lo Biancho, S.** Le pesche abissali eseguite da F. A. Krupp col Yacht Puritan nelle adiacenze di Capri ed in altre località del Mediterraneo. In: Mitth. Z. Stat. Neapel, xvi, p. 109–279, Pl. vii–ix. Vide pp. 243, 245, 246, 270.

1903 **Lundbeck, W.** Die Bezeichnung der Spongiennadeln und Anderes. Im Anschluss an Prof. v. Lendenfeld's Referat über meine Spongienarbeit. Von... In: Z. Anz. xxvi, p. 390–392.

1903 **Mandouli, H.** Recherches sur les colorations tégumentaires. In: Ann. Sc. natur. (8), xviii, p. 225–469, Pl. 3 and 4. Vide pp. 310–366, 367, 438.

1903 **Moore, J. E. S.** The Tanganyika Problem. An Account of the Researches undertaken concerning the existence of marine Animals in Central Africa. London. 1903. xxiii and 371 pp. Vide pp. 309–323, 331, 342, 353, 354.

1903 **Preiwisch, J.** Kalkschwämme aus dem Pacific, Ergebnisse einer Reise nach dem Pacific, Schauinsland 1896–97. Von...In: Z. Jahrb. Abth. Syst. xix, p. 9–26, Pl. 2–4.

1903 **Rousseau, E.** Note monographique sur les Spongiaires de Belgique. Par...In: Ann. Soc. malacol. Belg. xxxvii, p. 3–26, 17 figs. (woodcuts).

1903 **Schrammen, A.** Zur Systematik der Kieselspongien...In: Mitth. Römer. Mus. Hildesheim, No. 19, 21 pp.

* On the title is printed 1902; but at the end of the article we read: "Published 27th May, 1903."

1903 (α) **Schrammen, A.** Ueber den Horizont der Thecosiphonia nobilis, Roem...In: Centr. Min. Geol. Pal. p. 19–23.

1903 **Schulze, F. E.** Caulophacus arcticus (Armauer Hansen) und Calycosoma gracile, F. E. Sch. nov. spec. Von...In: Abh. K. Preuss. Akad. Wiss. (p. 1–22), Pl. I–II.

1903 **Sollas, I.** On Haddonella topsenti, gen. et sp. n., the Structure and Development of its Pithed Fibres. By...In: Ann. and Mag. N. H. (7), xii, p. 557–563, Pl. XXVIII–XXIX.

1903 **Thiele, J.** Kieselschwämme von Ternate. ii. Von...In: Abhandl. Senckenb. naturf. Ges. xxv, p. 933–968, Pl. XXVIII.

1903 (α) **Thiele, J.** Beschreibung einiger unzureichend bekannten monaxonen Spongien. Von...In: Arch. Naturgesch. lxix, p. 374–398, Pl. XXI, fig. 1–28 A.

1903 **Topsent, E.** Sur les larves cuirassées de Thoosa armata, par...In: Arch. Zool. Expérim. (4), i, p. i–iii.

1903 **Ungern-Sternberg, E. von.** Die Hexactinelliden der Senonen Diluvialgeschiebe in Ost- und Westpreussen...In: Schr. Physik. Oek. Ges. Königsberg, xliii, p. 132–151, Pl. IV–VI.

1903 **Urban, F.** Ueber das Dermalepithel der Kalkspongien...In: Verh. Ges. d. Naturf. u. Aerzte, Vers. 74 (2), i, p. 159.

1903 **Zacharias, O.** Eine zweite deutsche Fundstätte für Carterius stepanowi, Dyb. In: Biol. Centralbl. xxiii, p. 483–484.

1904 (α) **Cotte, J.** Des phénomènes de la nutrition chez les Spongiaires. In: C. R. Assoc. Franç. Av. Sc. xxxii, 2, p. 776–780.

1904 **Görich, W.** Zur Kenntniss der Spermatogenese bei den Poriferen und Coelenteraten nebst Bemerkungen über die Oogenese der ersteren. Von...In: Z. W. Z. lxxvi, p. 522–543, Pl. XXXI.

1904 **Henze, M.** Spongosterin, eine cholesterinartige Substanz aus Suberites domuncula, und seine angebliche Beziehung zum Lipochrom dieses Tieres. Von...In: Zeitschr. physiol. Chemie, xli, p. 109–124.

1904 **Hinde, G. J.** On the Structure and Affinities of the genus Porosphaera, Steinmann. By...In: Journ. R. Microsc. Soc. London, p. 1–25, Pl. I–II.

1904 **Ijima, I.** Studies on the Hexactinellida. Contribution iv. (Rossellidae.) By...In: Journ. Coll. Sc. Imp. Univ. Tokyo, xviii, p. 1–307, Pl. I–XXIII.

1904 **Lendenfeld, R. von.** Ueber die Herstellung von Nadelpräparaten von Kieselschwämmen. Von...In: Zeitschr. wiss. Mikrosk. xxi, p. 23–24.

1904 (**Maas, O.**) Ueber die Wirkung der Kalkentziehung auf die Entwicklung der Kalkschwämme...In: Sitz. Ber. Ges. f. Morphol. und Physiol. in München, xx, p. 4–21, 9 figs.

1904 (α) (**Maas, O.**) Ueber den Aufbau des Kalkskeletts der Spongien in normalem und in CaCO₃ freiem Seewasser. In: Verh. D. Zool. Ges. xiv, p. 190–199.

1904 **Schulze, F. E.** Hexactinellida. Bearbeitet von...In: Wiss. Ergebn. D. Tiefsee-Exp. Valdivia, iv, pp. i–viii, 1–226, Pl. I–LII.

1904 **Szymański, J. M.** Zur Anatomie und Systematik der Hornschwämme des Mittelmeers. Von...In: Zool. Anz. xxvii, p. 445–449.

1904 (α) **Szymański, J. M.** Hornschwämme von Aegina und Brioni bei Pola. Inaug. Diss....Breslau. 52 pp. and 9 woodcuts.

1904 **Thum, E.** Bericht über eine Sammlung trockener Chalineen-Skelete aus dem Brüsseler Museum. Von...In: Ann. Soc. R. zool. malacol. Belgique, xxxviii, p. 9–21, 26 figs.

1904 **Topsent, E.** Notes sur les éponges du Travailleur et du Talisman. In: Bull. Mus. Hist. Nat. Paris, x, p. 62–66.

1904 (α) **Topsent, E.** Spongiaires des Açores. In: Résult. Camp. Sc. Alb. Monaco, fasc. 25, p. 1–280, Pl. I–XVIII.

1904 (β) **Topsent, E.** Heteroclathria hallezi. Type d'un genre nouveau d'Ectyonines, par...In: Arch. Zool. Expérim. (4), ii, Notes et revue, p. xciii–xcviii. With 2 woodcuts.

1904 (γ) **Topsent, E.** Sarostegia oculata. Hexactinellide nouvelle des îles du Cap-Vert. Par...In: Bull. Mus. Océanogr. Monaco, No. 10, p. 1–8. With 3 woodcuts.

1904 **Ustinov, K.** (Russian.) (Einsturzseen des Kreises Laïschew, im Gouvernement Kasan.) In: Sitz. Protok. Nat. Ges. Kasan, Beil. 34, No. 219, 5 pp., 1 Karte.

1904 **Weltner, W.** Die Spongien im International Catalogue of Scientific Literature. Von...In: Zool. Anz. xxvii, p. 788–789.

1904 **Wilson, H. V.** The Sponges. (No. xxx of Reports on an Exploration off the West Coasts of Mexico, Central and South America, and off the Galapagos Islands, in charge of Alexander Agassiz, by the U. S. Fish. Commission Steamer "Albatross," during 1891, Lieut. Commander Z. L. Tanner, U. S. N., commanding.) In: Mem. Mus. Comp. Zoöl. Harvard Coll. xxx, No. 1. 164 pp. and 26 Pl.

1905* **Baer, L.** Silicospongien von Sansibar, Kapstadt und Papeete. Von ...In: Arch. Naturgesch. lxxii, p. 1–32, Pl. I–V.

1905 **Bartelletti, V.** Sulla posizione dei Poriferi nel regno animale. In: Boll. Natural. Siena, xxv, pp. 84–91, 105–106.

1905 **Bechhold, H.** Cf. 1906.

1905 **Bütschli, O.** Ueber die Einwirkung konzentrierter Kalilauge auf kohlensauren Kalk und das dabei sich bildende Doppelsalz. Von...In: Zool. Anz. xxix, p. 428–430.

---

* On the cover of Arch. Naturgesch. lxxii, Band i, Heft 1, is printed: "ausgegeben im Dezember, 1905." Consequently Maas mentioned the paper in the Zool. Jahresber. Neapel for 1905. For these reasons I bring the paper likewise under 1905. Cf. 1906, Baer.

203

1905 **Chapman, F.** New or Little-known Victorian Fossils in the National Museum, Melbourne. Part v, On the Genus Receptaculites. With a Note on R. australis from Queensland. In: Proc. R. Soc. Victoria N. S. xviii, p. 5–15.

1905 **Crossland, C.** The Oecology and Deposits of the Cape Verde Marine Fauna. In: Proc. Zool. Soc. London, ii, p. 170–186.

1905 **Dendy, A.** Report on the Sponges collected by Professor Herdman, at Ceylon, in 1902. By...In: Herdman, Rep. Pearl Oyster Fisheries, Suppl. xviii, p. 57–246, Pl. i–xvi.

1905 **Dragnewitsch, Pawla.** Spongien von Singapore. Inaug. Dissert. Bern. 36 pp.

1905 **Kirkpatrick, R.** On the Oscules of Cinachyra. By...In: Ann. and Mag. N. H. (7), xvi, p. 662–667, Pl. xiv.

1905 **Lambe, L. M.** A new recent marine Sponge (Esperella bellabellensis) from the Pacific coast of Canada. By...In: Ottawa Naturalist, xix, p. 14–15, Pl. i.

1905 **Lauterborn, R.** Die Ergebnisse einer biologischen Probeuntersuchung des Rheins. In: Arb. Kais. Gesundheitsamt. xxii, p. 630–652. (10 Pl.)

1905 **Lundbeck, W.** Porifera (Part ii). Desmacidonidae (pars). By...In: Danish Ingolf Exp. vi. 219 pp., Pl. i–xx.

1905 **Maas, O.** Entwicklungsmechanische Studien an Schwämmen. In: C. R. Congrès Internat. Zool. p. 238–239.

1905 (α) **Maas, O.** Zur Frage der Einwirkung von Kalilauge auf Kalkspat. Von...In: Zool. Anz. xxix, p. 558–559.

1905 **Minchin, E. A.** A Speculation on the Phylogeny of the Hexactinellid Sponges. By...In: Zool. Anz. xxviii, p. 439–448, 2 woodcuts.

1905 (α) **Minchin, E. A.** On the Sponge Leucosolenia contorta, Bowerbank, Ascandra contorta, Haeckel, and Ascetta spinosa, Lendenfeld. By...In: Proc. Zool. Soc. London, 1905, ii, p. 3–20, Pl. i.

1905 (β) **Minchin, E. A.** The Characters and Synonymy of the British Species of Sponges of the Genus Leucosolenia. By...In: Proc. Zool. Soc. London, 1904, ii, p. 349–396, 8 figs.

1905 **Neviani, A.** Spicole di tetractinellidi rinvenute nelle sabie post-plioceniche di Carrubare (Calabria)...In: Boll. Soc. geol. ital. xxiv, p. 265–274, 62 figs.

1905 **Pick, F. K.** Die Gattung Raspailia. Von...In: Arch. f. Naturgesch. p. 1–48, Pl. i–iv.

1905 **Swartschewsky, B.** (Beitrag zur Kenntniss der Schwamm-Fauna des Schwarzen Meeres.) In: Mém. Soc. Nat. Kiew, xx, p. 1–59, Pl. i–vii. (Russian; with German abstract.)

1905 **Thiele, J.** Die Kiesel- und Hornschwämme der Sammlung Plate. Von...In: Zool. Jahrb. Suppl. 6, p. 407–496, Pl. 27–33

204

1905 **Topsent, E.** Cliothosa seurati, Clionide nouvelle des Iles Gambier,
par...In: Bull. Mus. Hist. Nat. Paris, xi, p. 94–96.

1905 (α) **Topsent, E.** Étude sur les Dendroceratida par...In: Arch. Zool.
Expérim. (4), iii, p. clxxi–cxcii, 3 figs.

1905 (β) **Topsent, E.** Notes sur les Éponges du Travailleur et du Talisman
...In: Bull. Mus. H. N. Paris, x, p. 195–200, 372–378.

1905 (γ) **Topsent, E.** Note sur les Éponges recueillies par le Français dans
l'Antarctique; description d'une Dendrilla nouvelle. Par...In:
Bull. Mus. H. N. Paris, xi, p. 502–505.

1905 **Urban, F.** Kalifornische Kalkschwämme...Von...In: Arch. f.
Naturgesch. lxxii, p. 33–76, Pl. vi–ix.

1905 **Vosmaer, G. C. J. & H. P. Wijsman.** Over den bouw van sommige
Kiezelspicula bij Sponzen. i, De styli van Tethya lyncurium...
In: Versl. Gew. Vergad. W. en N. Afd. K. Akad. Wetensch.
Amsterdam, p. 733–748, 1 woodcut. Cf. 1905 (α).

1905 (α) **Vosmaer, G. C. J. & H. P. Wijsman.** On the Structure of some
Siliceous Spicules of Sponges. i, The styli of Tethya lyncurium.
By...In: Proceedings K. A. W. Meeting, May, p. 15–28, 1 wood-
cut. Transl. from 1905.

1905 **Weinschenk, E.** Ueber die Skeletteile der Kalkschwämme. Von...
In: Centralbl. f. Min. Geol. u. Pal. 1905, p. 581–588.

1905 **Whitelegge, Th.** Western Australian Prawns and Sponges. In: Rec.
Austral. Mus. vi, p. 119–120.

1905 **Whitfield, R. P.** Descriptions of New Fossil Sponges from the Hamilton
Group of Indiana. In: Bull. Amer. Mus. N. H. xxi, p. 297–300,
Pl. ix–xi.

1905 **Woodland, W.** Studies in Spicule Formation. i, The Development
and Structure of the Spicules in Sycons: with Remarks on the
Conformation, Modes of Disposition and Evolution of Spicules
in Calcareous Sponges generally. By...In: Quart. Journ.
Microsc. Sc. xlix, p. 231–282, Pl. 13–15.

1906 **Allemand, A.** La culture et l'acclimatation des éponges en Tunisie.
In: C. R. Assoc. Franç. Avanc. Sc. xxxv, pp. 117–118, 476–478.

1906 (α) **[Allemand, A.** (dit A. Allemand-Martin). Études de Physiologie
appliquée à la Spongiculture sur les côtes de Tunisie. Thèses...
pour obtenir le grade de docteur ès sciences naturelles. Lyon.
185 pp. and numerous illustrations.] Cf. Allemand, 1907.

1906 **Annandale, N.** Notes on the Freshwater Fauna of India. No. i. A
variety of Spongilla lacustris from Brackish Water in Bengal.
By...In: Journ. Asiatic Soc. Bengal (N. S.), ii, p. 55–58.

1906 **Arévalo, C.** Investigaciones ópticas sobre espiculas de algunas
especies de esponjas españolas. In: Bol. Soc. Españ. Hist. Nat.
vi, p. 368–375. (3 figs.)

1906 **Baer, L.** Silicispongien von Sansibar, Kapstadt und Papeete. Inaug. Dissert. von...Berlin, p. 5–36, Pl. ɪ–v. Reprint from 1905, Baer. N.B. different pagination.

1906 **Bechhold, H.** Strukturbildung in Gallerten...In: Zeitschr. Physik. Chem. lii, 1905, p. 185–199.

1906 **Bütschli, O.** Nochmals über die Einwirkung konzentrierter Kalilauge auf die Nadeln der Calcispongia. In: Zool. Anz. xxix, p. 640–643.

1906 (α) **Bütschli, O.** Ueber die Skeletnadeln der Kalkschwämme In: Centralbl. min. Geol. Pal. i, p. 12–15.

1906 (β) **Bütschli, O.** Über die Einwirkung koncentrirter Kalilauge und koncentrirter Lösung von kohlensaurem Kali auf kohlensauren Kalk. In: Verh. Nat. Med. Ver. Heidelberg (2), viii, p. 277–330.

1906 **Cotte, J.** La pêche des éponges en Tunisie. In: C. R. Assoc. Franç. Avanc. Sc. xxxiv, p. 587–593.

1906 **Darboux, G., P. Stéphan, J. Cotte & F. van Gaver.** L'Industrie des Pêches aux Colonies. Marseille.

1906 **Dragnewitsch, P.** Spongien von Singapore. In: Zool. Jahrb. Abt. Syst. xxiii, p. 439–448.

1906 **Dubois, R.** Le laboratoire maritime de biologie de Sfax (Tunisie). In: Ann. Soc. Linn. Lyon, lv.

1906 **Hammer, E.** Zur Kenntnis des feineren Baues und der Entwicklung der Calcispongien. Von...In: Sitz. Ber. Ges. Nat. Freunde, Berlin, p. 135–139.

1906 (α) (**Hammer, (E.).** Über Sycandra raphanus.) In: Verh. D. Zool. Ges. Marburg, p. 269–273.

1906 (β) **Hammer, E.** Zur Kenntnis von Hircinia variabilis. In: Sitz. Ber. Ges. Nat. Freunde, Berlin, p. 149–155, 1 Pl., fig. 1–3.

1906 **Kirkpatrick, R.** Zoological Results of the Third Tanganyika Expedition, conducted by Dr W. A. Cunnington, 1904–1905. Report on the Porifera, with Notes on Species from the Nile and Zambesi. By...In: Proc. Zool. Soc. London, i, p. 218–227, Pl. xv–xvii.

1906 **Maas, O.** Über die Einwirkung karbonatfreier und kalkfreier Salzlösungen auf erwachsene Kalkschwämme und auf Entwicklungsstadien derselben, von...In: Arch. Entw. Mechan. xxii, p. 581–599.

1906 **Miner, R. W.** A Guide to the Sponge Alcove in the American Museum of Natural History. In: Amer. Mus. Journ. vi, p. 219-250, fig. 1-31.

1906 **Richet, C.** De l'action toxique de la subéritine (extrait aqueux de Suberites domuncula). In: C. R. Soc. Biol. Paris, lxi, p. 598-600.

1906 (α) **Richet, C.** De la variabilité de la dose toxique de subéritine. In: C. R. Soc. Biol. Paris, lxi, p. 686–688.

1906 **Roemer, F.** Die Schwämme der neuen Schausammlung. In: Ber. Senckenb. nat. Ges. Frankfurt, p. 97–99.

1906 **Rousseau, E.** Note monographique sur les Spongiaires de Belgique. Par...ii, Les Spongilles. In: Ann. Soc. R. Zool. et Malac. xli, p. 119-127.

1906 **Sollas, Igerna B. J.** (Chapter on Porifera) in the Cambridge Natural History, edited by Harmer and Shipley. London. Vol. i. p. 165-242. Over 60 woodcuts.

1906 **Swartschewsky, B.** (Beiträge zur Spongien-Fauna des Weissen Meeres.) In: Mém. Soc. Nat. Kiew, xx, p. 307-371, Pl. x-xvi. (Russian, with German résumé.)

1906 **Thomson, J. A. & J. D. Fiddes.** Note on a Rare Sponge from the "Scotia" Collection. In: Proc. R. Phys. Soc. Edinburgh, xvi, p. 231-232.

1906 **Topsent, E.** Farrea occa (Bowerbank), var. foliascens, n. var. Par... In: Bull. Mus. Océanogr. Monaco, No. 83, p. (1-5).

1906 (α) **Topsent, E.** Éponges recueillies par M. Ch. Gravier dans la Mer Rouge, par...In: Bull. Mus. H. N. Paris, xii, p. 557-570.

1906 (β) **Topsent, E.** Les Clavulides purpurines. Par...In: Bull. Mus. Hist. Nat. Paris, p. 570-575.

1906 **Whitelegge, Th.** Scientific Results of the Trawling Expedition of H.M.C.S. "Thetis" off the coast of New South Wales in February and March, 1898. Part 9, Sponges. In: Austral. Mus. Mém. iv, p. 453-484, Pl. xliii-xliv.

1906 **Zeise, O.** Ueber die miocäne Spongienfauna Algeriens. In: Sitz. Ber Akad. Wiss. Berlin, p. 941-961.

1907 **Allemand, A.** Étude de physiologie appliquée à la spongiculture sur les côtes de Tunisie...Thèse, Lyon...195 pp., 17 Pl.

1907 **Annandale, N.** Notes on Freshwater Sponges...Part i. By...In: Rec. Indian Mus. i, pp. 267-273, 387-392, Pl. xiv.

1907 (α) **Annandale, N.** Notes on the Freshwater Fauna of India. No. ix. Descriptions of New Freshwater Sponges from Calcutta, with a Record of two known Species from the Himalayas and a List of the Indian Forms. In: Journ. Proc. Asiat. Soc. Bengal, iii, p. 15-26, 7 figs.

1907 **Boeke, J.** Sponsen en sponsenteelt...In: Rapport Visscherij Curaçao, i, p. 145-153, 1 Pl.

1907 **Chadenac, de.** La spongiculture. In: Revue générale des Sciences, Paris.

1907 **Cotte, J.** La Spongiculture peut-elle devenir une Industrie? Par... In: Bull. Enseignem. profession. Pêches maritimes, p. (1-12).

1907 (α) **Cotte, J.** La spongiculture. In: Rev. Sc. (5), viii, p. 332-338.

1907 **Giard, A.** La gastrula et les feuillets blastodermiques des Éponges. Par...In: Ann. Soc. R. Z. Malac. Belg. xlii, p. 199-202.

1907 **Ginestous, G.** L'industrie des éponges sur les côtes de Tunisie. In: Rev. Sc. (5), viii, p. 392-398. (7 figs.)

1907 **Kemna, A.** Les caractères et l'emplacement des Spongiaires. Par...
In: Ann. Soc. R. Z. Malac. Belg. xlii, pp. 72–97, 129–147.

1907 (α) **Kemna, A.** Réponse à la note de M. Giard sur la position systématique des Spongiaires. Par... In: Ann. Soc. R. Z. Malac. Belg. xlii, p. 228–230.

1907 **Kirkpatrick, R.** Preliminary Report on the Monaxonellida of the National Antarctic Expedition. By... In: Ann. and Mag. N. H. (7), xx, p. 271–291.

1907 (α) **Kirkpatrick, R.** Porifera. Hexactinellida. By... In: National Antarctic Expedition, Natural History, iii, p. 1–25, Pl. 1–7.

1907 (β) **Kirkpatrick, R.** Notes on Two Species of African Freshwater Sponges. By... In: Ann. and Mag. N. H. (7), xx, p. 523–525, 11 figs.

1907 **Lendenfeld, R. von.** Die Tetraxonia. Von... In: Wiss. Ergebn. "Valdivia," xi, pp. i–iv, 59–374, Pl. ix–xlvi.

1907 (α) **Lendenfeld, R. von.** Tetraxonia... In: D. Südpolar. Exped. 1901–1903, ix, p. 303–342, Pl. xxi–xxv.

1907 (β) **Lendenfeld, R. von.** (Review of Wilson's paper 1907.)

1907 **Maas, O.** Über die Wirkung des Hungers und der Kalkentziehung bei Kalkschwämmen und anderen kalkausscheidenden Organismen. In: Sitzungsber. Ges. Morph. Phys. München, xxiii, p. 82–89.

1907 **Minchin, E. A.** Spicule Formation. By... In: Rep. 76th Meet. Brit. Assoc. p. 605–606.

1907 **Théel, H.** Om utvecklingen af Sveriges zoologiska hafsstation Kristineberg och om djurlifvet i angränsande haf och fjordar. Af... In: Arkiv Zoologi, iv, p. 1–136, Pl. i–v and map.

1907 **Topsent, E.** Poecilosclérides nouvelles recueillies par le Français dans l'Antarctique, par... In: Bull. Mus. Hist. Nat. xiii, p. 69–76.

1907 (α) **Topsent, E.** Cliona purpurea, Hck., n'est pas une Clionide. Par... In: Arch. Zool. Expérim. (4), vii (notes et revues, p. xvi–xx, 1 fig.).

1907 **Weltner, W.** Spongillidenstudien. v, Zur Biologie von Ephydatia fluviatilis und die Bedeutung der Amoebocyten für die Spongilliden. Von... In: Arch. Naturgesch. lxxiii, p. 273–286, 2 figs.

1907 **Whitelegge, Th.** Scientific Results of the Trawling Expedition of H.M.C.S. "Thetis" off the Coast of New South Wales...Sponges. In: Austral. Mus. Mem. iv, p. 487–515, Pl. xlv–xlvi.

1907 **Wilson, H. V.** A new method by which sponges may be artificially reared. By... In: Science, N.S. xxv, p. 912–915. ? Also in: Journ. Elisha Mitchell Scient. Soc. xxiii, p. 91–97.

1907 (α) **Wilson, H. V.** On some phenomena of coalescence and regeneration in sponges. By... In: Journ. Experim. Zoölogy, v, p. 245–258, 4 figs.

208

1907 **Zimmermann, H.** Tierwelt am Strande der Blauen Adria. Von...
In: Zeitschr. Naturw. lxxviii (1905–1906), p. 293–322.

1908 **Annandale, N.** Notes on Freshwater Sponges. By...In: Rec. Indian
Mus. ii, p. 25–28, 5 woodcuts.

1908 (α) **Annandale, N.** Notes on some Freshwater Sponges collected in
Scotland. By...In: Journ. Linn. Soc. London, xxx, p. 244–250.

1908 (β) **Annandale, N.** Notes on the Freshwater Fauna of India, No. ix.
In: Journ. Proc. Assoc. Soc. Bengal, iii, p. 15–26, 7 figs.

1908 **Blin, H.** Nouvelles recherches sur la reproduction et la culture des
éponges. In: Suppl. La Nature, p. 181–182.

1908 **Chapman, F.** New or Little-Known Victorian Fossils in the National
Museum. Part ix. Some Tertiary Species. By...In: Proc. R.
Soc. Victoria, Melbourne, xx, p. 208–221, Pl. xvii–xix.

1908 **Chirica, C.** Spongillidae din România...In: Mem. Asoc. română
Inaintarea Rěspând. ii, p. 475.

1908 **Cotte, J.** Quelques observations de morphologie expérimentale sur
des Spongiaires. In: C. R. Soc. Biol. Paris, lxiv, p. 526–528.

1908 **Dubois, R. & A. Allemand Martin.** Contribution à l'étude de la biologie
des éponges et à la spongiculture sur les côtes de Tunisie. In:
Ann. Soc. Linn. Lyon (nouv. sér.), lv, p. 83–90.

1908 **Fages, E. de & Ponzevera.** Les pêches maritimes de la Tunisie. Tunis.

1908 **Hammer, E.** Neue Beiträge zur Kenntnis der Histologie und Entwicklung von Sycon raphanus. Von...In: Arch. Biontologie, ii,
p. 289–334, Pl. xxiii–xxviii.

1908 **Hanitsch, R.** Guide to the zoological collections of the Raffles Museum,
Singapore. By...Singapore. 112 pp. and 21 Pl.

1908 **Henze, M.** Ueber Spongosterin, das Cholesterin aus Suberites domuncula. Von...In: Zeitschr. physiol. Chem. lv, p. 427–432.

1908 **Jenkin, C. F.** Porifera. iii, Calcarea. In: National Antarctic
Expedition (Nat. Hist.), London, iv, 49 pp., 12 Pl.

1908 (α) **Jenkin, C. F.** The Marine Fauna of Zanzibar and British East
Africa, from Collections made by Cyril Crossland...in the Years
1901 and 1902. The Calcareous Sponges. By...In: Proc. Zool.
Soc. London, p. 434–456, fig. 81–104.

1908 **Kirkpatrick, R.** Description of a New Dictyonine Sponge from the
Indian Ocean. By...In: Rec. Indian Mus. ii, p. 21–26, Pl. i.

1908 (α) **Kirkpatrick, R.** Description of a new variety of Spongilla loricata.
By...In: Rec. Indian Mus. ii, p. 97–99, Pl. ix.

1908 (β) **Kirkpatrick, R.** Porifera (Sponge). ii, Tetraxonida...By...In:
National Antarctic Expedition (Nat. Hist.), London, iv, p.
1–56, Pl. 8–26.

1908 (γ) **Kirkpatrick, R.** On Two new Genera of Recent Pharetronid
Sponges. By...In: Ann. and Mag. Nat. Hist. (8), ii, p. 503–514,
Pl. xiii–xv.

1908 **Lameere, A.** Éponge et Polype. Par...In: Ann. Soc. R. Zool. et Malacol. Belgique, xliii, p. 107–124.

1908 **Minchin, E. A.** Materials for a Monograph of the Ascons. ii, The Formation of Spicules in the Genus Leucosolenia, with some Notes on the Histology of the Sponges. By...In: Quart. Journ. Micr. Sc. (2), lii, p. 301–355, Pl. 17–21.

1908 **Minchin, E. A. & D. J. Reid.** Observations on the Minute Structure of the Spicules of Calcareous Sponges. By...In: Proc. Zool. Soc. London, p. 661–676, Pl. xxxiv–xxxvii.

1908 **Sollas, I. B. J.** The Inclusion of Foreign Bodies by Sponges, with a Description of a new Genus and Species of Monaxonida. By... In: Ann. and Mag. N. H. (8), i, p. 395–401. With 5 woodcuts.

1908 **Thacker, A. G.** On Collections of the Cape Verde Islands Fauna made by Cyril Crossland...from July to September 1904. The Calcareous Sponges. By...In: Proc. Zool. Soc. London, p. 757–782, fig. 155–166, Pl. xl, fig. 1–9.

1908 **Topsent, E.** Spongiaires. Par...In: Expéd. antarctique Française (1903–1905)...Sciences natur. docum. scient. p. 1–37, Pl. i–v.

1908 (α) **Topsent, E.** Sur une variété de Clionopsis platei, Thiele. Par... In: Bull. Instit. Océanogr. No. 120, p. 1–3.

1908 **Urban, F.** Die Kalkschwämme der deutschen Tiefsee-Expedition. Von...In: Zool. Anz. xxxiii, p. 247–252.

1908 **(Vosmaer, G. C. J.)** Poterion een boorspons. In: Versl. Verg. Kon. Akad. Wet. Amsterdam, xvii, p. 16–22. See 1908 (α).

1908 (α) **Vosmaer, G. C. J.** Poterion a Boring Sponge. By...In: Proceedings K. A. W. Amsterdam, p. 37–41. Transl. from 1908.

1908 **Woodland, W.** Studies in Spicule-Formation. viii. Some Observations on the Scleroblastic Development of Hexactinellid and other Siliceous Sponge-Spicules. In: Quart. Journ. Microsc. Sc. (2), lii, p. 139–157, Pl. vii.

1909* **Annandale, N.** Report on a Collection of Freshwater Sponges from Japan. By...In: Annot. Zool. Japon. vii, p. 105–112, Pl. ii.

1909 (α) **Annandale, N.** Notes on Freshwater Sponges. No. x...In: Rec. Indian Mus. iii, p. 101–104, Pl. xii.

1909 (β) **Annandale, N.** Beiträge zur Kenntnis der Fauna von Süd-Afrika. Ergebnisse einer Reise von Prof. Max Weber im Jahre 1894. ix. Freshwater Sponges. By...In: Zool. Jahrb. Syst. xxvii, p. 559–568, 3 figs.

1909 (γ) **Annandale, N.** Freshwater Sponges in the Collection of the United States National Museum. Part i. Specimens from the Philippines and Australia. By...In: Proc. U. S. National Museum, xxxvi, p. 627–632.

1909 (δ) **Annandale, N.** Freshwater sponge from New Guinea. In: Nova Guinea, Rés. Expéd. sc. néerl. v, p. 421–422.

* In Jahresber. Neapel, f. 1910, erroneously placed in 1910; the number of the Annot. in which Annandale's paper appeared, is published June, 1909.

1909 (ε) **Annandale, N.** Description d'une nouvelle espèce d'Éponge d'eau douce du lac de Genève. Par...In: Rev. Suisse Zool. xvii, p. 367–369, Pl. 9.

1909 (ζ) **Annandale, N.** Fresh-Water Sponges Collected in the Philippines by the Albatross Expedition. In: Proc. U. S. National Museum, xxxvii, p. 131–132.

1909 (η) **Annandale, N.** Fresh-Water Sponges in the Collection of the United States National Museum. Part ii. Specimens from North and South America. By...In: Proc. U. S. National Museum, xxxvii, p. 401–406.

1909 **Blin, H.** La culture des éponges sur le littoral méditerranéen français. In: Suppl. La Nature, p. 53–54.

1909 **Burck, C.** Zur Kenntnis der Histologie einiger Hornschwämme, sowie Studien über einige Choanoflagellaten...von...Inaug. Diss. Heidelberg, p. 1–60, Pl. i–ii.

1909 **Dubois, R.** Rapport sur la spongiculture et sur la mytiliculture en France. In: C. R. du Congrès Internat. pêches marit. Sables d'Olonne.

1909 **Faurot, L.** Affinités des Tétracoralliaires et des Hexacoralliaires. In: Ann. Palaeont. iv, p. 69–108, 21 figs.

1909 **Gerth, H.** Timorella permica, n. g. n. sp. eine neue Lithistide aus dem Perm von Timor. In: Centralbl. Min. Geol. Pal. p. 695–700.

1909 **Harold Row, R. W.** Vide Row.

1909 **Hentschel, E.** Tetraxonida. 1. Teil. Von...In: Fauna Südwest-Australiens (Michaelsen und Hartmeyer), Bd. ii, Lfrg. 21, p. 347–402, Pl. xxii–xxiii.

1909 **Kirk, H. B.** Two Sponges from Campbell Island...In: Subantarct. Isl. N. Zealand, ii, p. 539–540, 1 Pl.

1909 **Kirkpatrick, R.** Notes on Merlia normani, Kirkp. By...In: Ann. and Mag. N. H. (8), iv, p. 42–48.

1909 (α) **Kirkpatrick, R.** On the Regular Hexactine Spicule of Hexactinellida. By...In: Ann. and Mag. N. H. (8), iv, p. 505–509.

1909 (β) **Kirkpatrick, R.** On the Phylogeny of the Amphidiscophora. In: Ann. and Mag. N. H. (8), iv, p. 479–484.

1909 **Kirsch, A. M.** Fresh Water Sponges and Particularly Those of the United States. In: Amer. Midland Naturalist, i, p. 29–38.

1909 **Kolb, R.** Die Kieselspongien des schwäbischen Weissen Jura. Von ...In: Palaeontogr. lvii, p. 141–256, 27 figs., Pl. xi–xxi.

1909 **Korschelt, E. & K. Heider.** Lehrbuch der vergleichenden Entwicklungsgeschichte der wirbellosen Thiere, von...(Allgem. Theil., dritte Lief.)

1909 **Lo Bianco, S.** Notizie biologiche riguardanti specialmente il periodo di maturità sessuale degli animale del golfo di Napoli, del...In: Mittheil. Zool. Stat. Neapel, xix, p. 513–763.

1909 **Lühe, M.** Unsere einheimischen Süsswasserschwämme. In: Schrift. Physik. Oek. Ges. Königsberg, xlix, pp. 309–312, 409.

1909 **Lundbeck, W.** The Porifera of East Greenland. By...In: Meddelelser om Grönland, xxix, p. 423–464, Pl. xiv.

1909 **Maas, O.** Zur Entwicklung der Tetractinelliden...In: Verh. D. Zool. Ges...p. 183–200, 11 woodcuts.

1909 **Maire, V.** Les Spongiaires oxfordiens. Recueillis dans le département du Jura...In: Bull. Soc. Grayloise Émul. No. 12, p. 249–263.

1909 **Minchin, E. A.** Sponge-Spicules. A summary of present Knowledge. By...In: Ergebn. und Fortschr. Zoologie, ii, p. 171–274, fig. 1–26.

1909 (α) **Minchin, E. A.** The relation of the flagellum to the nucleus in the collar-cells of Calcareous Sponges. By...In: Zool. Anz. xxxv, p. 227–231. Woodcuts.

1909 **Neviani, A.** Nuova specie di "Psammophyllum," Haeck. In: Boll. Soc. zool. ital. (2), x, p. 265–266.

1909 **Reynolds, J. E.** Recent Advances in our Knowledge of Silicon and its Relations to Organized Structures. In: Nature, lxxxi, p. 206–208, 2 woodcuts.

1909 **Row, R. W. H.** Report on the Sponges collected by Mr Cyril Crossland in 1904–5. Part i. Calcarea. By...(xiiith Report on the Marine Biology of the Sudanese Red Sea.) In: Journ. Linn. Soc. xxxi, p. 182–214, Pl. 19–20.

1909 **Seale, A.** The Fishery Resources of the Philippine Islands: ii, Sponges and Sponge Fisheries. By...In: Philippine Journ. Sc. iv, p. 57–64, Pl. i–ix, 4 figs.

1909 **Topsent, E.** La coupe de Neptune, Cliona patera, par...In: Arch. Zool. Expérim. (4), ix, p. lxix–lxxii.

1909 (α) **Topsent, E.** Étude sur quelques Cladorhiza et sur Euchelipluma pristina, n. g. et n. sp. Par...In: Bull. Instit. Océanogr. No. 151, p. 1–23, Pl. i–ii.

1909 (β) **Topsent, E.** Description d'une Variété nouvelle d'Éponge d'eau douce (Ephydatia fluviatilis, Auct. var. syriaca, Tops.)...Par... In: Bull. Soc. Amis Sc. nat. Rouen, p. 1–5.

1909 **Urban, F.** Die Calcarea. In: Wiss. Ergebn. D. Tiefsee-Exped. xix, p. 1–40, Pl. i–vi.

1909 **(Vosmaer, G. C. J.)** Over de spinispirae van Spirastrella bistellata (O. S.) Ldfd. In: Versl.Verg. Kon. Akad. Wet. Amsterdam, xvii, p. 637–644, 1 Pl.

1909 (*a*) **Vosmaer, G. C. J.** On the spinispirae of Spirastrella bistellata
(O. S.) Lhfd. By...In: Kon. Akad. Wetensch. Amsterdam,
Proceedings Meeting February, p. 642–648, 1 Pl.

1909 **Weltner, W.** Ist Merlia normani Kirkp. ein Schwamm? Von...In:
Arch. f. Naturgesch. lxxv, p. 139–141.

1909 (*a*) **Weltner, W.** Spongillidae, Süsswasserschwämme. Von...In:
Die Süsswasserfauna Deutschlands, herausg. v. Brauer, xix,
p. 177–190, figs. 293–332.

1909 **Wilson, H. V.** A Further Contribution on the Regenerative Power of
the Somatic Cells of Sponges after Removal from the Parent.
In: Science (N. S.), xxix, p. 430.

1910 **Annandale, N.** Notes on Freshwater Sponges. By...xii, Description
of a new species from Cape Comorin. In: Rec. Indian Mus.
Calcutta, v, p. 31.

1910 (*a*) **Annandale, N.** Fresh-water Sponges in the Collection of the
United States National Museum. Part iii–iv...By...In: Proc.
U. S. National Mus. xxxviii, pp. 183, 649–650.

1910 (*β*) **Annandale, N.** Contributions to the Fauna of Yunnan based on
Collections made by J. Coggin Brown, B.Sc., 1909–1910. Part i.
Sponges and Polyzoa. In: Rec. Indian Mus. Calcutta, v, p. 197–
199, 2 figs.

1910 (**Biedermann, W.** Die Ernährung der Spongien.) In: Winterstein,
Handb. Vergl. Physiol. Bd. ii, Erste Hälfte, p. 426–444, 5 figs.

1910 (**Burian, R.** Die Exkretion...B. Coelenteraten.) In: Winterstein,
Handb. Vergl. Physiol. Bd. ii, Zweite Hälfte, p. 280–287,
1 fig.

1910 **Ceccaty, R. de.** La Dépêche Sfaxienne, 29. Septembre 1908, 5. Dé-
cembre 1908, 1ᵉʳ, 8, 16 Septembre, 1910.

1910 **Chadenac, de.** Le laboratoire de Sfax, la station maritime de Tamaris-sur-Mer, la spongiculture et l'industrie des éponges en Tunisie.
In: Ligue maritime.

1910 **Fernández, G. E.** Consideraciones acera de la posición de las esponjas
en el reino animal...In: Bol. Soc. españ. H. N. x, p. 75–80.

1910 **Flegel, C.** The Abuse of the Scaphander in the Sponge Fisheries.
By...In: Bull. Bureau Fisheries, xxviii*, p. 513–543.

1910 **Fredericq, L.** Die Sekretion von Schutz- und Nutzstoffen...ii, Coelenterata (inkl. Porifera). In: Winterstein, Handb. Vergl.
Physiol. Bd. ii, Zweite Hälfte, p. 3.

1910 **Hentschel, E.** Das Skelett der Schwämme...In: Verh. nat. Ver.
Hamburg (3), xvii, 1909, p. lvii–lviii.

1910 **Jörgensen, M.** Beiträge zur Kenntnis der Eibildung, Reifung, Befruchtung und Furchung bei Schwämmen (Syconen). In: Arch.
f. Zellforschung, iv, p. 163–242, Pl. xi–xv.

* No. 668, in which the article occurs, "issued March, 1910."

1910 **Kemna, A.** Sur la position systématique des Spongiaires. Par...In: Ann. Soc. Z. Malac. Belg. xlv, p. 13–26.

1910 **Kirkpatrick, R.** On Hexactinellid Sponge Spicules and their Names. By...In: Ann. and Mag. N. H. (8), v, p. 208–213, Pl. viii.

1910 (α) **Kirkpatrick, R.** On Hexactinellid Spicules and their Names. Part ii. Supplementary. By...In: Ann. and Mag. N. H. (8), v, p. 347–350, 5 figs.

1910 (β) **Kirkpatrick, R.** On the Affinities of Astrosclera willeyana, Lister. By...In: Ann. and Mag. N. H. (8), v, p. 380–383, Pl. xi.

1910 (γ) **Kirkpatrick, R.** A Sponge with a Siliceous and Calcareous Skeleton. In: Nature, lxxxiii, p. 338, 3 figs.

1910 (δ) **Kirkpatrick, R.** On new Species of Hydrozoa and Porifera of St Helena. By...In: Proc. Zool. Soc. London, p. 127–131, Pl. vii.

1910 (ε) **Kirkpatrick, R.** On a Remarkable Pharetronid Sponge from Christmas Island. By...In: Proc. R. Soc. London, lxxxiii, p. 124–133, Pl. 10–11.

1910 (ζ) **Kirkpatrick, R.** Further Notes on Merlia normani, Kirkp....In: Ann. and Mag. N. H. (8), v, p. 288–291.

1910 **Korschelt, E. & K. Heider.** Lehrbuch der vergleichenden Entwicklungsgeschichte der wirbellosen Thiere. Von...(Erste und zweite Aufl. allgem. Theil.) Jena, p. 167–896.

1910 **Lendenfeld, R. von.** ...The Sponges. i. The Geodidae. By...In: Mem. Mus. Comp. Zoölogy Harvard Coll. xli, No. 1, p. 1–259, Pl. 1–48.

1910 (α) **Lendenfeld, R. von.** ...The Sponges. 2. The Erylidae. By...In: Mem. Mus. Comp. Zoöl. Harvard Coll. xli, No. 2, p. 267–323, Pl. 1–8.

1910 **Lundbeck, W.** Porifera. (Part iii.) Desmacidonidae (Pars). By... In: Danish Ingolf-Exped. vi, p. 1–124, Pl. i–xi.

1910 **Maas, O.** Ueber Nichtregeneration bei Spongien. Von...In: Arch. f. Entw. Mechan. xxx, p. 356–378, 4 figs.

1910 (α) **Maas, O.** Ueber Involutionserscheinungen bei Schwämmen und ihre Bedeutung für die Auffassung des Spongienkörpers. Von... In: Festschrift f. R. Hertwig, iii, p. 93–130, 2 figs, Pl. 8–10.

1910 **Minchin.** See Robertson.

1910* **Moore, H. F.** The Commercial Sponges and the Sponge Fisheries. By...In: Bull. Bureau Fisheries, xxviii, p. 399–511, 4 figs., Pl. xxviii–lxvi.

1910 (α)† **Moore, H. F.** A Practical Method of Sponge Culture. By...In: Bull. Bureau Fisheries, xxviii, p. 545–585, 7 figs., Pl. lxvii–lxxvi.

---

* Paper presented before 4th Intern. Fish. Congress, Washington, 1908; "issued March, 1910."
† Issued 1910.

214

1910 **Parker, G. H.** The Reactions of Sponges, with a consideration of the origin of the Nervous System. By...In: Journ. Experim. Zoölogy, viii, p. 765–805, 3 figs.*

1910 **Robertson, M. & E. A. Minchin.** The Division of the Collar-Cells of Clathrina coriacea (Montagu): A Contribution to the Theory of the Centrosome and Blepharoplast. By...In: Quart. Journ. Microsc. Sc. lv, p. 611–640, Pl. 25–26.

1910 **Schrammen, A.** Die Kieselspongien der oberen Kreide von Nordwestdeutschland. 1 Theil. Tetraxonia, Monaxonia und Silicea incert. sedis von...In: Palaeontogr. Suppl. v, p. 1–175, Pl. I–XXIV.

1910 **Schulze, F. E. & R. Kirkpatrick.** Preliminary notice on Hexactinellida of the Gauss-Expedition. By...In: Zool. Anz. xxxv, p. 293–302.

1910 (α) **Schulze, F. E. & R. Kirkpatrick.** Die Hexactinelliden der Deutschen Südpolar-Expedition 1901–1903. Von...In: D. Südpolar-Expedition, xii, Zool. iv, p. 1–62, Pl. I–x.

1910 **Topsent, E.** Les Hexasterophora recueillies par la Scotia dans l'Antarctique (note préliminaire). Par...In: Bull. Inst. Océanogr. Monaco, No. 166, p. 1–18.

1910 **Urban, F.** Zur Kenntnis der Biologie und Cytologie der Kalkschwämme (Familie Clathrinidae Minch.). In: Internat. Rev. ges. Hydrobiol. iii, p. 37–43, 6 figs.

1910 **Walther, J.** Die Sedimente der Taubenbank im Golfe von Neapel. Von...In: Abh. K. Pr. Akad. Wiss. Phys.-math. Cl. Anhang, p. 1–49, 2 Pl.

1910 **Weltner, W.** ...Spongiae für 1907. Von...In: Arch. f. Naturgesch. 74. Jahrg. 1908, Bd. ii (p. 1–35).

1910 (α) **Weltner, W.** Ist Astrosclera willeyana, Lister, eine Spongie? Von ...In: Arch. f. Naturgesch. lxvi, p. 128–134.

1910 (β) **Weltner, W.** Spongillidae. In: Michaelsen und Hartmeyer, Die Fauna Südwest-Australiens, iii, p. 137–144, 22 figs.

1910 (γ) **Weltner, W.** Spongiae für 1908. Von...In: Arch. f. Naturgesch. lxxv, Bd. ii, p. 1–26.

1910 **Wilson, H. V.** Development of Sponges from tissue cells outside the Body of the Parent. By...In: Bull. Bur. Fish. xxviii, p. 1267–1271.

1910 (α) **Wilson, H. V.** A Study of some Epithelioid Membranes in Monaxonid Sponges. By...In: Journ. Experim. Zoöl. ix, p. 537–577, 21 figs.

1910 (β) **Wilson, H. V.** On the Structure and Regeneration of the Epidermal Layer in some Siliceous Sponges. In: Science (2), xxxi, p. 469 (Preliminary account).

1910 **Winterstein, H.** Handbuch der vergleichenden Physiologie. Cf. 1910, Biedermann; 1910, Fredericq; 1910, Burian.

* Also printed with separate paging 1–41.

1911 **Annandale, N.** Some Sponges associated with gregarious Molluscs of the Family Vermetidae. By...In: Rec. Indian Mus. Calcutta, vi, p. 47–55, Pl. VIII–IX.

1911 (α) **Annandale, N.** Notes on Freshwater Sponges. By...xiii. Specimens collected in the Poona District, Bombay Presidency, by S. P. Agharkar. In: Rec. Indian Mus. Calcutta, vi, p. 225–226.

1911 (β) **Annandale, N.** Note on a Freshwater Sponge and Polyzoon from Ceylon. In: Spolia Zeylan. Colombo, vi, p. 63–64, Pl. I.

1911 **Breitfuss, L. L.** (Russian, with German abstract, entitled Zur Kenntnis der Spongio-Fauna des Kola-Fjords.) In: Trav. Soc. Imp. Natural. St Pétersbourg, xlii, p. 209–226.

1911 **Cotte, J.** Sponge Culture. By...In: Bull. Bureau Fish. Washington, xxviii, p. 587–614.

1911 **Delage, Yves.** La Spongiculture à Tamaris, par...In: Bull. Inst. Océanogr. No. 198, p. 1–4.

1911 **Dubois, R.** Nouveaux essais de spongiculture au Laboratoire Maritime de Biologie de Tamaris-sur-Mer...par...In: Bull. Inst. Océanogr. No. 191, p. 1–19.

1911 **Fraipont, C.** Une Hexactinellide nouvelle du Dévonien belge (Calcaire Frasnien). Par...In: Ann. Soc. Géol. Belgique, xxxviii, p. 197–208, Pl. XIII and XV.

1911 **Hentschel, E.** Tetraxonida. 2 Teil. Von...In: Fauna Südwest-Australiens (Michaelsen und Hartmeyer), iii, p. 279–393, 54 figs.

1911 (α) **Hentschel, E.** Über den Ursprung der Microsclere der Desmacidoniden. Von...In: Zool. Anz. xxxviii, p. 148–155.

1911 **Holst, N. O.** Beskrifning till kartbladet Börringe kloster. In: Sverig. geol. Undersökn. (ser. Aa) No. 138.

1911 **Huxley, J. S.** Some Phenomena of Regeneration in Sycon; with a Note on the Structure of its Collar-cells. By...In: Phil. Trans. R. Soc. London, Series B, ccii, p. 165–189, Pl. 8.

1911 **Kirk, H. B.** Sponges collected at the Kermadec Islands by Mr W. R. B. Oliver. In: Trans. N. Zealand Inst. Wellington, xliii, p. 574–581, Pl. XXVII and 5 figs. in text.

1911 **Kirkpatrick, R.** On Merlia normani, a Sponge with a Siliceous and Calcareous Skeleton. By...In: Quart. Journ. Microsc. Sc. lvi, p. 657–702, Pl. 32–38.

1911 (α) **Kirkpatrick, R.** On a new Lithonine Sponge from Christmas Island. By...In: Ann. and Mag. N. H. (8), viii, p. 177–179, 10 figs.

1911 **Lendenfeld, R. von.** Bemerkungen über die technische Ausführung und biologische Verwertung mikroskopischer Messungen. Von ...In: Zeitschr. wiss. Mikrosk. xxviii, p. 27–34.

1911 (**Maas, O.**) Ueber das Ausbleiben der Regeneration und Regulation bei niederen Tieren. In: Sitz. Ber. Ges. Morphol. u. Physiol. München. (5 pp.)

1911 **Müller, K.** Versuche über die Regenerationsfähigkeit der Süsswasserschwämme. Von...In: Zool. Anz. xxxvii, p. 83–88,

1911 (α) **Müller, K.** Beobachtungen über Reduktionsvorgänge bei Spongilliden, nebst Bemerkungen zu deren äusserer Morphologie und Biologie. Von...In: Zool. Anz. xxxvii, p. 114–121 (3 woodcuts).

1911 (β) **Müller, K.** Ueber eine vermutliche Varietät von Ephydatia fluviatilis. Von...In: Zool. Anz. xxxviii, p. 495–500, 5 figs.

1911 (γ) **Müller, K.** Das Regenerationsvermögen der Süsswasserschwämme, insbesondere Untersuchungen über die bei ihnen vorkommende Regeneration nach Dissociation und Reunition. Von...In: Arch. Entw. Mech. xxxii, p. 397–446 (28 figs.).

1911 (δ) **Müller, K.** Reductionserscheinungen bei Süsswasserschwämmen von...In: Arch. Entw. Mech. p. 557–607 (16 figs.).

1911 **Robertson, M.** The Division of the Collar-Cells of the Calcarea Heterocoela. By...In: Quart. Journ. M. Sc. lvii, p. 129–139, Pl. 19.

1911 **Row, R. W. Harold.** Report on the Sponges collected by Mr Cyril Crossland in 1904–5. Part ii. Non-Calcarea. By...(xixth Rep. of Reports Marine Biol. Sudanese Red Sea.) In: Journ. Linn. Soc. xxxi, p. 287-400, Pl. 35–41 (and 26 text-figures).

1911 **Smith, J.** Carboniferous Limestone Rocks of the Isle of Man. In: Trans. Geol. Soc. Glasgow, xiv, p. 119–164.

1911 **Thomas, A. O.** A Fossil Burrowing Sponge from the Iowa Devonian. By...In: Bull. State Univ. Iowa, New Series, vi, No. 2, p. 165–168, 1 Pl.

1911 **Topsent, E.** Sur une magnifique Geodia megastrella Carter du Muséum de la Rochelle. Par...La Rochelle. 7 pp. With illustr.

1911 (α) **Topsent, E.** Sur les affinités des Halichondria et la classification des Halichondrines d'après leurs formes larvaires, par...In: Arch. Zool. Expérim. (5) vii, p. i–xv, 3 figs.

1911 (**Vosmaer, G. C. J.**) Opmerkingen omtrent het geslacht Spirastrella. In: Verslag Gew. Verg. W. en N. Afd. Kon. Akad. v. Wetensch. Amsterdam, p. 1243–1250.

1911 (α) **Vosmaer, G. C. J.** Observations on the genus Spirastrella. By... In: Kon. Akad. Wetensch. Amsterdam, Proc. p. 1139–1146.

1911 (β) **Vosmaer, G. C. J.** The Porifera of the Siboga-Expedition. ii. The Genus Spirastrella. By...In: Siboga-Expeditie, vi A¹, p. 1–69, Pl. i–xiv.

1911 **Welter, O. A.** Die Pharetronen aus dem Essener Grünsand. In: Verh. Nat. Ver. Bonn, lxvii, p. 1–82, 10 figs., Pl. i–iii.

1911 **Weltner, W.** (Beiträge Fauna Turkestans.) viii. Spongillidae des Issyk-Kul-Sees und des Baches bei Dschety-Ogus von...In: Trav. Soc. Imp. Natural. St Pétersbourg, xlii, p. 59–88, Pl. i.

1911 **Wilson, H. V.** Development of Sponges from dissociated tissue cells. By...In: Bull. Bureau Fisheries, xxx (1910*), p. 1-30, Pl. i–v.

* "Issued June 16, 1911."

1911 **Zschokke, F.** Die Tiefseefauna der Seen Mitteleuropas...von...
Leipzig. 246 pp., 2 Pl. 4º.

1912 **Annandale, N.** Notes on Freshwater Sponges. xiv. The generic
position of "Spongilla ultima." In: Rec. Indian Mus. Calcutta,
vii, p. 99.

1912 (α) **Annandale, N.** Systematic and geographical notes on the Sponges.
In: Rec. Indian Mus. Calcutta, vii, p. 137–139.

1912 (β) **Annandale, N.** The Freshwater Sponges of the Malabar Zone.
In: Rec. Indian Mus. Calcutta, vii, p. 383–397.

1912 (γ) **Annandale, N.** Porifera. In: Rec. Indian Mus. Calcutta, viii,
p. 67.

1912 (δ) **Annandale, N.** Fresh-Water Sponges in the Collection of the
United States National Museum. Part v. A new Genus pro-
posed with Heteromeyenia radiospiculata, Mills, as type. By...
In: Proc. U. S. National Mus. xl, p. 593–594.

1912 **Breitfuss, L. L.** Zur Kenntnis der Spongio-Fauna des Kola-Fjords...
In: Trav. Soc. Imp. Natural. St Pétersbourg, xli, p. 61–80, Pl. I–II.

1912 **Crawshay, L. R.** On the Fauna of the Outer Western Area of the
English Channel. By...In: Journ. Mar. Biol. Assoc. Plymouth,
ix, p. 292–393, Pl. VI.

1912 **Dubois, R.** Sur la spongiculture par fragmentation au laboratoire de
Tamaris-sur-Mer. In: C. R. Assoc. Franç. Avanc. Sc. p. 141.

1912 (α) **Dubois, R.** Essais de spongiculture par fixation des larves
d'éponges. In: C. R. Assoc. Franç. Avanc. Sc. p. 141–142.

1912 **Emery, C.** Alcune riflessioni sulla classificazione zoologica. Del...
In: Monitore Zool. Ital. xxii, p. 224–231.

1912 **Hallman, E. F.** Report on the Sponges obtained by the F.I.S.
"Endeavour" on the Coasts of New South Wales, Victoria,
South Australia, Queensland, and Tasmania. Part i. By...In:
Zool. Results Fish. Experim. "Endeavour," Part ii, p. 117–300,
Pl. XXI–XXXVI.

1912 **Hentschel, E.** Kiesel- und Hornschwämme der Aru- und Kei-Inseln.
Von...In: Abh. Senckenb. Ges. xxxiv, p. 295–448, Pl. 13–21.

1912 **Hernández, F. F.** Notas sobre alcunas esponjas de Santander con
una introducción sobre sistemática...Por...Madrid. 18 pp.
Also in: Bol. Soc. españ. Hist. Nat. xii, p. 573–589 (?).

1912 **Jaffé, G.** Die Entwicklung von Spongilla lacustris, L., und Ephydatia
fluviatilis, L., aus den Gemmulae. Inaug. Diss. Berlin. 38 pp.

1912 (α) **Jaffé, G.** Die Entwicklung von Spongilla lacustris, L., und Ephydatia fluviatilis, L., aus der Gemmula. Von...In: Zool. Anz.
xxxix, p. 705–719, 21 figs. Abstr. from Jaffé, 1912 (?).

1912 (β) **Jaffé, G.** Bemerkungen über die Gemmulae von Spongilla lacustris, L., und Ephydatia fluviatilis, L. Von...In: Zool. Anz.
xxxix, p. 657–667.

1912 **Kirkpatrick, R.** Note on Astrosclera willeyana, Lister. By...In: Proc. R. Soc. B. lxxxiv, p. 579–580.

1912 (a) **Kirkpatrick, R.** On the Nature of Stromatoporoids. In: Nature, lxxxix, p. 607.

1912 (β) **(Kirkpatrick, R.** Merlia normani.) In: Nature, lxxxix, p. 352–353.

1912 (γ) **Kirkpatrick, R.** Note on Merlia normani and the "Monticuliporas." By...In: Proc. R. Soc. London, lxxxv, p. 562–563.

1912 (δ) **Kirkpatrick, R.** Merlia normani and its Relation to Certain Palaeozoic Fossils. In: Nature, lxxxix, p. 502–503.

1912 **Lendenfeld, R. von.** Spinosella infundibulum, n. sp. Von...In: Mitt. Osterlande, xxxiv, p. 14–18, Pl. i.

1912 (a) **Lendenfeld, R. von.** Untersuchungen über die Skelettbildungen der Kieselschwämme von...i. Die Mikrosklere der Caminus-Arten...In: Denkschr. math.-naturw. Kl. Akad. Wien, lxxxviii, p. 693–709, Pl. i–vi.

1912 **Maas, O.** Vitalfärbung bei Kalkschwämmen im normalen und in Involutionszustand. Von...In: Zool. Jahrb. Suppl. xv, p. 253–268, Pl. 9–10.

1912 (a) **Maas, O.** Porifera. In: Handwörterb. Naturw. Jena, vii, p. 1028–1047.

1912 **Ortmann, P.** Die Mikroscleren der Kieselspongien in Schwammgesteinen der senonen Kreide. Von...In: N. Jahrb. Mineral. ii, p. 127–149.

1912 **Ravn, J. P. J.** Om de Saakaldte Bløddyraeg fra vore Kridtaflejringer. (Aphrocallistes sp. in Danish cretaceous deposits.) In: Medd. Dansk. Geol. Kjøbenhavn, iv, i, p. 55–60.

1912 **Schrammen, A.** Porifera. Palaeontologie. In: Handwörterb. Naturw. Jena, vii, p. 1047–1053.

1912 (a) **Schrammen, A.** Die Kieselspongien der oberen Kreide von Nordwestdeutschland. ii. Theil. Triaxonia (Hexactinellida), von... In: Palaeontogr. Suppl. v, pp. i–ii, 177–385, Pl. xxv–xlv.

1912 **Sella, M.** La pesca delle Spugne nella Libia per...In: R. Comitato talassogr. Italiano, Mem. xiii, 154 pp., 2 maps and 12 Pl.

1912 **Stephens, Jane.** Clare Island Survey. Part 59. Marine Porifera by... In: Proc. R. Irish Acad. xxxi, p. (1–42), 1 Pl.

1912 (a) **Stephens, Jane.** Clare Island Survey. Part 60. Fresh-water Porifera by...In: Proc. R. Irish Acad. xxxi, p. (1–18), 1 Pl.

1912 **Thomas, A. O.** Fossil burrowing Sponge from the Iowa Devonian. In: Bull. Lab. N. H. Iowa (2), vi, 2, p. 165–168.

1912 **Topsent, E.** Sur une grande Tedania abyssale des Açores (Tedania phacellina, n. sp.). Par...In: Bull. Inst. Océanogr. Monaco, No. 252, p. 1–5, 2 figs.

1912 (α) **Topsent, E.** Description de Spongilla (Stratospongilla) Gilsoni, n. sp., Éponge d'eau douce des Iles Fidji. Par...In: Ann. Biol. Lac. Bruxelles, v, p. 187–191, Pl. i, fig. 1–8.

1912 (β) **Topsent, E.** Sur la contribution apportée par les explorations scientifiques dans l'antarctique à la connaissance des Euplectellinae. Par...In: C. R. Assoc. Franç. Avanc. Sc. p. 518–520.

1912 **Vosmaer, G. C. J.** On the Distinction between the genera Axinella, Phakellia, Acanthella a. o. By...In: Zool. Jahrb. Suppl. xv (Festschr. Spengel), p. 307–322, Pl. 15–16.

1912 **Wierzejski, A.** Über Abnormalitäten bei Spongilliden. Von...In: Zool. Anz. xxxix, p. 290–295, 4 figs.

1913 **Annandale, N.** Notes on some sponges from Lake Baikal in the collection of the Imperial Academy of Sciences, St Petersburg. In: Ann. Mus. Zool. Acad. Sc. St Pétersbourg, xviii, p. 96–101.

1913 (α) **Annandale, N.** Introduction to a Report on the Biology of the Lake of Tiberias. In: Calcutta Journ. As. Soc. Beng. ix, p. 17–23, Pl. i.

1913 (β) **Annandale, N.** An Account of the Sponges of the Lake of Tiberias, with observations on certain genera of Spongillidae. In: Calcutta Journ. and Proc. As. Soc. Beng. ix, p. 57–87, Pl. ii–v.

1913 (γ) **Annandale, N.** Note on a Sponge Larva from the Lake of Tiberias. In: Calcutta Journ. As. Soc. Beng. ix, p. 221–222, Pl. vii, fig. 3.

1913 (δ) **Annandale, N.** Notes on Freshwater Sponges. xv. Sponges from shells of the genus Aetheria. In: Rec. Indian Mus. Calcutta, ix, p. 237–240.

1913 **Arndt, W.** Zoologische Ergebnisse der ersten Lehr-Expedition der Dr P. Schottländerschen Jubiläums-Stiftung. In: Jahresber. Schles. Ges. vaterl. Cultur, 1912, p. 110–136.

1913 **Bidder, G. P.** Amoebocytes in Calcareous Sponges. In: Nature, xcii, p. 479.

1913 **Bösraug, E.** Die Tetractinelliden. In: Reise Ostafrika A. Voeltzkow, iii, p. 231–251, 4 Pl.

1913 **Boyer, J.** Les Éponges. In: Cosmos, Paris (N.S.), lxviii, p. 322–325, 2 figs.

1913 **Cirvinsky, P. N.** (Russian.) (Geol. Structure, Seims.) Part ii. In: Kiev Zap. Obšč. jest. 23 (1), p. 1–132, Germ. transl., p. 133–141, 3 Pl.

1913 **Clark, J. M.** Fosseis devonianos do Paraná. In: Serviço Geol. Min. Brasil, Vol. i, pp. xx and 353, Pl. 1–27.

1913 **Collins, J. H.** Addenda to the Working List of Cornish Palaeozoic Fossils. In: Penzance Trans. R. Geol. Soc. Cornwall, xiii, p. 385–427.

1913 **Dendy, A.** By-products of Organic Evolution. By...In: Journ. Quekett Microsc. Club (2), xii, p. 65–82, Pl. 7.

1913 (α) **Dendy, A.** Report on the Calcareous Sponges collected by H.M.S. "Sealark" in the Indian Ocean. By...In: Trans. Linn. Soc. London, xvi, 1, p. 1–29, Pl. 1–5.

1913 (β) **Dendy, A.** Amoebocytes in Calcareous Sponges. In: Nature, xcii, p. 399.

1913 (γ) **Dendy, A.** (Reply to Mr Bidder on Amoebocytes in Calcareous Sponges.) In: Nature, xcii, p. 479.

1913 **Dendy, A. & R. W. H. Row.** The Classification and Phylogeny of the Calcareous Sponges; with a Reference List of all the described Species, systematically arranged. By...In: Proc. Zool. Soc. London, p. 704–813.

1913 **Felix, J.** Ueber ein cretaceïsches Geschiebe mit Rhizocorallium gläseli, n. sp. aus dem Diluvium bei Leipzig. In: Sitz. ber. nat. Ges. Leipzig, xxxix, p. 19–25, Pl. i.

1913 **Flegel, C.** Sulla questione dei pescatori di spugne del Mediterraneo e del golfo del Messico. In: Atti 5. Congresso internaz. pesca (1911), Roma, p. 345–355, 3 figs.

1913 **Hentschel, E.** Über einen Fall von Orthogenese bei den Spongien. Von... In: Zool. Anz. xlii, p. 255–267.

1913 (α) **Hentschel, E.** Über die Anwendung der funktionalen Betrachtungsweise auf die biologische Systematik. Von...In Biol. Centralbl. xxxiii, p. 644–649.

1913 **Hovasse, R.** Nos Éponges d'eau douce. In: Bull. Soc. Sc. Nat. Chalons-sur-Saône, xix, p. 12–16.

1913 **Jordan, H.** Vergleichende Physiologie wirbelloser Tiere. Von...i. Die Ernährung...Jena, xxii and 738 pp., 277 figs.

1913 **Kirkpatrick, R.** Note on the Occurrence of the Euplectellid Sponge. Regadrella phoenix, O. Schm., off the South African Coast. In: Ann. S. Afric. Mus. Cape Town, xiii, p. 63–64, 1 Pl.

1913 **Kudelin, N.** (Russian.) ("Zur Frage der Spongien des Schwarzen Meeres.") In: Mem. Soc. Nat. Odessa, xxxv (1910), p. 1–40.

1913 **Leriche, M.** Deuxième Note sur les fossiles de la Craie phosphatée de la Picardie. In: Bull. Soc. Géol. Bruxelles, xxv, 1911, p. 297–312, Pl. i.

1913 **Orton, J. H.** On a Habitat of a Marine Amoeba. In: Nature, xcii, p. 371–372.

1913 **Parker, W. N.** Sponges in Waterworks. In: Proc. Zool. Soc. London, p. 973–976.

1913 **Rauff, H.** Barroisia und die Pharetronenfrage. In: Palaeontol. Zeitschr. i, p. 74–144, Pl. i and ii.

1913 **Schwan, A.** Über die Funktion des Hexactinelliden-skelets und seine Vergleichbarkeit mit dem Radiolarien-skelet. In: Zool. Jahrb. Abth. Allgem. Zool. xxxiii, p. 603–616.

1913 **Sella, M.** La pesca delle spugne nella Libia. In: Atti Soc. Ital. Progr. Sc. Roma, vi, p. 585–606.

1913 **Shimer, Hervey W. & Sidney Powers.** A New Sponge from the New Jersey Cretaceous. In: Proc. U. S. National Mus. xlvi, p. 155–156, 1 Pl.

1913 **Siemiradzki, J. von.** Die Spongien der polnischen Juraformation. In: Beitr. Palaeont. Geol. Oesterr. Ungarn, xxvi, p. 163–211, Pl. viii–xiii.

1913 **Silvestri, A.** Spicole di Tetractinellidi rinvenute da Ambrogio Soldani nei Sedimenti del Mediterraneo. In: Mem. Accad. Nuovi Lincei, Roma, xxx, p. 125–146.

1913 **Smith, W. D.** Contributions to the stratigraphy and fossil invertebrate fauna of the Philippine Islands. In: Journ. Sc. Manilla, Sect. A, viii, p. 235–300.

1913 **Swartz, C. K.** Porifera. In: Maryland Geol. Survey Lower Devonian, Baltimore, p. 195–198, 1 Pl., 4 figs.

1913 **Topsent, E.** Spongiaires provenant des Campagnes scientifiques de la Princesse Alice dans les Mers du Nord (1898–1899, 1906–1907). Par...In: Résult. Camp. Sc. Monaco, xlv, p. 1–67, Pl. i–v.

1913 (α) **Topsent, E.** Spongiaires de l'Expédition Antarctique Nationale Ecossaise. Par...In: Trans. R. Soc. Edinburgh, xlix (Part iii, No. ix), p. 579–643, Pl. i–vi.

1913 **Toula, F.** Die Kalke vom Jägerhause unweit Baden...In: Jahrb. Geol. Reichs-Anst. Wien, lxiii, p. 77–126, Pl. iv–vii.

1913 **Walcott, C. D.** Notes on Fossils from Limestone of Steep-rock Series, Ontario, Canada. In: Geol. Survey, Ottawa Mem. xxviii, p. 16–19, Pl. i and ii.

1913 (α) **Walcott, C. D.** The Cambrian faunas of China. In: Publ. Carnegie Inst. No. liv, p. 1–276, 24 Pl.

1913 **Walton, C. W.** The Shore Fauna of Cardigan Bay. In: Journ. Mar. Biol. Assoc. x, 2 (1), p. 102–113.

1913 **Weltner, W.** Süsswasserschwämme (Spongillidae) der Deutschen Zentral-Afrika-Expedition, 1907–1908...In: Wiss. Ergebn. D. Zentr. Afr. Exped. iv, p. 475–485, 53 figs.

1913 **Zahálka, B.** (Kridovy utvar v zapadnim Povltavi.) ("Kreideformation im westlichen Moldaugebiet," Zone iii, iv and v.) In: Sitz. ber. Böhm. Ges. Wiss. Prag, 1912, vii, p. 1–80.

INDEX OF AUTHORS

Oswald, F., 1847, 1850
Ott, Ch., 1892
Otto, E. von, 1852, 1854
Owen, D. D., 1844, 1852, 1857
Owen, R., 1835, 1841, 1843, 1857

Pacht, R., 1858
Packard, A. S., 1875, 1881
Pagenstecher, H. A., 1860, 1872
Pallas, P. S., 1766, 1771, 1776
Pantanelli, D., 1880
Parfitt, E., 1868, 1870, 1871, 1872, 1878
Parker, G. H., 1910
Parker, W. N., 1913
Parkinson, J. [botanist], 1640
Parkinson, J. [geologist], 1808, 1818, 1822, 1833
Pasquier, A. du, 1843
Passy, A., 1832
Pavesi, P., 1881
Payer, J., 1876
Peach, C. W., 1843, 1868, 1878
Pearcey, F. G., 1892, 1893
Pekelharing, C. A. See Vosmaer, 1893, 1894, 1898
Pélissier, A. J. J., 1861
Pengelly, W., 1861, 1871
Pérez, J., 1895
Péron, F., 1810
Pérot, F., 1890
Perrier, E., 1881, 1898
Perty, J. A. M., 1841, 1852, 1870
Peters, K. F., 1855
Petiver, J., 1704, 1712
Petr, F., 1885, 1886, 1887, 1894, 1895, 1898, 1899
Peyssonel, J. A., 1752, 1758
Pfeiffer, L., 1873
Philippi, R. A., 1843
Phillips, J., 1829, 1836, 1837, 1841, 1875
Phillips, J. & J. W. Salter, 1848
Phillips, W., 1822. See Conybeare, 1822
Pichler, A., 1857
Pick, F. K., 1905
Picot de Lapeygrouse, P., 1781
Pictet, F. J., 1857
Pictet, F. J. & E. Renevier, 1858
Pidgeon, E., 1830
Pillet, L. & E. de Fromentel, 1875
Piso, G. (W.), 1658
Plancus, J. See Bianchi
Plant, J., 1866
Plot, R., 1677
Plukenet, L., 1691, 1696, 1705
Počta, P., 1882, 1883, 1884, 1885, 1886, 1887, 1890, 1892
Poey, F., 1854
Poiret, J. L. M., 1789
Poléjaeff, N., 1882, 1883, 1884, 1885, 1889
Poli, I. X., 1791

Pomel, A., 1867, 1872
Pomet, P., 1694
Pontoppidan, E., 1752
Portlock, J. E., 1843
Posselt, L., 1843
Potts, E., 1880–1887, 1889
Pourtales, L. F. de, 1868
Preiwisch, J., 1903
Prestwich, J., 1871
Preuss, 1837
Price, F. G. H., 1874, 1877
Prichard, J. C., 1813
Priest, B. W., 1881, 1883, 1884, 1886, 1888
Pusch, G. G., 1836
Pylaye, B. de la, 1826

Quekett, J. E., 1848, 1854
Quenstedt, F. A., 1836, 1838, 1842, 1843, 1848, 1852, 1857, 1866, 1867, 1871, 1872, 1876, 1877, 1878
Quoy, J. R. C. & G. Gaimard, 1824, 1833

Rafinesque (Schmaltz), C. S., 1814, 1818
Ramond de Carbonnière, L. F. E., 1801
Raspail, F. V., 1827, 1828, 1832
Rath, G. vom, 1869
Rathbun, R., 1882, 1883
Rauff, H., 1886, 1891, 1892, 1893, 1894, 1895, 1913
Raulin, F. V. & A. Leymerie, 1858
Ravn, J. P. L., 1899, 1912
Ray, J., 1686, 1688, 1690, 1704
Reeker, R., 1895
Reichert, K. B., 1869
Rein, J. J., 1881
Reinsch, P. F., 1877, 1879
Renard, A., 1878
Reneaume, M. L., 1717
Renevier, E., 1867
Renier, S. A., 1793, 1807
Rennes, A. J. M., 1868
Renou, F., 1857
Retzer, W., 1883
Reuss, A. E., 1844, 1845, 1846, 1854, 1861, 1866, 1867
Reynolds, J. C., 1909
Rice, H. J., 1885
Richards, J., 1889
Richet, C., 1906
Richter, R., 1853
Ridley, S. O., 1880, 1881, 1883, 1884, 1885
Ridley & Carter. See Carter, 1890
Ridley, S. O. & A. Dendy, 1886, 1887
Ringueberg, E. N. S., 1885
Risso, A., 1826
Ritgen, F. A. von, 1831
Robertson, M., 1911
Robertson, M. & E. A. Minchin, 1910
Roemer, C. F., 1844, 1848, 1855, 1857–1863, 1866, 1870, 1872, 1876, 1880

Solger, B., 1883
Sollas, Igerna B. J., 1902, 1903, 1906, 1908
Sollas, W. J., 1872, 1873, 1876-1889
Sommer, P. J., 1834
Sorby, H. C., 1875
Soubeiran, J. L., 1861
Sovinsky, W., 1890
Sowerby, J., 1806, 1811, 1817
Spada Lavini, A. & A. Orsini, 1855
Spallanzani, L., 1784
Speyer, O., 1875
Spix, J., 1811
Spratt, T. A. B., 1865
Städeler, G., 1859
Staring, W. C. H., 1856, 1860
Stark, J., 1828
Stebbing, F. R. R., 1875, 1878
Stedman, J. M., 1893
Steiniger, J., 1831, 1840
Steinmann, G., 1877, 1880, 1881, 1882, 1883
Stephens, J., 1912
Sternberg, K. M. von, 1833
Sterzel, J. F., 1875
Stewart, C., 1802, 1818
Stewart, C., 1870, 1880, 1881
Stimpson, W., 1863
Stoliezka, F., 1873
Stoppani, A., 1860
Strange, J., 1770
Strickland, H. E., 1854
Ström, H., 1762
Strombeck, A. von, 1849
Struckmann, C., 1880
Studer, B., 1853
Studer, T., 1879, 1882
Stutchbury, S., 1841
Stuxberg, A., 1880
Suess, E., 1866
Sukatschoff, B., 1899
Swartschewsky, B., 1902, 1905, 1906
Swartz, C. K., 1913
Syen, B. W., 1879
Szymanski, M., 1904

Tate, R., 1864, 1865
Taylor, R. C., 1824, 1829, 1830
Tchihatcheff, P. von, 1854
Tempère, J., 1897
Templeton, J., 1836
Terquem, O., 1855
Terquem, O. & E. Piette, 1862, 1865
Thacker, A. G., 1908
Théel, H., 1907
Thiele, J., 1891, 1898, 1899, 1900, 1903, 1905
Thomas, A. O., 1911, 1912
Thompson, W., 1840, 1841, 1844, 1846, 1856
Thompson, W. d'Arcy, 1885

Thomson, C. Wyville, 1867-1871, 1873-1877, 1880
Thomson, J. A., 1887
Thomson, J. A. & J. D. Fiddes, 1906
Thoulet, J., 1884
Thum, E., 1904
Thurmann, J., 1851, 1863
Tietze, E., 1870
Tilenau. See Tilesius
Tilesius, A. von, 1826
Tilesius von Tilenau, W. G., 1803
Tizard & J. Murray, 1882
Topsent, E., 1887-1913
Toucas, A., 1874
Toula, F., 1875, 1877, 1879, 1913
Tournefort, J. P. de, 1694, 1700
Townsend, J., 1813
Trautschold, H., 1861, 1870, 1876, 1877, 1878, 1879, 1881
Traxler, L., 1894, 1895, 1896, 1898
Treviranus, G. R., 1803
Tribolet, M. de, 1873, 1876
Tromsdorff, J. B., 1805
Troost, G., 1838, 1840
Tschernychew, Th., 1898, 1899
Turgot, E. F., 1758
Turpin, P. J. F., 1837, 1838
Turton, W., 1806, 1807
Tyler, C., 1878

Uhle, A. F. (see also Ludwig, 1819), 1819
Ulex, G. L., 1864
Uljanin, W., 1872
Ulrich, E. O., 1879, 1890
Ungern-Sternberg, E. von, 1903
Urban, F., 1902, 1903, 1905, 1908, 1909, 1910
Urban, W. S. M. D', 1880
Ustinov, K., 1904

Vahl, M., 1793
Vaillant, L., 1869, 1870
Valenciennes, A., 1860
Valle, A. della, 1893
Vallisnieri de Vallisnera, A., 1722
Vallon, J. N., 1833
Vanuxem, L., 1842
Vasseur, G., 1880
Vaughan Jennings. See Jennings, 1891
Vauquelin, 1811 (Fourcroy)
Vejdovský, F., 1883, 1884, 1886
Verneuil. See Archiac & Verneuil, 1842, Murchison & Verneuil, 1844
Verneuil, P. E. P. de, 1847, 1864
Vernhout, J. H. See Vosmaer, 1902
Verrill, A. E., 1869, 1871, 1873, 1874, 1878, 1879
Vincent. See Bory
Vio, G., 1792
Vogel, H. A. von, 1824